W. Schiefer · E. Kazner · St. Kunze

Clinical Echo-Encephalography

With an Introduction into the Fundamental Principles of Acoustics by
W. Güttner

Translation by Margareta Klinger, M. D., Saskatoon, Saskatchewan, Canada

With 195 Figures

Springer-Verlag Berlin · Heidelberg · New York 1968

Dr. med. WOLFGANG SCHIEFER, o. Professor für Neurochirurgie, Direktor der Neurochirurgischen Klinik der Universität Erlangen-Nürnberg

Dr. med. EKKEHARD KAZNER, Oberarzt an der Neurochirurgischen Klinik der Universität Erlangen-Nürnberg

Dr. med. STEFAN KUNZE, wissenschaftlicher Assistent an der Neurochirurgischen Klinik der Universität Erlangen-Nürnberg

Dr. phil. WERNER GÜTTNER, Abteilungsleiter des Wernerwerks für Medizinische Technik, Siemens AG, Erlangen

German Edition Published in 1967 with the Title "Klinische Echo-Encephalographie"

ISBN-13: 978-3-642-48723-1 e-ISBN-13: 978-3-642-48721-7
DOI: 10.1007/978-3-642-48721-7

Softcover reprint of the hardcover 1st edition 1968

Library of Congress

Catalog Card Number 68—56950. Title No. 1543

Preface

Echo-encephalography, introduced by LEKSELL in 1955, has gained increasing importance for the early detection of numerous intracranial lesions in the last decade. The main advantage of this diagnostic procedure lies in the fact that it permits a rapid orientation about the spatial relationships within the skull without stressing or endangering the patient. Although this method alone only rarely allows a complete diagnosis, the echo-encephalographic findings always indicate which further diagnostic measures are most suitable for establishing the diagnosis with the greatest accuracy in every case. However, the correct interpretation of an echo-encephalogram is possible only, if the findings which are assumed to be pathological are evaluated in the light of the clinical symptomatology.

Since JEPPSSON's excellent monograph on the origin of the midline echo and its importance for the diagnosis of intracranial expansivities, published in 1961, a great deal of work has gone into the development of echo-encephalography all over the world. For this reason the possibilities of this procedure today go far beyond the mere demonstration of a supratentorial shift. Now we can frequently outline the width of the ventricles exactly and localize tumors or hematomas by means of abnormal reflections. Since a detailed description of the technique, application and present-day diagnostic uses of echo-encephalography has not been available as yet, we undertook to fill this gap in the German literature in 1967 with a monograph summarizing the hitherto existing experience as well as our own extensive case material.

The good reception of this book and the wish for an English edition of "Klinische Echo-Encephalographie" which many colleagues expressed both to the publisher and to us, motivated us to translate this monograph. In doing so, it seemed feasible to include the experience we had gained since the appearance of the German edition as well as numerous new publications on echo-encephalography. For this reason, large sections of the book were revised and brought up to date.

The book is now based on 10,478 echo-encephalograms carried out on 3140 patients from the Neurosurgical Clinic of the University of Erlangen-Nuremberg; the period of study extends from 1962 to 1968. The skilful evaluation of these echo-encephalograms enabled us to describe numerous ultrasonic findings which had been hitherto unknown or barely known and which increased the value of the method considerably. The book is based chiefly on the ultrasonic trace. Typical abnormalities of the oscilloscopic picture which could be observed again and again were worked out and in each case we attempted to correlate these with the clinical, neuroradiological and if possible, with the pathological-anatomical findings. The neurosurgeon is in the unique position of being able to examine the accuracy of the echo-encephalogram immediately by contrast medium studies and biopsy. Our own results were compared with those reported in the literature. The papers presented at the International Symposium on Echo-Encephalography in April 1967 in Erlangen, organized by the authors together with Prof. Dr. H.-J. ZÜLCH, Cologne, were also drawn into consideration. Special attention was devoted to the unfortunately inavoidable misinterpretations of echo-encephalograms. In the appendix on two-dimensional echo-encephalography (B-scan), we have evaluated the most recent reports in the literature and presented our own results up to the present illustrated with several examples. The list of references includes the most important publications up to July 1, 1968.

A certain amount of knowledge of ultrasonic physics must precede the clinical use of the procedure; therefore we deemed it necessary to begin the book with a short but adequate introduction into the fundamentals of acoustics. Dr. phil. WERNER GÜTTNER, who has been acquainted with the problems of ultrasonic diagnosis on the skull for decades, has done this in an admirable fashion.

At this point we want to express our thanks to the Wernerwerk für Medizinische Technik, Siemens AG, Erlangen, and especially to Dr. rer. nat. J. SPERLING for his support and constant readiness to help. His collaborators also met our suggestions with enthusiasm at every opportunity and advised us on all technical problems.

The pathological anatomical specimens were kindly placed at our disposal by Priv. Doz. Dr. med. H. SCHMIDT from the Pathology Institute and the scintigrams by Priv. Doz. Dr. med. F. WOLF from the Dept. of Nuclear Medicine, Medical Clinic, University of Erlangen-Nuremberg. Special credit goes to Mrs. MARIE-LUISE IMGRUND for her painstaking work on the photographic material and to Mr. W. AUER for the well-drawn schematic illustrations.

We wish to express our gratitude to Dr. MARGARETA KLINGER, Saskatoon, Saskatchewan, Canada, for the translation which was completed so rapidly; we believe that the spirit of the book has been transmitted.

The cooperation, advice and insistence on high publication standards and especially the excellent reproduction of the pictorial material by the Springer-Verlag, Berlin-Heidelberg-New York, are gratefully acknowledged.

Erlangen, Germany, November 1968 W. SCHIEFER · E. KAZNER · ST. KUNZE

Contents

Part I. Fundamental Principles of Acoustics and its Methodology

A. Introduction

Again and again we find that nature has solved problems millions of years ago which we face today as technical problems and which we can approach only in a very simplified form. Everyone knows of the bats who hunt their prey in flight at night and is aware of their ability to fly in complete darkness without hitting an obstacle. More careful observation has shown that these animals emit cries at an ultrasonic level inaudible for us and interpret the echoes returning from impending obstacles for the direction and distance of their flight. The apparently giddy flight through the air is a very successful method for hunting insects who must be localized and caught. Apparently these animals are in a position to differentiate between objects which can be eaten and must therefore be flown at directly, and other objects which must be circumvented in order to avoid dangerous collisions.

The character of the cries and their source varies with the different types of bats but the basic principle is the same for all them: the cry is a mixture of frequencies in the range of 30 to 100 kHz, so that the wave lengths of these cries in the air ranges from 1 to 0.3 cm. These can be radiated from the mouth cavity which lies in an acoustic baffle and the sound is directional. In one species the nose serves as the sound generator. A swelling around the nose increases the directional effect because alterations in the muscular tension affect the degree of bundling. To obtain information on an object in close proximity, a wide beam width is radiated while more distant objects require a very narrow beam width. Thereby adjustment to the energy relations which occur with distant objects is achieved and in addition bearings are taken on close-lying, lateral objects so that the batwings can circumnavigate them in safety. The intensity of the orientation cry is exceptionally great, being about 120 dB at a distance of 10 cm from the animal. The duration of this signal is about 2 ms and is adjusted to the expected echo from a near object. For superficial orientation the cry is repeated about 25 times per second. To detect an object in the immediate vicinity, the frequency is increased to about 100 per second.

Objects in the operational radius of the bat reflect the emitted cry. The animal perceives the echo with its ears. The directional effect of the individual ear exerts an exceptionally strong bundling effect on the ultrasonic echoes which are measured by the cone-shaped opening of the ears. The reception is facilitated by the fact that the ears are rapidly directed towards the object of reflection. In certain species not only is the highest echo value taken bearing on by the movement of the ears, but the time difference between the two ears is also evaluated in localizing the origin of the echo. At any rate the echo reception is improved in this way. It probably explains the fact that the animals can still perceive an echo even if the sound level has fallen to 0 dB. The remarkable thing is that the bat can still localize perfectly even when there is a high interference echo level. Obviously the powerful directional effect of its hearing filters out the echo from the diffuse sound field. Furthermore the bat must have the ability to adapt to enormous variations in sound volume between its own high intensity cry and the low intensity echo which follows it.

The efficiency of echo orientation is surprising for us considering that insects with a total body length of 2 mm can be detected at a distance of 50 cm. Even a thin wire only 0.2 mm in diameter can be recognized as an obstacle at a distance of 1 to 2 m. Only in the vicinity of moths is the procedure not quite satisfactory for the bats because nature has given these insects good protection: a thick fur cover absorbs most of the orientation cry so that no echo can betray the location of the moth if it is far enough eway from the bat. As an additional precaution, moths drop away by letting their wings fall whenever they perceive the ultrasonic pulse. They escape out of the sounding area so as to avoid being detected by a bat who might yet localize them at a closer range.

In discussing the flight of bats, we have already touched the phenomena necessary for understanding echo-encephalography.

The physician is used to demonstrating the internal structures of the human body with the help of x-rays. This radiation penetrates the various parts of the body and the absorption of the rays which depends on the density (atomic weight) of each body medium is depicted as a two-dimensional picture perpendicular to the direction of the beam. From this picture we can deduce the internal construction of that part of the body. However, normal x-ray technique is useless in the demonstration of cerebral structures which are of interest in the diagnosis of intracranial disorders, because the difference in attenuation between CSF and brain tissue for example, is too slight for x-rays to pick up. But if the cerebrospinal fluid is exchanged for air, satisfactory pictures of the ventricular system can be obtained because air has a lower atomic weight than the brain tissue around it. Angiography rests on the same principle. Both investigations have certain side-effects however, and can therefore be carried out only on hospitalized patients. Consequently the need arose for another physical, in our case acoustic method of investigation for diagnosing pathological processes of the skull contents.

An ultrasonic transmission method which produces a two-dimensional picture similar to that obtained with x-rays, has been used for industrial purposes in order to detect cavities and malpositionings within certain materials. This is based on work by Sokolow and Mühlhäuser. Here a plane-wave beam is produced by a source and is sent through the material to be investigated. A receiving transducer fixed to the opposite boundary surface picks up the sound which has passed through the examined medium. By co-ordinated movements of the two transducers, one obtains a view of the inside of the material. A detailed report may be found in the monograph "Werkstoffprüfung mit Ultraschall" by the brothers J. and K. Krautkrämer.

To consider the applicability of the transmission procedure for the demonstration of structures within the head as well, was an obvious step. It was thought that the attenuation of sound within the different tissue media similar to the absorption of x-rays would result in a picture adequate for the diagnosis of lesions within the skull. Dussik (1942) believed he had realized this idea, for he showed ultrasonic pictures which resembled x-rays of an air-filled ventricular system on superficial examination (see Fig. 15). However, it turned out that this procedure could not be employed for diagnostic purposes because of a series of physical interference effects. The most important of these is the variation in the thickness of the skull bone, which envelops the head. For although ultrasound is attenuated less in CSF than in brain tissue, the bony substance absorbs an unusually large amount of sound. A picture obtained by through-transmission sounding of the skull with plane waves is due to the non-uniform thickness of the skull; the slight differences in the absorption of sound by CSF and by brain are not detected (see Fig. 16).

An entirely different principle is employed when the ultrasound method is used. Just as navigators used to lower a weighted line into the water and recognized that it had reached the bottom when the resistance to the plumb-line changed, in the same way the echo procedure used a sound beam which is reflected when it hits the interface between two media, thereby reporting the presence of this interface to the transducer. The locations of the interfaces can be calculated from the directions from which the sonic beam is emitted and from the echo.

While the transmission procedure can depict only the sum of all information obtained on the path of the sound beam, the echo-pulse-method records every alteration in the resistance at the interfaces between media and depicts it in time sequence, i. e. separately.

The echo-pulse-method was initially used in navigation sounding in order to detect the presence of underwater obstacles and their distance from the ship. This development was stimulated by the sinking of the "Titanic" in a collision with an ice-berg in the year 1912 and was pursued by RICHARDSON and LANGEVIN. Now echo sounding can be employed even for fishing in order to proceed in an optimal economic way, for the net is thrown out only after a swarm of fish has been detected by echo sounding. The resolution of the pictures obtained in this manner can be perfected to such a degree that individual fish can be recognized. The depth of the water over important navigation routes and a more exact profile of the ocean floor were also mapped out by echo sounding. Continuous knowledge of the water depth below a moving ship is a necessary prerequisite for safe navigation today. And finally the localization of mines and submerged submarines was a matter of life-and-death importance for a ship, especially during the last world war.

FIRESTONE was the first to utilize the echo method for flaw detection in raw materials. Here it serves to detect cavities and weaknesses in metals and synthetics and today it is used whenever the materials must satisfy high industrial safety standards.

In the meantime it has become clear that the procedure can be used successfully for the detection of pathological changes in the brain as well. The credit for the first experiments on the intact skull goes to LEKSELL (1955). An ultrasonic pulse is projected into the skull. The reflections from the boundaries between different media in the brain, e. g. between brain tissue and CSF are then observed. If one knows the velocity of sound propagation within the various media, then the distance from the interface to the surface of the head can be calculated from the time difference between emission of the ultrasonic pulse and reception of the echo. A reproducible signal is obtained if it appears at the receiving site with great intensity. Only then does it rise above the interference level so that it can be recognized. For the normally built skull, a typical normal structure will be found for the distances from the surface; pathological changes can then be related to these. In order to evaluate these traces correctly, a thorough knowledge of the anatomy within the skull and experience with the acoustic relations is necessary.

The conditions necessary for the production of sufficient echo energy are different for different application purposes such as the bat's search for food, brain diagnostics, water sounding and flaw detection.

The signal of the bat is attenuated very little even in air of high humidity, especially since the objects which are to be localized are only a few meters away. The surfaces of objects reflect practically everything. However, insects for example are so small that the sound energy reflected from them is very scattered. The echo intensity heard by the bat is therefore usually slight and permits good localization only of the objects in the immediate neighborhood.

In order to ascertain the water depth below a ship by means of vertical sounding, a pulse with a frequency of approximately 20 kHz is used. At this frequency the absorption in water is small as yet. The ocean floor can reflect ultrasound quite variably. However, it is often so even that scatter is slight. The screw of the ship and the movement of the waves due to the wind on the surface of the water produce noise which the echo must supersede. Therefore the maximal intensity of the sound pulse is increased up to the cavitation limit of sea water. On the whole the system functions satisfactorily then. Fish are detected by the reflection from their air-bladders, not from the external skin. Air reflects sound in water completely, but here too the single, relatively small air-bladder is a strongly scattering reflection surface. A swarm of fish on the other hand, acts as a wall of air and therefore reflects a great deal of sound. Mines and submerged submarines are also large, air-filled bodies. Horizontal sounding which is used in the detection of the latter would probably always give very good pictures if it were not for the occasional appearance of water stratifications which deflect the ultrasonic front and veils of air bubbles which make the water impenetrable.

The search for cavities, weaknesses, errors in the crystal structure and similar faults with the pulse-echo-method for the purpose of evaluating materials is generally carried out in media whose absorption is very small in a given range of frequencies. However, the flaws which are to be detected are occasionally minute in relation to the other dimensions so that they scatter sound. Nevertheless the energy which returns to the reception site is enough for an evaluation. Because several types of waves can occur within solid bodies — not only longitudinal waves but also transverse, torsional, lateral, sheer waves etc. — special knowledge is necessary for interpreting the echoes.

When the echo method is applied to the human skull, the ultrasonic wave passes through a series of media which absorb a great deal of energy in the range of frequencies used and whose interfaces reflect little sound in contrast to other applications of ultrasound. But the radiation path is only a few centimeters thick and some of the interfaces lying in the path of the sound beam have adequately large surfaces, perpendicular to the beam. Therefore the intensity of the ultrasonic beam which is radiated into the skull is enough to produce echoes; its biological mean value is considerably lower than that which becomes effective in therapeutic ultrasonics. Although the interfaces within the brain reflect relatively little sound, the penetrability for sound energy however, is great. Therefore so much energy floods through the interfaces between the individual media that echoes are reflected from other boundaries and from the opposite side of the head as well. Only these values put us in a position to give satisfactory answers about the site of interfaces, i. e. about the inner structure of the brain.

B. Fundamental Principles of Acoustics

In order to understand echo-encephalography, the physical principles of this procedure must be brought to mind:

1. Sound, Infrasound, Ultrasound

By sound we mean all natural processes which evoke the sensation of "hearing" in us, e. g. the rustling of leaves in a tree, thunder during a strom, the song of a bird, the cries of animals and the talk of human beings. But the noises which are troublesome accompaniments of modern technology are also perceived.

Our ears can analyze sound according to pitch, volume and time sequence. When both ears are used, deductions can be made about the site and size of a source, making it possible to recognize sound procedures if they are repeated and to classify them with certain sound sources. Mechanical oscillations can be a source of sound in a manner resembling wind passing over certain objects. The transmission medium is the surrounding atmosphere which our ears are best adjusted to. The human ear can pick up only a limited range of pitch and volume. Very low frequencies outside the limit of human hearing, e. g. gusts of wind or alterations in pressure during weather changes from high to low pressure do not leave auditory impressions. We describe this range of frequencies as infrasound.

The proportion of sound frequencies which lies above the limit of human hearing, e. g. the bat's cry, is termed ultrasound (above 15 kHz approx.). Borderline values of human hearing are determined in young people with normal hearing because the sensitivity for high frequencies decreases with age.

In the range of audible frequencies, the subjective interpretation depends on the intensity of sound. We hear sound only above a certain threshold. This intensity is greater than the rushing of air molecules and greater than the noise produced by the pulse. Very great sound intensities may be experienced as being unpleasant, they produce pain and may result in permanent damage to the organ of hearing. Such sensations can be produced in the range of ultrasound also, without however, leaving an auditory impression.

2. Sound Propagation, Sound Velocity, Sound Fields

The propagation of sound is dependent on the substance, and the velocity with which mechanical oscillation energy is passed on to an adjacent area is characteristic for each transmission medium. Sound cannot be propagated in vacuum. The velocity of sound in the surrounding air is about 340 m/s. Immediately after a flash of lightning appears, we count, "thousand-and-one, thousand-and-two, thousand-and-three" and if we then hear a clap of thunder, we calculate that lightning has struck about 1 km away by taking the time difference of 3 seconds and the sound velocity. In water the sound velocity is greater, being about 1500 m/s here. In the solid stony layers of the earth, the velocity increases, e. g. in granite it is about 4000 m/s, while the highest values are found in metals, e. g. in steel it is over 5000 m/s.

The sound velocity c is related to the wave-length λ of a frequency f as follows:

$$c = f \cdot \lambda.$$

In air the wave-length for a signal with a frequency of approximately 300 Hz is around 1 m, for a frequency of 3000 Hz it is about 10 cm and for a frequency in the ultrasonic range e. g. 3 MHz, the wave-length is approximately only 0.1 mm.

In water where the sound velocities are higher, the wavelengths for the same frequencies are also greater, e. g. for 3 MHz the wave-length is about 0.5 mm.

A source of sound, whose dimensions are small in comparison to the wave-length it produces, may be regarded as a spherical source, i. e. the sound spreads out evenly in all directions, losing power at the same time. A spherical sound field is thus formed. All particles of matter at an equal distance from the source are in the same energy state or are in the same phase. A pulsating ball is the ideal spherical sound source. When we enunciate vowels, the mouth must be regarded as such a spherical transducer because its dimensions are small for both basic forms; for the vowel a, in the word father, for example they lie close together at about 800 Hz while the corresponding wave length is around 40 cm.

At a sufficiently great distance from a sound source, part of the space through which sound energy travels may be regarded as a plane sound field; just as the surface of a lake seems to us to be a plane surface although it is part of the round surface of the earth, so the wave-fronts are practically plane here (see Fig. 1).

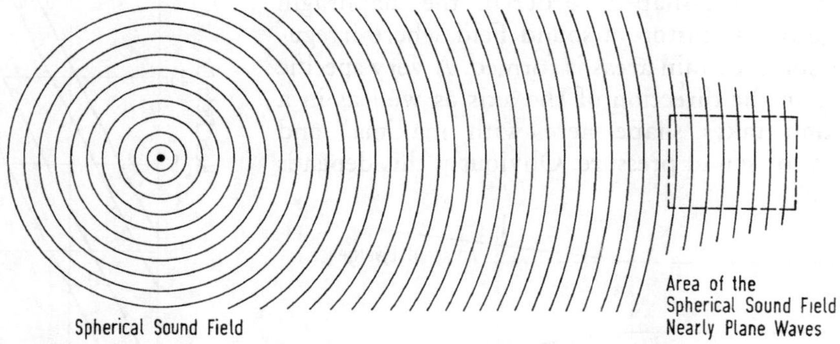

Fig. 1. Sound field of a circular piston: if a small part of the sound field is far enough away from the sound source, it is practically a plane sound field

In a medium with negligible absorption, the sound intensity is uniformly the same throughout a plane sound field. In a spherical sound field it depends on the site of observation.

Since we are dealing with echo-encephalography, we will limit our observations to the plane sound field. It is the ideal sound field for acoustic examinations.

3. The Production of Plane Sound Waves

A plane sound field can be produced by an oscillating wall from which the waves are propagated perpendicular to the wall. However, in practice waves can be produced satisfactorily by means of a piston.

By this we mean a rigid plate located in an infinitely large wall which can swing back and forth as a piston in a medium, thereby producing sound. If its diameter is very much greater than the wave length in the medium corresponding to its excitation frequency and if we let it swing as a plane piston, we obtain the same sound field as behind a baffle with an aperture the diameter of the piston on to which the plane wave falls (Fig. 2).

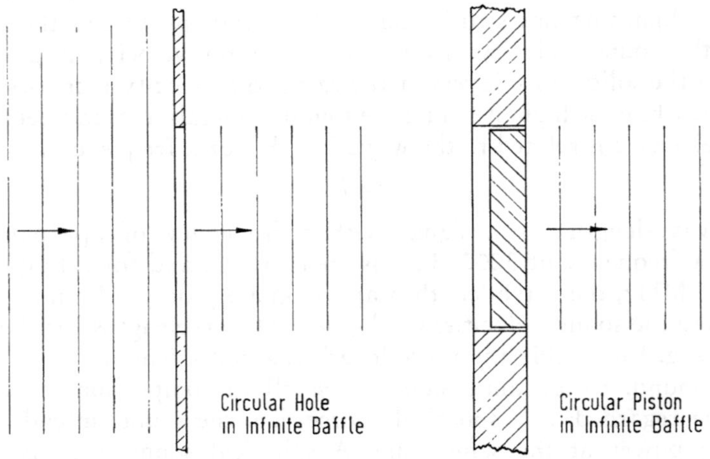

Fig. 2. Baffle and piston as identical radiators of plane sound waves

If we regard this from a geometric point of view, we find a cylindrical sound beam which looks just like the sunbeams falling through a hole in a baffle. However, signs of refraction appear at the edge of the baffle and the piston thereby altering certain sound field parameters, for example the sound pressure within the sound beam. If the plane sound waves are overlapped by basic waves developing at the ring-shaped border of the diaphragm or at the edge of the piston, a sound field where irregularities can occur in certain areas is formed. A very specific field structure in the direction of the axis as well as in a radial direction takes shape here with maximal and minimal values of sound pressure. Obviously this depends

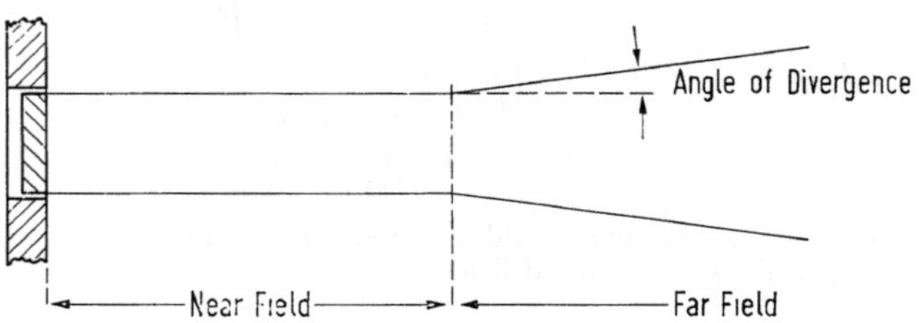

Fig. 3

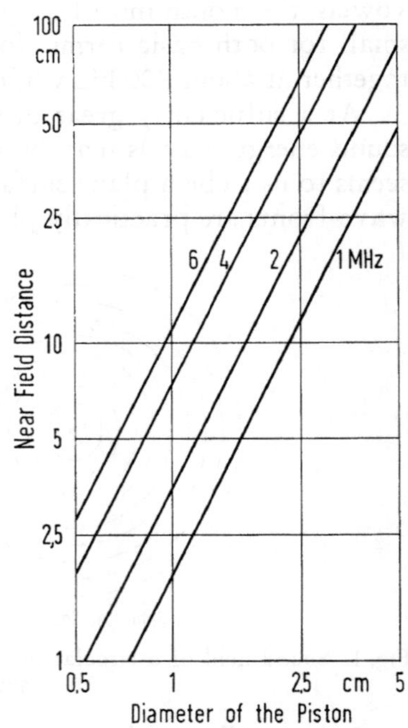

Fig. 4

Fig. 3. Extent of the near and far field in front of a piston

Fig. 4. Nomogram on the dimensions of the near field in front of pistons of various diameters in water

on the diameter of the spherical piston and the wavelength of the emitted oscillations. This area is called the near field. Fig. 3 shows a diagrammatic representation of this sound beam. The near field is the domain of the cylindrical sound beam. Its dimensions for different diameters of the piston and for the frequencies employed in echo-encephalography can be read directly off Fig. 4. The basic sound velocity for this diagram is $1.5 \cdot 10^3$ m/s.

In contrast to the structured field in front of the piston membrane, the sound field then progresses into an area which is more uniform. This is termed the far field (Fig. 3) and has a divergence corresponding to the diameter of the piston and the wave-length of the radiation. However, it is extraordinarily slight for the size of the pistons utilized in the pulse-echo-method. At a sufficiently great distance from the piston, a sound field corresponding to that of a spherical wave is found.

The structure of the near field depends on the distribution of stimulation on the piston. If we change the uniform stimulation into a bell-shaped distribution, i. e. to a Gauss function, a near field is created which is practically a plane sound field. Therefore the probes applied in echo-encephalography have contact surfaces which can oscillate like bells.

4. Elements of the Sound Field

In all physical experiments, it is simplest to proceed from the energy flow or to set up an energy balance. The energy E passed on to a medium by a sound generator is

$$E = P \cdot t$$

P = acoustic power in watts (W), t the time in seconds (s), E in Ws.

The total power P is the product of the sound intensity I (W/m²) and the area S (m²) through which the sound energy is transmitted.

$$P = I \cdot S.$$

The sound intensity is defined as the acoustic power through a unit area perpendicular to the specific direction at the point considered; it is frequently used in ultrasound therapy with the dimension W/cm².

$$I = \frac{P}{S}.$$

In plane sound fields Ohm's Law of acoustics is applicable: the relationship of sound pressure p (in N/m² where N is the abbreviation for a unit of force, namely a Newton) to particle velocity v (m/s) is a constant one for a given medium and is termed specific acoustic impedance Z (Ns/m³)

$$\frac{p}{v} = Z.$$

On the other hand, this acoustic impedance is proportional to sound velocity c and density ϱ

$$Z = \varrho \cdot c.$$

The density of air under normal conditions is $\varrho = 1.2 \cdot \text{kg/m}^3$. If $c = 340$ m/s and the conversion constant is 1 N = 1 kg m/s², we find

$$Z_{\text{air}} = 4.1 \cdot 10^2 \text{ Ns/m}^3.$$

In water where $\varrho = 10^3$ kg/m³ and $c = 1.4 \cdot 10^3$ m/s, the acoustic impedance is higher by several orders of magnitude

$$Z_{\text{water}} = 1.4 \cdot 10^6 \text{ Ns/m}^3.$$

Thus the particle velocity at the same sound pressure is much lower than it is in air.

The intensity of sound is related to sound pressure and particle velocity according to the following formula:

$$I = p \cdot v.$$

If we use

$$\frac{p}{v} = Z$$

then

$$I = \frac{p^2}{Z} = v^2 \cdot Z.$$

The energy can be calculated from the sound intensity. In practice only the sound pressure is measured because our ears work like sound pressure receptors: the effective sound receptor surface of our ear drums have a size of the order of 1 cm² and react to sound pressure. Therefore most measuring devices for sound are based on sound pressure measurements.

The dimension of sound pressure N/m² is frequently replaced by μN/m² where 1 μN = 10^{-6} N. Previously the unit microbar (μb) was used. The conversion constants are as follows:

$$1\ \mu b = 0.1\ N/m^2 = 10^5\ \mu N/m^2.$$

This is approximately the sound pressure we pick up when we listen to someone talking 0.5 m away.

The particle velocity for this sound pressure calculated according to the formula

$$v = \frac{p}{Z_{air}}$$

is $2.4 \cdot 10^{-4}$ m/s = 0.24 mm/s.

At these values the sound intensity becomes

$$I = 0.1\ N/m^2 \cdot 1.4 \cdot 10^{-4}\ m/s$$
$$= 2.4 \cdot 10^{-5}\ N/ms = 2.4 \cdot 10^{-5}\ W/m^2 = 2.4 \cdot 10^{-9}\ W/cm^2.$$

The lowest sound intensity that we can just hear is about 10^{-16} W/cm² or 10^{-12} W/m² for middle frequencies. This corresponds to a sound pressure of approximately 20 μN/m².

The greatest sound intensity that the human ear can endure without registering a painful sensation is about 1 W/m² or 10^{-4} W/cm² which corresponds to a sound pressure of approximately 10 N/m².

In auditory acoustics the sound pressures thus lie in a range of values extending over 6 powers. Therefore a logarithmic scale was chosen for expressing the intensities of sound. Logarithms can only be numbers not quantities. For this reason we use the relationship between two sound pressures or two sound intensities for this calculation so that no dimensions are involved. The sound intensity $I_0 = 10^{-12}$ W/m² serves as the standard value. In practice we proceed from a sound pressure $p_0 = 20$ μN/m² and relate other sound pressures to it. We then no longer speak of sound pressures and sound intensities but rather of sound pressure levels and acoustic intensity levels. Thus we calculate

$$L = 20 \log \frac{p}{p_0}$$

or

$$L = 10 \log \frac{I}{I_0}$$

and apply the term dB to these value levels.

The sound pressure level of a sound pressure of 1 μb or 10^5 μN/m² in air is

$$L_p = 20 \log \frac{10^5}{20} = 20 \log 5 \cdot 10^3$$
$$= 20 \cdot 3.7 = 74\ dB.$$

The sound pressure $10^5\,\mu N/m^2$ which is just enough to evoke an auditory impression corresponds to a sound pressure level of 0 dB. Painful sensations occur at sound pressures greater than $10\,N/m^2$ which is a sound pressure level of more than 120 dB.

The acoustic intensities employed in ultrasonic therapy for frequencies near 1 MHz lie at about $1\,W/cm^2 = 10^4\,W/m^2$. The sound pressures produced in human tissue at this acoustic intensity are of the order of $1.2\cdot 10^5\,N/m^2$, the particle velocity approximately $8\cdot 10^{-2}\,m/s = 8\,cm/s$. If we proceed from a standard value of $I_0 = 10^{-16}\,W/cm^2 = 10^{-12}\,W/m^2$, the acoustic intensity levels lie at about 160 dB. The specific acoustic impedance value of ca. $1.4\cdot 10^6\,Ns/m^3$ was used for human tissue. Therefore the corresponding standard sound pressure level in air is not $20\,\mu N/m^2$ as with air sound, but rather it is

$$p_0 = 1.2\cdot 10^{-3}\,N/m^2 = 1.2\cdot 10^3\,\mu N/m^2.$$

Then the sound pressure level which corresponds to an acoustic intensity of $10^4\,W/m^2$ is

$$L = 20 \log 1.2\cdot 10^5/1.2\cdot 10^{-3}$$
$$= 20 \log 10^8 = 160\,dB.$$

As a result of absorption in tissue, these values are valid only at the point where sound enters but not any deeper.

In echo-encephalography, sound pulses with a duration of approximately $1\,\mu s$ are used at a repetition rate of 500 times per second. If the effective acoustic intensity appearing during the pulse is $1\,W/cm^2$, then the effective mean value of the entire intensity radiated at the surface of the head will be

$$1\,\mu s\cdot 500\,Hz\cdot 1\,W/cm^2 = 5\cdot 10^{-4}\,W/cm^2.$$

The acoustic intensity level is then

$$L = 10 \log \frac{5\cdot 10^{-4}}{1\cdot 10^{-16}} = 127\,dB.$$

We shall see later that the bony skull attenuates sound markedly. The passage of a sonic pulse with a middle intensity level of 127 dB and a frequency of 2 MHz through bone 5 mm in thickness reduces the intensity level by 30 dB because of absorption. The sound reaches the surface of the brain with an acoustic intensity level of only 97 dB. This corresponds to an intensity of $5\cdot 10^{-7}\,W/cm^2$ which is 7 powers below that used in ultrasonic therapy. Incidentally this is practically the same intensity level as when the eardrum is hit by the sound waves from a person speaking right into the ear.

5. Echoes

If the sound wave passes through a medium such as air for example and then encounters an interface, the propagation of sound is altered at this interface. We know that echoes appear when sound hits solid bodies with large surfaces. More careful observation shows that reflections are obtained not only at this site but anywhere in the atmosphere where the density of the air is altered by physical influences, for example. Thus clouds as well as temperature stratifications in the atmosphere can give rise to reflections. Sound is not necessarily reflected completely here as is the case with solid bodies, rather we are dealing with interfaces which permit the transmission of part of the sound from one medium to another one with different properties. If we examine the transition we recognize that it is due to the difference in the specific acoustic impedance. At an incidence of 90° the part of the plane wave energy which is reflected into the original medium (impedance $Z_1 = \varrho_1\cdot c_1$) from a plane interface behind which there is a medium with an impedance of $Z_2 = \varrho_2\cdot c_2$, can be calculated according to

$$R = \left(\frac{Z_1 - Z_2}{Z_1 + Z_2}\right)^2.$$

Thus if a plane sound wave passing through air hits the surface of the water, then R is nearly equal to 1 providing the commonly employed values of impedances are used, in other words almost all the energy is reflected by the water. That is why a thin layer of air in the water reflects sound so completely. For the same reason a coupling medium must be utilized between the probe and the human head in order to exclude a layer of air which is impenetrable to sound and which would prevent the transmission of ultrasound.

The smaller the difference in the impedance of two media lying one behind the other, the less sound is reflected and the more energy will penetrate the interface.

The impedances of human body tissues differ only very little from each other so that the reflections values between the layers which are to be demonstrated are very small. For example, the value for brain tissue is

$$Z_{\text{brain}} = 1.59 \cdot 10^6 \text{ Ns/m}^3$$

and the value for CSF is

$$Z_{\text{CSF}} = 1.51 \cdot 10^6 \text{ Ns/m}^3.$$

If a sound wave hits a CSF-filled ventricle in the brain, R is found to be about $7 \cdot 10^{-4}$.

In other words less than 1% of the energy is reflected at the interface, the rest passes through the ventricle (see Fig. 5). Here the energy balance at the boundary between two media is split into two parts: one part penetrates the layer, the other is reflected at the interface.

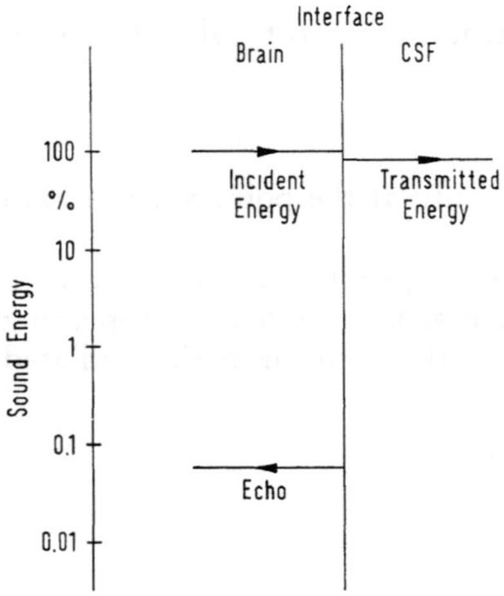

Fig. 5. Energy balance at the interface between two media of different specific impedances

If the ventricle is filled with air however, the transmission of sound is blocked completely and all of the energy is thrown back.

In practical usage it was found that the unusually small reflections make it possible to obtain echoes of several interfaces, one behind the other, within the skull, so that the procedure is entirely feasible.

With this method only sound returning to the generator provides information, i. e. only the surfaces which are at right angles to the direction of propagation of plane sound waves at the interface between two media return the signal to the place from whence it came Diagonal surfaces lead to echoes which are not picked up immediately. They can give rise to interference and become disturbances.

The energy balance will be influenced not only by the reflection conditions of the layer in question but also by the dimensions of the surface which is perpendicular to the sound beam. Similarly the transmission from probe to the surface of the head must be assured by the careful application of coupling material.

6. The Propagation of Plane Sound Waves in an Absorbing Medium

In echo-encephalography we use a plane sound wave which is radiated into a medium having absorbing properties. Along its distance sound energy is transformed into another form of energy, namely into heat.

If we observe the sound wave, which begins its transmission path at a plane $x = 0$ with an energy value E_0 and [after a distance at x (Fig. 6)] we find the energy E_x

$$E_x = E_0 \, e^{-2\alpha x}$$

where $2\,\alpha$ is the absorption coefficient.

Fig. 7 shows the decrease in energy as a function of the distance of a plane wave. A logarithmic scale is used for the energy coordinates here.

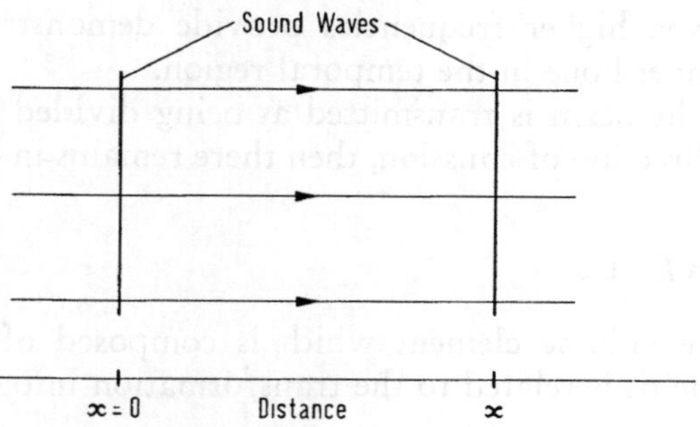

Fig. 6. Propagation of a plane sound wave front in a geometric fashion

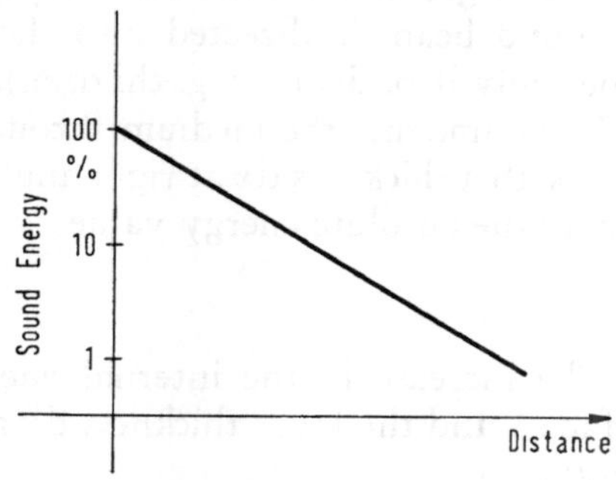

Fig. 7. Attenuation of a plane sound wave in an absorbing medium

If we relate the energy at a distance x to the entrance site ($x = 0$) then

$$\frac{E_x}{E_0} = \frac{I_x}{I_0} = e^{-2\,\alpha x}\,.$$

By using logarithms we arrive at the sound energy level or sound intensity level which we relate to the incident energy level

$$10\log\frac{E_x}{E_0} = 10\log\frac{I_x}{I_0} = -10\cdot 2\,\alpha x\cdot\log e.$$

For each unit distance

$$\frac{10\log\frac{E_x}{E_0}}{x} = \frac{10\log\frac{I_x}{I_0}}{x} = -10\cdot 2\,\alpha\cdot\log e.$$

This value is given in dB/cm. In this way it can be estimated at what gauge level the echo can be expected after it has passed through one stratum of a medium. If, for example, brain substance produces an attenuation of 2 dB/cm at a frequency of 2 MHz the intensity of the sound pulse after one passage through the brain and its return in a skull with a diameter of 10 cm is decreased by 40 dB.

Experimentally it has been shown that the absorption coefficient of human body tissue depends on the frequency of sound used. For frequencies of 1 to 6 MHz which are employed for the echo-pulse-method, the absorption rises linearly with the frequency. The absorption in bone is proportional to the square of the frequency used in contrast to the absorption in tissues enclosed within the skull. The following survey indicates the decrease in the sound intensity level per centimeter for a frequency of 1 MHz:

Brain	0.9 dB/cm
Muscle	1.2 dB/cm
Bone	15.0 dB/cm
CSF	0.01 dB/cm
Blood	0.17 dB/cm

The attenuation of energy in brain tissue is about 1 dB/cm at 1 MHz rising to about 2 dB/cm at 2 MHz and to about 4 dB/cm with 4 MHz, etc. By contrast the absorption in CSF is negligible and in blood it is low. In bone, on the other hand, it is exceptionally high, e. g. 15 dB/cm at 1 MHz. In order to have as little energy loss in bone as possible we choose a very thin bony area on the side of the head as the application point of the probe from which the transmitter pulse is projected (the bone thickness here varies from 3 to 6 mm). A bone thickness of 5 mm attenuates the intensity of a sound pulse passing through the brain and its return by 15 dB. If the test frequency of 1 MHz is raised to 2 MHz, the attenuation is 4 times as great. Thus the bony layer reduces the echo intensity by 60 dB at the site where the sound beam is directed into the skull. Even higher frequencies provide demonstrable echoes only if patients (e. g. children), have thinner bone in the temporal region.

If we imagine the medium through which the beam is transmitted as being divided into slices with a thickness dx at right angles to the direction of emission, then there remains in each slice dx the absolute energy value

$$dE_x = 2\,\alpha\,E_x\,dx.$$

The increase in the internal energy of the volume element which is composed of the surface S and the layer thickness dx after a time dt is related to the transformation into heat energy:

$$dE_x = \varrho\,C_p\,S\,dx\,dT_x\,,$$

where ϱ is the density in kg/m^3, C_p the mean specific heat in Ws/kg and dT_x the temperature increase in degrees. This results in a temperature rise at a distance x during one time unit

$$\frac{dT_x}{dt} = \frac{2\,\alpha}{\varrho\,C_p}\cdot I_x\,,$$

where I_x is the sound intensity at the distance x.

In the following section we see a summary of the quotient $2\,\alpha/\varrho\,C_p$ for a frequency of 1 MHz in different media. It is dependent on frequency in the same way as the absorption coefficient itself

Brain	0.054 cm^2 degrees/Ws
Muscle	0.073 cm^2 degrees/Ws
Bone	1.5 cm^2 degrees/Ws
Blood	0.0083 cm^2 degrees/Ws

We make use of this increase in heating in ultrasonic therapy. Above all, these effects which appear at the outer layer of bone and which are caused by unusually great absorption by the bone, are made use of. In the echo-pulse-method we saw that the sound intensity radiated into the skull at the application point is approximately $5\cdot10^{-4}$ W/cm^2. The increase in temperature per second at the entrance point into the skull at a frequency of 2 MHz would be

$$\frac{dT}{dt} = 6\ \text{cm}^2\ \text{degrees/Ws}\cdot 5\cdot10^{-4}\ \text{W/cm}^2$$
$$= 3\cdot10^{-3}\ \text{degrees/s}$$

or would lead to an increase in temperature of about 1 °C at a radiation time of 5 minutes, if the heat were not conducted away as in a calorimeter, for example. Thus it is of almost no practical importance in echo-encephalography at the most absorptive site already. After penetrating through a bony layer 5 mm thick, the sound intensity entering the brain at the same frequency would be about $5\cdot10^{-7}$ W/cm^2, and the increase in temperature per second in the brain at the same frequency would be

$$\frac{dT}{dt} = 0.1\cdot5\cdot10^{-7}\ \text{degrees/s}$$
$$= 5\cdot10^{-8}\ \text{degrees/s}$$

i. e. the surface of the brain would practically not be heated up at all and the deeper areas of the brain even less so. Therefore we can absolutely exclude any possible damage to the brain.

C. Detection of Structures within the Skull by Means of Echoes

In order to be able to give some information on the site of the reflection from the echo duration, we need the sound velocity value of the tissue through which the ultrasound has travelled. After measuring the time difference between the time of sound emission and time of the echo reception, we can calculate the distance and thereby the site from the product of the sound velocity and half the time difference.

The sound velocity of the different media we need in echo-encephalography are taken from Table 4:

Brain	$1.53 \cdot 10^3$ m/s
Compact bone	$3.38 \cdot 10^3$ m/s
CSF	$1.50 \cdot 10^3$ m/s
Blood	$1.57 \cdot 10^3$ m/s

Most of the tissues of the human body, have a sound velocity which differs only slightly from that of water. CSF and blood have almost the same acoustic behavior as water. Only in bone is sound propagated somewhat more than twice as fast.

The differences between the various sound velocities lead us to difficulties which must be considered in the localization of ventricles and accumulations of blood or CSF in the brain.

1. Inaccuracy in Localization

In echo-encephalography the time difference between the generation of the transmitter pulse and the reception of the echo, is transformed into a longitudinal display. For this purpose the light point moves across the screen of the cathode ray tube from left to right at a constant speed. The horizontal light line is deflected vertically by the transmitter pulse and the received echoes. The demarcations on the light line produced in this manner are a direct measure of the depth of the medium if the propagation of sound in the investigation medium and the velocity of the light point are considered. If we examine stratified substances whose sound velocities are not equal, with the same apparatus, we find deformation and inaccuracies in the localization of the stratified structure.

We will examine this process on the typical layers:

Bone / Brain / CSF

If we depict the time difference on a longitudinal axis by a light point which moves horizontally across the screen at a constant velocity (the sound velocity in brain being the basic velocity for all layers) we obtain an apparent thickness compression of the bony layer (H. Kresse, 1968) because the velocity in bone is about 2.4 times greater than in brain substance. Because of this, the thickness of the bone is portrayed as being too small and the interface of the ventricle too close to the surface of the head — see Fig. 8 a. Only after the velocity of sound in bone and in brain are multiplied by the right conversion factor do we obtain a true picture of the actual thickness of the bone and the true site of the ventricular outlines.

Brain / CSF / Brain

The CSF stratum is depicted here as being somewhat wider because the sound velocity in CSF is about 3% smaller than in brain tissue — see Fig. 8 b. This error is normally irrelevant for the diagnosis for it lies within the range of resolutions. However it can become important if the CSF layer becomes very large (at death, for example). If we see a hydrocephalus in a child for example, then the normal picture exaggerates the degree of hydrocephalus (see Fig. 9).

Therefore we can evaluate the echo reflections for such structures accurately only if we know what they are. It is absolutely necessary to know normal and pathological anatomy of the brain for the interpretation of the A-scan picture obtained during an examination.

Fortunately the errors in localization are normally small enough especially if we are adequately informed on the thickness compression of the bony layer.

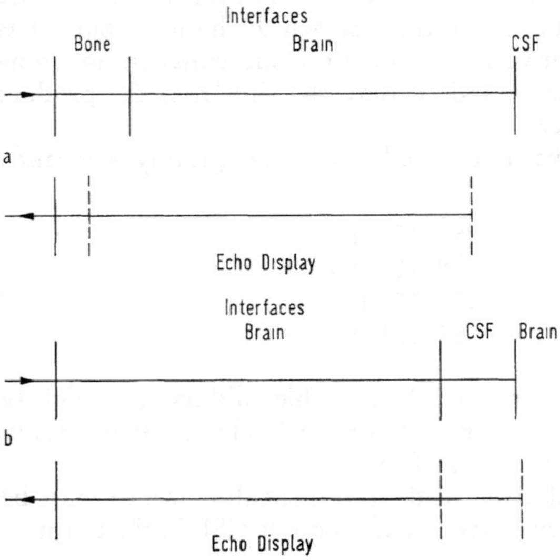

Fig. 8

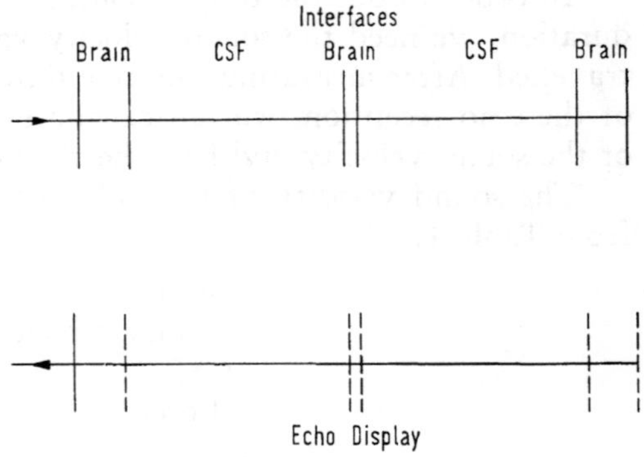

Fig. 9. Thickness distortion in a case of infantile hydrocephalus using the pulse echo technique

Fig. 8. Thickness distortion in the diagram with the pulse echo technique

2. Inaccuracy in Resolution

Boundary resolution of two closely adjoining layers in the direction of the beam is dependent on the geometric dimensions of the pulse used. In order to see how far the layers must be separated from each other, before they can be distinguished we will follow the path of a square pulse whose duration is 1 µs. For simplicity's sake, we will disregard energy changes at the penetration site and at the reflection site, as well as alterations in the sound velocity of different layers as much as possible. In Fig. 10 a, the pulse is approaching the edge of a layer and will reach it in 2 μs. In Fig. 10 b a time of 1 μs has passed since the interface was crossed. The pulse has thus penetrated the boundary of the layer completely and its echo (which we want to take a picture of for demonstration purposes) is just leaving it. If the pulse which is moving in the direction of radiation were to hit a second interface which corresponds to the longitudinal distance of the pulse in the medium, then we could expect Fig. 10 c after a period of 1 μs, which is the time necessary to cross the second interface. Here the echo from the first interface precedes the echo from the second interface by the length of one pulse. In order to just distinguish the two echoes from each other, the distance from the boundaries can be smaller. If the two interfaces are exactly one longitudinal distance of half a pulse width apart, then we

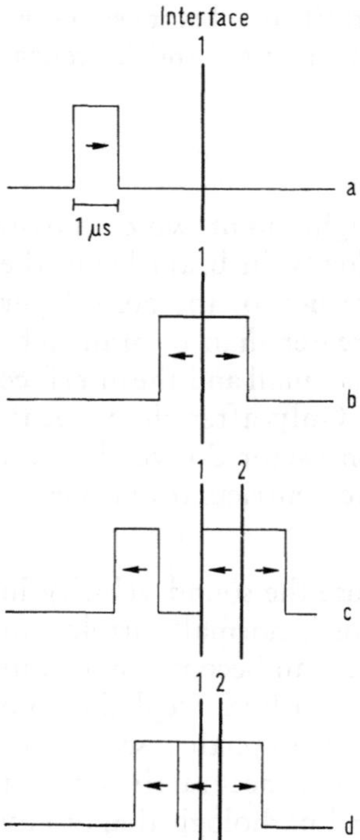

Fig. 10. Interface resolution of the pulse echo technique: pulse duration 1 μs. a) pulse 2 μs before reaching an interface; b) pulse and echo 1 μs after passing through the interface; c) pulse and 2 echoes, 1 μs after penetrating the interface 2, distance from the interface = length of pulse in medium; d) as in c), however the distance from the interface = $^1/_2$ the length of the pulse in the medium

find Fig. 10 d after a further interval of 1 μs following the crossing of the second interface. The echoes from the first and second interfaces follow each other immediately. This distance corresponds to the boundary resolution: it has the geometrical dimension of half a pulse length. In reality it is not a square pulse but rather a sinus wave. For this reason we do not receive the sharp sides of the square pulse and can resolve the two boundaries over an even smaller distance. If the pulse has a duration of 1 μs and we assume a sound velocity of $1.5 \cdot 10^3$ m/s, then the smallest distance between two surfaces which can still be distinguished from each other is 0.75 mm. If we must have details of the bone structure, we can expect a resolution of half a pulse length even there. Since the sound velocity in bone is 2.4 times higher, we will detect only structural layers which are more than 1.8 mm apart at a pulse duration of 1 μs. The greater pulse dimensions in bone with a constant pulse duration allows us to recognize 2.4 times fewer acoustic details in bone per unit length than in brain tissue. However, since a simultaneous thickness compression by the same factor occurs when the pulse-echo method is used, this decrease in the amount of information is not immediately recognizable.

D. Energy Balance of Echoes from Structures within the Skull

If we want to obtain a picture of the expected echo intensities originating from structures in the head, we can undertake an estimate with the help of some acoustic data. The result provides us with information on the necessary technical equipment, its limitations and on the frequencies which are optimal under the existing anatomical conditions.

The echoes depend on the reflection and penetrability of the interfaces and on the sound absorption of the various strata.

The differences in the specific acoustic impedance of two adjoining media are decisive for the reflection and penetrability of a plane sound wave which hits an interface vertically.

The impedance of a single medium is proportional to its sound velocity and its density. In order to estimate the echo intensities, the following specific acoustic impedance values may be used (see Tables 4 and 5):

Brain	$1.59 \cdot 10^6$ Ns/m³
CSF	$1.51 \cdot 10^6$ Ns/m³
Blood	$1.66 \cdot 10^6$ Ns/m³
Bone	
compact	$6.10 \cdot 10^6$ Ns/m³
porous	2.2 to $2.9 \cdot 10^6$ Ns/m³

On the basis of these values the following reflection quotients which express the relation between the reflected energy E_R and energy E_I incident to the interface may be calculated for typical stratifications within the head:

Blood/Brain	$6.9 \cdot 10^{-4}$
Bone/Blood	0.35
Bone/Brain	0.36
Brain/CSF	$9.7 \cdot 10^{-4}$

In order to simplify this survey it is feasible not to form the relationship E_R/E_I but rather to observe the sound energy level which we relate to the energy E_I arriving at the interface. This is

$$10 \log E_R/E_I \text{ (dB)}.$$

For typical stratifications, the following values have been found by which the sound level of the echo intensity is decreased in relation to the sound energy level arriving at the interface:

Brain/CSF	– 32 dB
Bone/Brain	– 4.5 dB
Bone/Blood	– 4.4 dB
Blood/Brain	– 30 dB

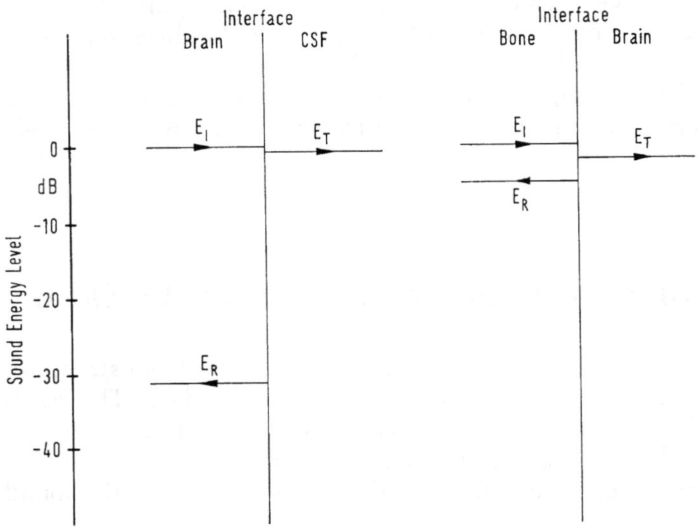

Fig. 11. Examples of the energy balance at typical interfaces in the skull

Instead of the ordinate with the % scale seen in Fig. 5, we now use the difference in sound level (dB) (Fig. 11).

Through the interface floods the energy E_T, which is the difference between the arriving energy and the echo energy:

$$E_T = E_I - E_R$$

$$\frac{E_T}{E_I} = 1 - \frac{E_R}{E_I}$$

Here too we can find the sound energy level

$$10 \log \frac{E_T}{E_I}$$

and then the transmitted sound level is decreased from the sound energy level arriving at the interface by the following values which have been calculated for the typical stratifications:

Brain/CSF	≈0 dB
Bone/Brain	– 1.9 dB
Bone/Blood	– 1.9 dB
Blood/Brain	≈0 dB

For the estimation of echo intensities, we must continue to consider the attenuation in the various media. The absorption occurring in bone is of unusually great importance here. Because we are assuming an intact skull and want to transmit the sound pulse through the outer bony layer, we must count on considerable attenuation before the inside of the skull is ever reached. This attenuation is partly dependent on the anatomy, i. e. on the bony stratum which varies from 3 to 8 mm in thickness and partly on the frequency used because the attenuation increases proportional to the square of the frequency.

Let us study the diagram in Fig. 12 whose coordinates are attenuation in dB and bone thickness in mm. The curves depicted here are valid for the frequencies 1, 2, 4 and 6 MHz which are commonly used in echo-encephalography. As the thickness of the bone increases, the attenuation also increases at twice the frequency, for example, the attenuation is four times as great while the bone density remains the same.

Effective attenuation which can be compensated by the amplifier of an echo-encephalograph is also indicated in this diagram. The limit of this effective attenuation is about 120 dB and is drawn here at 60 dB because the sound pulse travels back and forth through the thickness of the bone before it triggers a reception signal.

We see that the curves which indicate that the attenuation is a function of bone thickness at 4 and 6 MHz cross this limiting curve at a frequency of 4 MHz where the thickness is greater than 3 mm and at a frequency of 6 MHz already, if the thickness is 1.5 mm. No echo reception can therefore be expected because the returning signals are smaller than the noise of the amplifier itself.

Therefore the use of the pulse-echo method at high frequencies has chances only when the skull bone is relatively thin or in children where the bone is not fully developed yet. In old people with skulls of high calcium content and great thickness, we will have to use a frequency range of 1 to 2 MHz if we want to obtain intelligible echoes.

A frequency of 2 MHz is optimal as long as the thickness of the skull is not too great.

However, the attenuation of sound in the layers enclosed within the skull and the penetration loss at the interfaces have not been considered here.

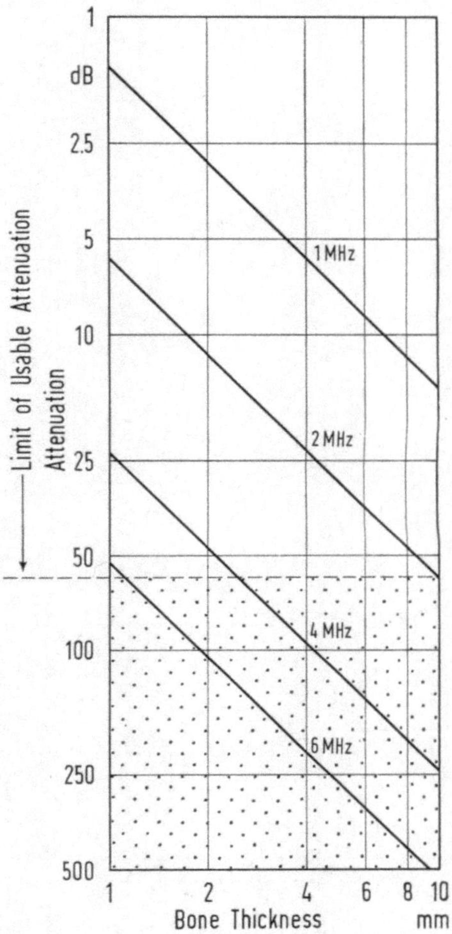

Fig. 12. Attenuation of sound as a function of bone thickness for the typical test frequencies 1, 2, 4 and 6 MHz

In order to obtain a mental picture of the energy balance of echoes from the structures within the head, we will observe a sound wave with a frequency of 1 MHz travelling successively through the strata of bone, brain, CSF, brain and bone (Fig. 13), and record the energy losses on the way. The choice of the frequency affects only sound attenuation, not however, the reflections and permeability of their strata. In this diagram, the interfaces are numbered and the echoes grouped with interfaces have the same marking: an echo from interface 2 (brain—CSF) also bears the number ②. The sound pulse begins on the left of the diagram. A bony layer 5 mm thick immediately attenuates the energy travelling through it. The echo originating from interface 1 (bone—brain) — indicated as ① — reaches the receiver at 20 dB below the sender level. On the other hand, the wave which has passed through interface 1 (bone—brain) is hardly attenuated at all and reaches the border of the ventricle after a distance of 5 cm. This interface reflects only a very small part — ②. The amount of energy returning from there reaches the surface of the skull having lost about 60 dB. The echo ③ from interface 3 (CSF—brain) has practically the same energy as the echo at interface 2 because the attenuation within the CSF over a distance of 1.5 cm, is negligible.

The sound beam continues its way through a layer of brain 3.5 cm in depth and is reflected at the bone (interface 4). The amount of sound reflected from there — ④ —reaches the receiver having lost barely 40 dB.

Finally there is total reflection at the external bony layer (interface 5), which triggers an echo — ⑤. This is attenuated by 50 dB at the receiving end when compared to the original energy. Assuming this stratification, the following echo intensities appear in time sequence

from 1	– 20 dB	from 4	– 37 dB
from 2	– 58 dB	from 5	– 49 dB
from 3	– 58 dB		

Naturally the energy returning to the generator results in multiple echoes as well. In the picture the first ones are indicated by broken lines: the first one which originated from the bone/brain interface and which is indicated with marker ①', reaches the receiver attenuated by 40 dB; in the time sequence it appears immediately after the echo ①. The echo complex originated from the interface brain/CSF, marked by ②' is notably weaker than all the others already. It is depicted between ② and ③ in the time sequence but its intensity is hardly detectable.

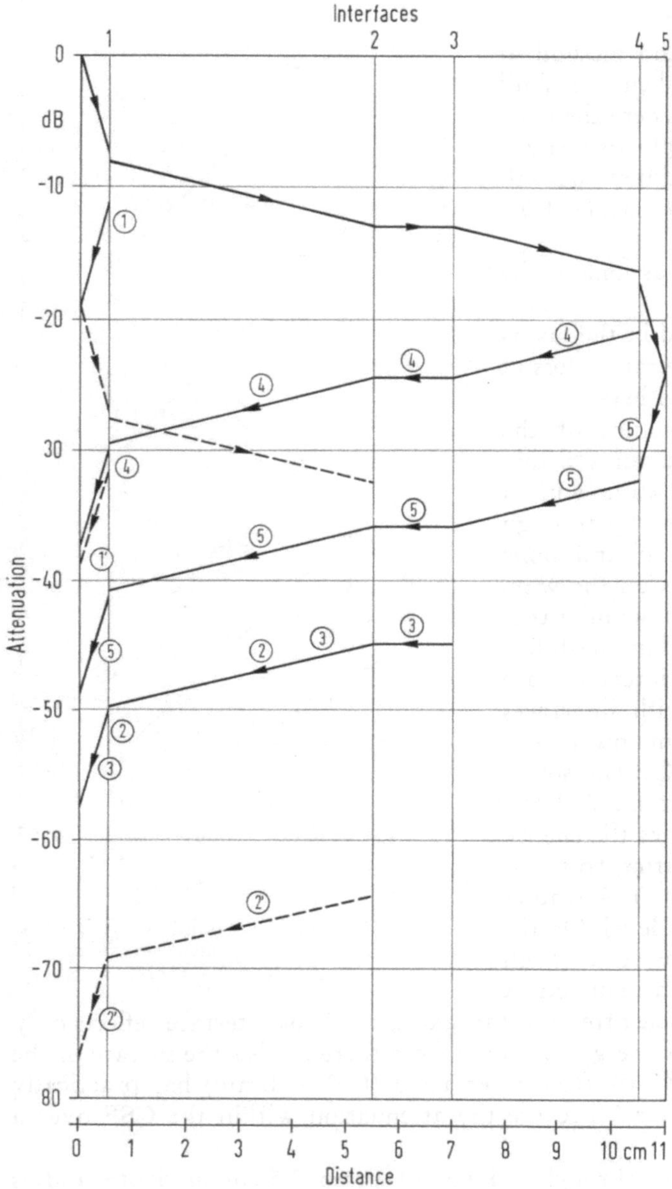

Fig. 13

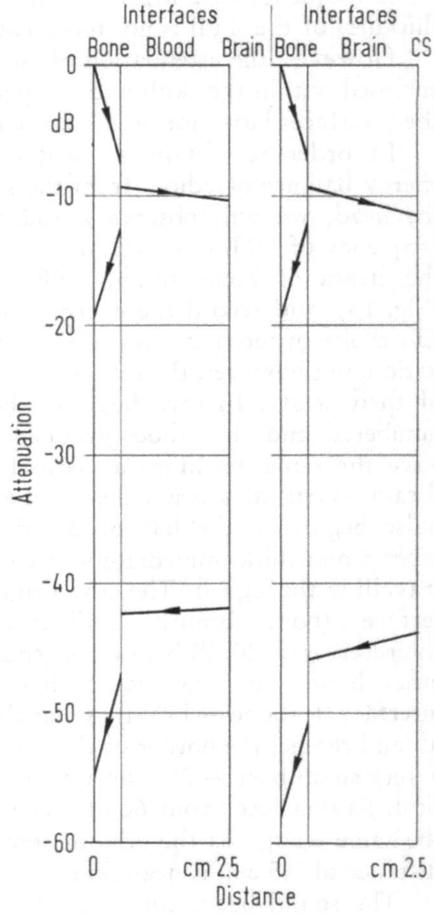

Fig. 14

Fig. 13. Energy balance of the echoes (frequency 1 MHz) with typical stratification of the structures within the skull

Fig. 14. Energy balance of the echoes (frequency 1 MHz) with typical stratification hematoma and brain/CSF

It is obvious that we are dealing with effective echo levels in this example which can lie as much as 60 dB below the original energy level. Since the amplifier can compensate attenuations up to 120 dB, this frequency delivers a high signal to noise ratio. At higher frequencies the echo intensities are smaller because absorption is increased. The use of a higher frequency can be successful only if the echo signals rise above the noise of the amplifier.

The values which actually appear can be even lower if the coupling between probe and surface of the head is faulty and if the reflecting surface assumed for this calculation is not smooth so that the echo returns unbundled or scattered. In children on the other hand, whose skull bones are not yet solid, the sound attenuation is considerably less.

Finally we want to estimate the echo energy in the presence of a hematoma: the stratification here is bone/blood/brain and we compare it with the stratification bone/brain/CSF.

Fig. 14 shows the details: ultrasound attenuation over the thickness of the layer. A comparison between the two adjoining stratifications shows that the echo intensity from the interface blood/brain, i. e. the depth of the hematoma at a frequency of 1 MHz is greater than from the stratification brain/CSF, the spatial geometry being the same. The differences become even greater if we choose a higher test frequency. Here then it is possible to use the level of intensity for diagnostic purposes.

E. Tissue Reactions to Ultrasonic Radiation

Ever since the introduction of ultrasound to medicine for therapeutic purposes, many investigations of the *damaging effects* of ultrasound have been made. These effects are of three types: thermal (due to the absorption of ultrasound), mechanical (due to cavitation) and chemical. The therapeutic ultrasonic apparatus works at a maximal energy of 3 W/cm². At this intensity no tissue damage occurs under normal conditions (see also p. 13).

LYNN and PUTNAM (1944) carried out the first experimental investigations and attempted to use energy doses sufficient to produce demonstrable damage. The authors focussed a beam with a maximum dose on the cerebrum, cerebellum and spinal cord of 3 dogs, 30 cats and 4 monkeys for periods from 5 to 15 minutes producing partly irreversible damage in these experimental animals. All these cases showed clinical defects after a more or less prolonged period of time following the application of the ultrasonic beam. In some of these cases, the clinical changes returned to normal within a few days. On autopsy, wedge-shaped areas of necrosis in which the peak of the wedge pointed towards the white matter, were found in every case. Here there was also total necrosis of the parenchyma. The nerve cells had become necrotic but the glia were still viable.

BARTH and BÜLOW (1949) undertook a series of animals experiments on young dogs using continuous sound at a frequency of 0.8 MHz. Growing bone was exposed to an intensity of 2.5 W/cm² for time periods up to 15 minutes. Aside from some swelling in the soft tissues, no tissue lesion was observed. A distinct pain reaction occurred at an intensity of 3.25 W/cm². A radiation period of 5 minutes led to slight tissue damage only, while a duration of 15 minutes resulted in massive bone damage. The authors emphasized that the pain reaction appears at a dose which does not lead to tissue damage yet. The tissue destruction threshold for continuous sound lies at 2.5 W/cm² for 5 minutes according to these authors.

BRÜSCHKE (1955) undertook similar investigations with rats, exposing organs to prolonged ultrasonic radiation generally regarded as being particularly sensitive to ultrasound namely testes, ovaries and pregnant uteri. He used continuous sound at a frequency of 1 MHz and an intensity of 3.5 W/cm² for an exposure time of 5 minutes; this dose was repeated in equal strength 5 to 6 times every second day. In several cases the radiation time was extended to 15 minutes. The function of the above-mentioned organs was not to be interrupted. A spotty degeneration was observed in the testes without unequivocal limitation of function. A pregnancy was damaged only when unusually high doses were used.

PETERS (1949) carried out extensive investigations regarding damage to the central nervous system by ultrasound. The experimental animals were guinea pigs. The frequency used was 1 MHz. Ultrasound was applied either by moving the probe back and forth on an oily film in the sagittal area of the head or by holding the animal's head in the cone of a water stream. If the oily film was used, the intensity ranged from 4 to 6 W/cm². If contact was brought about by a stream of water, it was 3 to 6 W/cm². The duration of radiation was 5 to 12.5 minutes. The most minor changes produced in this manner consisted of a dilatation of the subarachnoid vessels and cortical capillaries in a circumscribed area corresponding to the irradiated area. At the next level of change, subarachnoid hemorrhages leading to a tamponade of the basal cisterns were observed in some cases. All animals radiated with an intensity of more than 4.8 W/cm² for a period of 5 to 12.5 minutes, showed distinct morphological changes in the brain tissue itself. Depending on the survival time, the changes consisted of hemorrhages, necrosis of brain tissue or early scarring. The necrotic lesions were always wedge-shaped. PETERS found a general correlation between the brain tissue lesions caused by the effects of ultrasound and those of head injuries; he suggests a similar mechanism for the development of both types of lesion, „Wir können uns die besondere Form der traumatischen Herde nur durch den Verlauf der Stoßwellen, also rein mechanisch bedingt, vorstellen. Dafür spricht auch das momentane Eintreten gewisser Gewebsläsionen (Gefäßzerreißungen) noch während der Gewalteinwirkung. Eine gleiche mechanische Wirkungsweise nehmen auch zahlreiche Autoren, vor allem Physiker (PÄTZOLD, POHLMAN, u. a.) beim Ultraschall an. Die beiden Autoren haben nachgewiesen, daß bei einer Schalldichte von 4 W/cm² durch Schwingung im menschlichen Gewebe infolge der nur geringen Kompressibilität erhebliche Druck- und Zugwirkungen auftreten“ (PETERS).

BALLANTINE, BOLT, HUETER and LUDWIG (1951) exposed two dogs to continuous sound at a frequency of 2.4 MHz. One dog was radiated 11.5 minutes with an intensity of 3 W/cm², the other for 15 minutes at an intensity of 1.5 W/cm². Neither of the two animals showed brain tissue lesions on histological examination. Similarly a cat was exposed to a radiation intensity of 15 W/cm² for 15 minutes using a frequency of 0.8 MHz while an EEG was being performed simultaneously. No EEG changes were observed. Two human subjects were also exposed to continuous sound applied directly to the head at a frequency of 0.8 MHz and at an intensity of 2 W/cm² for 9 seconds with simultaneous EEG control. Although no alterations could be detected in the EEG, both subjects complained of slight headache in the main area of radiation.

Further investigations include those by ADAM (1949), PEZOLD (1952), WACHSMANN (1952), HUETER (1954), FRENCH, WILD and REID (1951, 1952) as well as those by BALDES, HERRICK and STROEBEL (1958).

The only investigation which is especially devoted to the question of ultrasonic tissue damage through the energy emitted by echo-encephalography was performed by GARG and TAYLOR (1967). The authors employed a Siemens Echo-encephalograph USM 1 und exposed 6 patients (children and adults) to an ultrasound dose of 1 mW/cm² for over an hour. The clinical, electroencephalographic, biochemical and histopathological findings fail to reveal any traces of tissue damage.

The results of the investigations described above indicate that continuous sound of an intensity of several watts for a period of several minutes must be applied in order to produce tissue lesions. Interposed bone raises the threshold of tissue damage.

However, the intensities of ultrasound employed for diagnosis in the form of echo-encephalography is several orders lower than the values given above. The maximum ultrasound intensity with a 24 mm probe is $1 \cdot 10^{-3}$ W/cm² and with a 10 mm probe, $5 \cdot 10^{-3}$ W/cm². Furthermore the absorption of ultrasound in the bone decreases the intensity of ultrasound entering the brain by another order of magnitude so that only a fraction of the $1 \cdot 10^{-3}$ W/cm² is effective, i. e. practically less than one ten-thousandths of the experimentally determined threshold of tissue destruction. Therefore it may be assumed with certainty *that tissue damage is not possible even with prolonged ultrasonic investigations* (see also GORDON, 1959; JEPPSSON, 1961; TAYLOR et al., 1961). Although the absorption of ultrasound in the thinner and less calcified skull of children is distinctly less, the intensity of ultrasound is considerably lower here, so that sufficiently high echoes can still be determined.

F. The Use of Ultrasound in Medical Diagnostics

The diagnostic use of ultrasound in medicine originates with to K. TH. DUSSIK. In the year 1937 the brothers DUSSIK performed experiments in order to obtain a picture of the cerebral ventricles by depicting absorption. From 1941 to 1947 they developed this into a technique, the so-called "hyperphonography of the brain" from this. This was the *through-*

transmission method with which they attempted to demonstrate the cerebral ventricles using grid-shaped scanning of the skull from the side and from the front (Fig. 15). This method proceeded from the assumption that ultrasound is only slightly attenuated in CSF and damped much more in brain tissue. The differences in attenuation obtained in this manner were interpreted by the DUSSIK brothers to be pictures of the ventricles. In 1952 DUSSIK reported his results with this procedure for the last time.

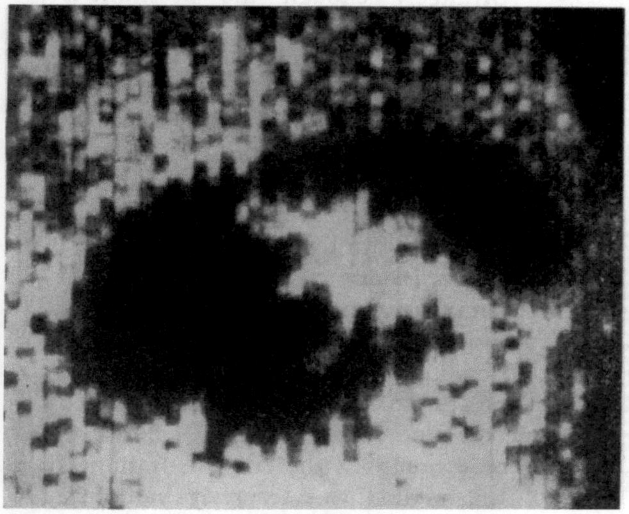

Fig. 15

Fig. 16

Fig. 15. Hyperphonogram according to DUSSIK (from K. TH. DUSSIK in: Der Ultraschall in der Medizin, Zürich: Hirzel, 1949).

Fig. 16. Ultrasonic picture of the human skull. The picture shows that the outline which DUSSIK interpreted as being the lateral ventricles caused by differences in sound attenuation is in reality due to variations in the thickness of the bony skull (from GÜTTNER, FIEDLER and PÄTZOLD, Acustica 2, 148—156, 1958)

In the United States of America, BALLANTINE, BOLT, HUETER, LUDWIG and others have carried out detailed investigations with the DUSSIK procedure since 1950. Initially they arrived at the conclusion that a demonstration of the ventricular system by the transmission method was possible even without pneumoencephalography. In the opinion of these authors, the attenuation differences between the ventricular system and surrounding brain tissue was enough to be registered. Certain types of brain tumors were also said to be demonstrable with this method. In the year 1952, HUETER reported on ultrasonic investigations in a sagittal direction where ventricular displacements were found which corresponded very well to the operative site of the tumor. According to this author the ultrasonograms in an AP direction were easier to interpret than the lateral pictures because of their symmetry.

On testing DUSSIK's ultrasonic method for demonstrating ventricles in live brain, GÜTTNER, FIEDLER and PÄTZOLD (1952) arrived at the conclusion that variations in the thickness of the skull bone alone are sufficient to produce a markedly structured picture (Fig. 16). Thus the attenuations in ultrasound caused by the ventricles are no longer recognizable making a faithful reproduction of the ventricles by means of the DUSSIK shadow method impossible. Of particular importance in this misinterpretation is the fact that the skull bone is thinner in the temporal area than elsewhere and that this simulates the picture of the lateral ventricles. Initially an attempt was made to eliminate this source of error electronically, leaving the actual attenuation differences between CSF spaces and brain tissue to be registered. However, the results were unsatisfactory and in 1954 GÜTTNER's objections finally led to the cessation of the experiments with through-transmission sounding carried out under American researchers such as BALLANTINE, HUETER and BOLT. The most recent evidence for the validity of GÜTTNER's objections was published by WHITE et al. (1968) who investigated the absorption of ultrasound by the bony skull as part of their experimental work with two-dimensional echo-encephalography. The pictures obtained in this manner are amazingly similar to the lateral pneumoencephalograms (Fig. 17).

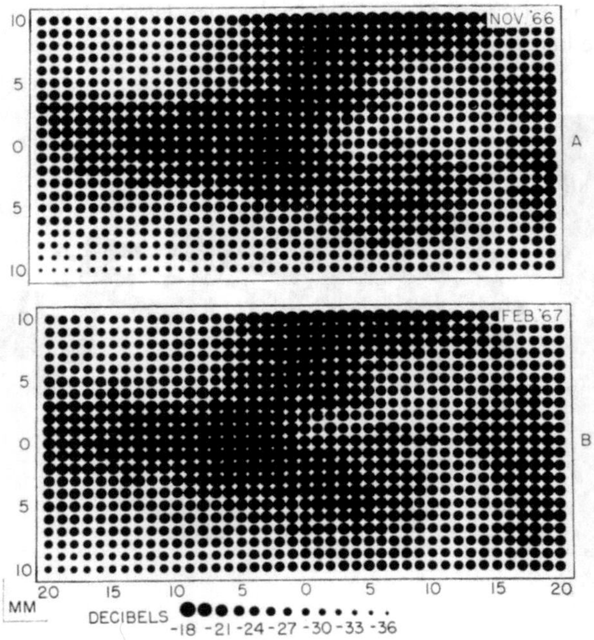

Fig. 17. Ultrasound attenuation by the temporal bone using punctate scanning. In this way, a picture is produced which bears a remarkable resemblance to a lateral pneumoencephalogram (from WHITE et al., Proceedings in Echo-encephalography, New York: Springer 1968)

Strangely enough the possibilities of the *ultrasound-echo-method* were regarded as unpromising for a long time. Although the principles of echo sounding in this connection were discussed by SCHLIEPHAKE (1940), GOHR and WEDEKIND (1940), diagnostic results did not appear. The fact that the acoustic impedance of brain tissue and CSF differed very little from each other and that only minimal quantities are reflected seemed to be points against the applicability of this method. In addition, the complicated shape and layer relations in the skull led one to expect such involved conditions of reflection that the clinical usefulness of the method was seriously in doubt. KEIDEL (1947) did not deny the clinical applicability of echo sounding completely, but he believed that the severe loss of energy and the corresponding technical expenditure were hardly worthwhile in view of the diagnostic implications. GÜTTNER et al. (1952) also felt that the prospects of applying the echo-pulse-method on the intact skull were very slim, since echoes from the admissable sizes of transducers would not rise above the noise level. Finally in 1955 the *United States Atomic Energy Commission* published a report in which they arrived at the conclusion that the ultrasound-echo-method was unsuitable for the detection of tumors in the human head and recommended testing the method on other parts of the body.

The difficulties which were suspected only in part can naturally be circumvented by applying the echo-pulse-method to the operatively exposed brain where the bone no longer presents a barrier (FRENCH, WILD, NEAL, REID, 1950, 1952; KIKUCHI, UCHIDA et al., 1957). The extent of such investigations however, is extremely limited from the outset and the diagnostic results affect only a few cases. Today this method is used primarily by Japanese and American neurosurgeons for the localization of deeply situated tumors (TANAKA et al., 1961—1967; WALKER and UEMATSU, 1967; DYCK et al., 1967) or of massive hypertensive hemorrhages (MITSUNO et al., 1966).

The credit for developing a method of ultrasonic investigation on the intact skull goes to LEKSELL; his method brought useful diagnostic results in cases of space-occupying intracranial lesions for the first time. In the same year as the *US Atomic Energy Commission* declared that the echo-pulse-method on the skull was worthless, LEKSELL (1955/56) published his first paper on "*Echo-Encephalography*" whereby it was possible to detect complications of head injuries. He was able to show that certain echoes originating from the midline structures of the skull regularly appeared on lateral sounding of the skull. He has been using this "*midline echo*" to determine on which side posttraumatic intracranial hemorrhages are located since 1953. Further publications by LEKSELL (1958) and his collaborator JEPPSSON (1960, 1961) concerned the origin of the midline echo and clinical results. The findings reported here were confirmed by a number of authors, some of whom reported further diagnostic advances in echo-encephalography (de Vlieger, Ridder, Grandia, Ter Braak, Crezée, 1959, 1961; Gordon, 1959; Lithander, 1960, 1961; Jefferson, 1959, 1961; Taylor, Newell and Karvounis, 1961; Ford and Ambrose, 1963; Dreese and Netsky, 1963; Schiefer, Kazner, Brückner, Kunze, 1963, 1964, 1965, 1966). In the meantime the method has been adopted all over the world and already belongs to the routine diagnostic procedures.

Besides one-dimensional echo-encephalography, *two-dimensional echo-encephalography* was also tested on the skull. However, this method is still in the experimental stage and the results obtained thus far do not diminish the value of the simple time-amplitude method in any way (see Appendix).

The pulse-echo-technique has taken a place in the diagnostic arsenal of other fields of medicine as well. Wild and Reid (1952, 1954) as well as Howry, Stott and Bliss (1954) were the first to draw ultrasound into the diagnosis of breast tumors. The introduction of so-called ultrasonic cardiography by Edler and Hertz (1954) represents a step forward in the diagnosis of cardiac lesions, particularly in the recognition and evaluation of mitral stenosis. Other papers on these special applications were published by Effert, Erkens and Grosse-Brockhoff (1957), Effert and Domanig (1959) as well as by Schmitt and Braun (1960). In recent years the method has obviously assumed increasing importance. The American Journal of Cardiology, for example, devoted one entire instalment (Vol. 19, No. 1, 1967) to the ultrasonic diagnosis of the heart just a short time ago. The most recent progress was achieved in the evaluation of artificial cardiac valves whose function may be observed through ultrasound (Gimenez, Winters, Davila, Connell and Klein, 1965).

Pericardial effusions may also be diagnosed by means of echo-encephalography (Edler, 1955; Hertz, 1965; Effert et al., 1965; Feigenbaum et al., 1965, 1966; Soulen et al., 1966; Rothman et al., 1967). In cases of aortic aneurysms, it is possible to determine the thickness of the intravascular thrombus and the amount of lumen as well as the diameter of the entire vessel (Segal, Likoff, Asperger and Kingsley, 1966).

Investigations on the abdomen were first carried out by Donald, MacVicar and Brown in 1958. In obstetrics, ultrasonic diagnosis acquired great value in the determination of fetal head size (Thompson, Holmes, Gottesfeld and Taylor, 1964, 1965). Not long ago Hofmann, Holländer and Weiser (1966) as well as Donald and Abdulla (1967) showed very impressive ultrasonotomograms of ovarian tumors, cysts, normal fetal heads and multiple pregnancies. Even pathological alterations of the fetus' skull may be recognized in ultrasonic tomograms (Hofmann, Holländer and Weiser, 1967). The localization of the placenta by means of ultrasound is also assuming greater importance (Kratochwil, 1966; Gottesfeld et al., 1966; Donald and Abdulla, 1967; Hofmann et al., 1967.

Just recently Kratochwil (1968) summarized the diagnostic possibilities of ultrasound in obstetrics and gynecology in a monograph.

In the diagnostic investigation of eye diseases, ultrasound was first used by Mundt and Hughes (1956) and by Oksala and Lehtinen (1957). The eye is particularly suitable for ultrasonic examinations because its functional parts are separated from each other by distinct interfaces which are in themselves however, relatively homogeneous. Intraocular foreign bodies and retinal detachments are detected especially well with this method (echo-ophthalmography). Tumors can also be diagnosed. Futhermore the length of the ocular axis can be measured (see Oksala and Lehtinen, 1957—1962; Baum and Greenwood, 1958, 1960; Nover and Stallkamp, 1960, 1962; Löpping, 1962; Nover, 1963 and Buschmann, 1964 among others). Buschmann (1966) published an introduction to ultrasound diagnosis in ophthalmology where the method is described precisely and the diagnostic possibilities of the echo technique are discussed in detail.

The pulse echo technique is also an important diagnostic tool in detecting orbital lesions (Baum and Greenwood, 1960). Ultrasonotomography of the orbit enables us to discover and localize retrobulbar tumors with greater accuracy than any other method (Ossoinig, 1968).

Aside from these, Japanese and American authors have been especially concerned with the ultrasonic diagnosis of diseases of the liver, kidneys, thyroid and prostate gland. Renal calculi are much easier to detect by means of ultrasound on the exposed kidneys, thereby simplifying the operation (SCHLEGEL, DIGGDON and CUELLAR, 1961). Cysts, tumors and cirrhotic processes of the liver may also be localized exactly. The investigation of the spleen by ultrasound has been undertaken with success as well (LEHMAN, EVANS and BRADY, 1966). The present status of ultrasonic diagnostic techniques in internal medicine has been described by HOLMES in 1967.

G. The Pulse-Echo Method Applied to the Skull

1. Principles of the Method

We can understand the principle of this procedure if we imagine a needle passing from one side of the skull through to the other side; the resistance which is felt when the needle penetrates the tissue surface is indicated on it by a mark corresponding to the amount of resistance. When the needle is pulled out, it is a one-dimensional report on the state of the path which the needle has followed through the head. This principle is realized when a sound pulse instead of a needle is sent into the head as a plane wave packet. If the pulse hits interfaces these offer resistance to the sound pulse and partly reflect it. The returning sound echoes are conducted to an indicator apparatus which converts the time differences between sender pulse and echo received so that they appear on a longitudinal axis, thereby creating a one-dimensional picture (A-scan).

2. Construction and Mechanism of the Pulse-Echo Apparatus

A modern echo-encephalograph looks practically the same as an oscillograph which is familiar to us from other techniques of measurement (Fig. 18). All construction units which are necessary for this procedure are included in it. The probes are coupled with the apparatus by means of a cable and can be exchanged in order to achieve optimal adjustment to the diagnostic investigation.

A probe with an electroacoustic transducer which is stimulated electrically, serves to generate the sound pulse. The same probe is used for receiving the echo after the sender pulse has been generated. The intensity of the echo and the distance from the probe to the different boundaries within the skull and back is measured. The result of this measurement is depicted as a curve on the screen of the cathode ray tube, where the abscissa is the distance and the ordinate the amplitude of the echo. The distance on the time axis can be calibrated in longitudinal units corresponding to the velocity of sound in brain, so that the depth of the interface can be read directly from the screen. In order to facilitate the observation the measurement is rhythmically repeated until a stationary picture appears on the screen, the point at which the sender pulse is produced being on the left edge of the screen. The following requirements must be met here: the repetition rate must be so slow that the last expected echo for one specific investigation arrives with certainty before the subsequent transducer pulse does. The transmitter pulse and the lateral deviation on the screen must be synchronized. In order to determine the depth of the interface, the lateral deviation on the screen must remain constant.

The construction of such an pulse-echo apparatus will be illustrated by means of a circuit diagram (Fig. 19).

Two generators are necessary for the production of the sound pulse. One produces the high frequency oscillation of the ultrasound pulse and the other the low frequency repetition rate.

The high frequency generator [1] can produce a series of different frequencies from 1 to 6 MHz which are necessary for an optimal adjustment of the ultrasonic frequency to the absorption and to the different pathological disturbances within the head. The energy for this can be regulated and passed on to the probe [15]. It can be increased until the middle acoustic output reaches an order of 10^{-3} W/cm^2. The low frequency repetition rate which is produced in another generator [2] steers the high frequency pulse. The time between two pulses at the commonly used repetition rate of 500 Hz is 2 ms. This corresponds to a distance of 2.8 m in brain tissue, i. e. an investigation depth of 1.4 m. The echoes of a transmitter which develop at an interface have entirely disappeared at this repetition rate before the next transmitter pulse is sent out.

The low frequency generator [2] also furnishes the electrical voltage

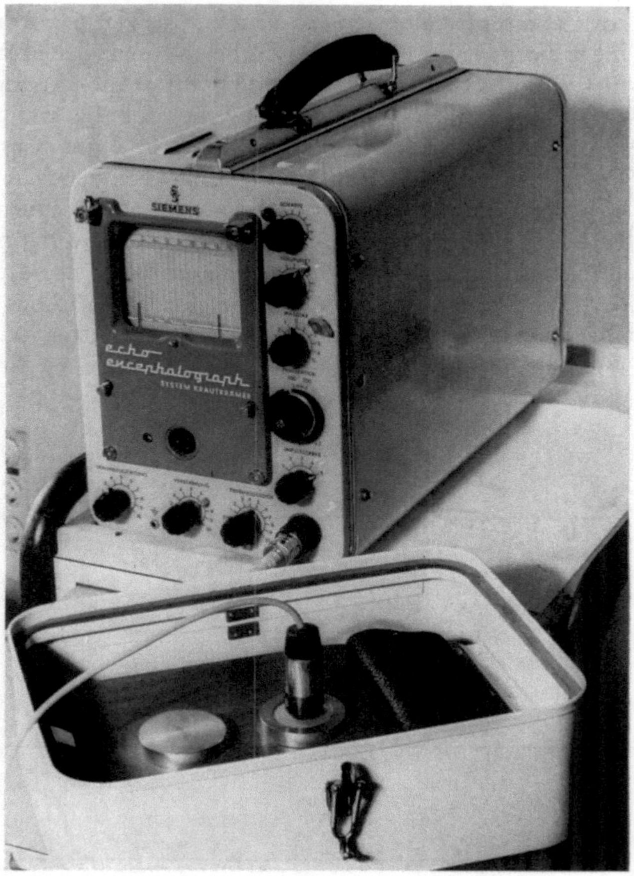

Fig. 18. Portable ultrasound equipment (echo-encephalograph System Krautkrämer, Siemens AG). In the foreground, the probe has been applied to the phantom in order to calibrate the echo-encephalograph

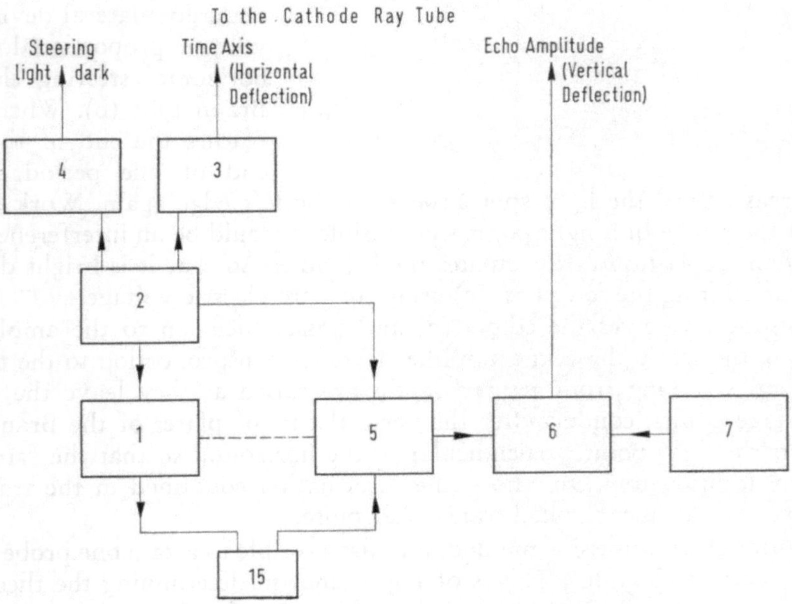

Fig. 19. Circuit diagram of a pulse echo apparatus

for the lateral deviation [3] on the screen of the Braun tube which is due to the distance. In the same way, the electron ray is bright during the receptor phase and dark during the interval in which it returns to the starting point at the left edge of the screen [4].

The sound echo is picked up by the probe [15] and passed on to the amplifier [5] as an electric pulse. This apparatus is a compromise of 2 qualities: broad beam width and amplifier noise. For if the amplifier transmits a broad frequency it can be used for all frequencies and the pulses which are picked up undergo almost no linear distortion. However, the disadvantage here is that the echo signals will be masked by the noise of the amplifier thus narrowing the amplifier dynamics. Therefore the amplifier is adjusted to a frequency range from 0.5 to 6 MHz. The amplification can be regulated. It has been found useful to let the sensitivity increase logarithmically after the transmitter pulse has been generated. Thereby it is possible to achieve depth resolution (sensitivity time control) based on the absorption of attenuated echo pulses from greater tissue depths. The repetition rate generator is responsible for initiating the rise sensitivity as well.

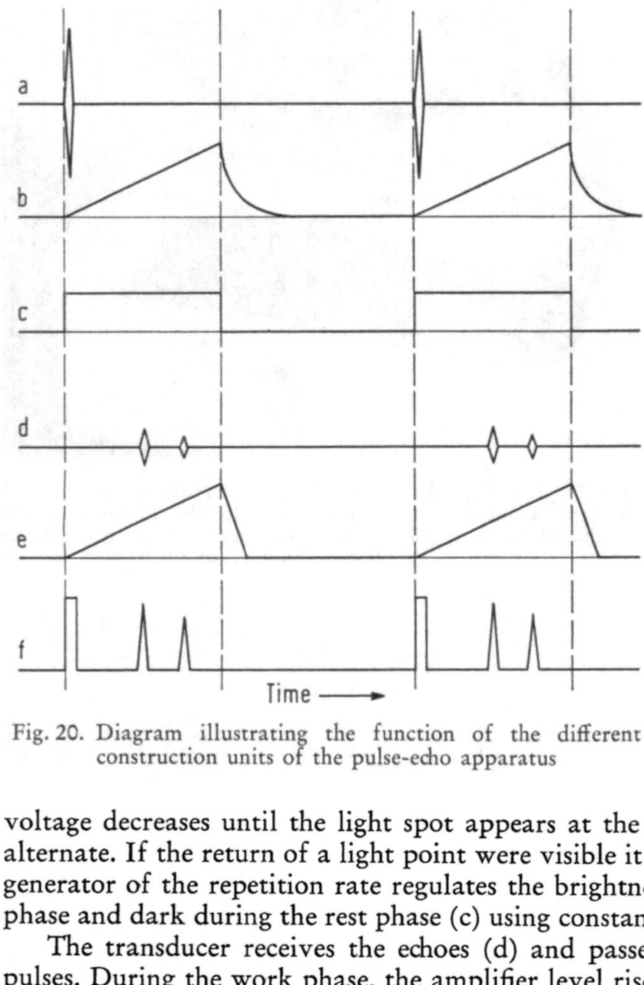

Fig. 20. Diagram illustrating the function of the different construction units of the pulse-echo apparatus

It is advisable to rectify all echo pulses [6]. On the screen of the apparatus the transmitter pulse and the echo have the shape of peaks. The distance between the peaks corresponds to the distance the ultrasound has travelled through the brain.

Echo suppression [7] serves to eliminate smaller interfering echoes.

The function of the component parts of the echo-pulse apparatus are shown as a time diagram in Fig. 20.

The transmitter pulse recurs in the rhythm of the repetition rate (a). Just before the generation of the transmitter pulse, the electron beam which appears on the screen as a light spot, undergoes lateral deviation through a voltage proportional to time at the horizontal steering electrodes of the Braun tube (b). When the light spot reaches the end of the screen at the end of one period of observation, voltage decreases until the light spot appears at the left edge again. Work and rest phases alternate. If the return of a light point were visible it would be an interference, therefore the generator of the repetition rate regulates the brightness so that it is bright during the work phase and dark during the rest phase (c) using constant electric voltage.

The transducer receives the echoes (d) and passes them on to the amplifier as electric pulses. During the work phase, the amplifier level rises in proportion to the time (e), so that the echo energies coming from greater depths are raised as they leave the apparatus. The rectified voltage is then conducted to the vertical pair of plates of the Braun tube (f). The latter deflects the light point perpendicular to the horizontal so that the familiar peaks are produced. In addition to these echoes, the information contained in the trace includes the srongly suppressed or uncontrolled transmitter pulse.

For the through-transmission method, it is also possible to attach one probe as a sender and the other as receiver (Fig. 26). This is of importance in determining the theoretical midline of the head.

3. Probes

The electroacoustic transducers built into the probes work according to the piezoelectric principle. The input of electric energy produces power which transforms the substance used in the rhythm of the induced oscillation. This mechanical deformation triggers the acoustic

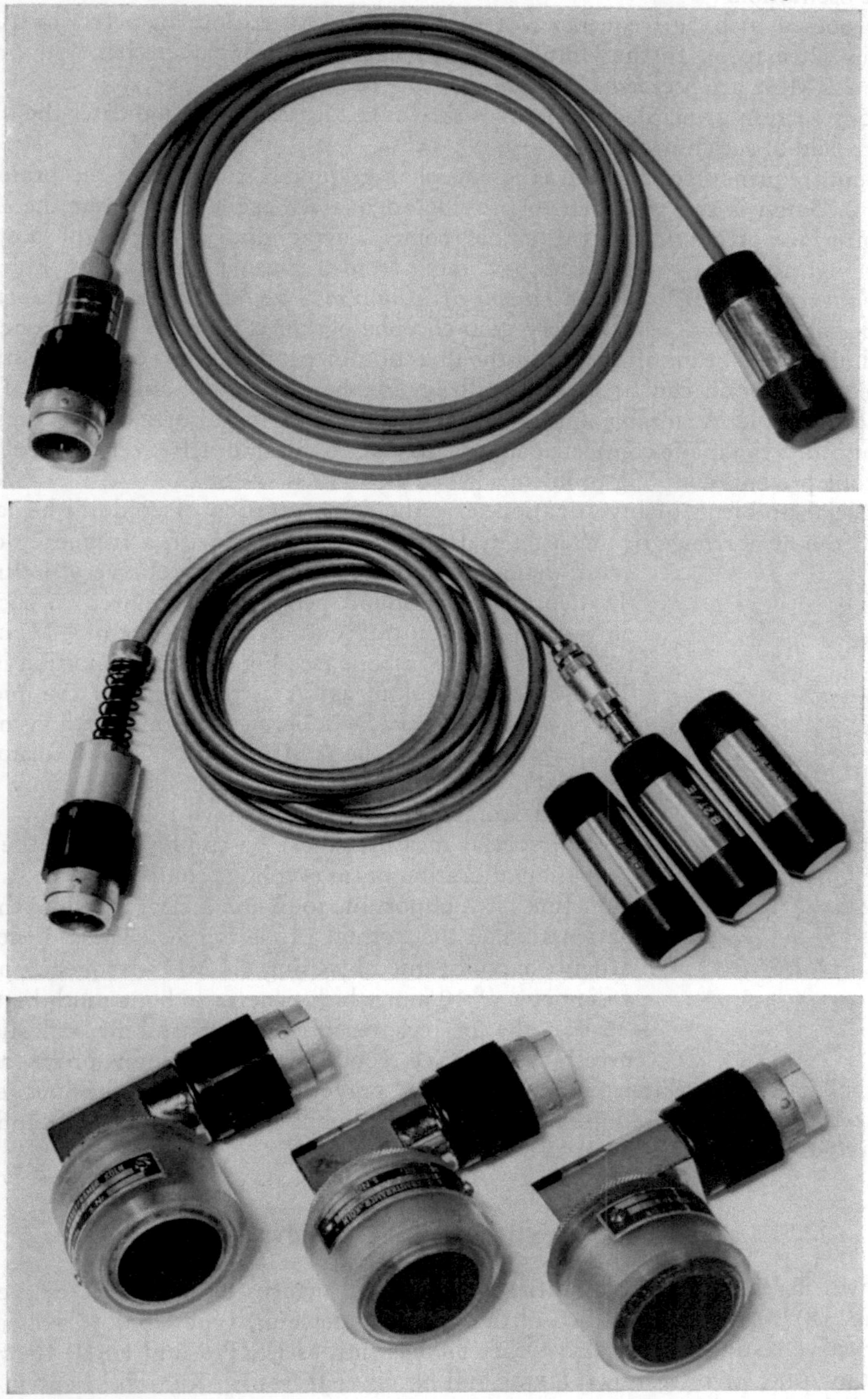

Fig. 21. a) Universal probe with an electroacoustic transducer 15 mm in diameter with a frequency of 2 MHz. b) Interchangeable probes with frequencies of 1, 2 and 4 MHz and a diameter of 15 mm. c) Interchangeable probes with frequencies of 1, 2 and 4 MHz and a diameter of 24 mm

radiation of ultrasound. The same transducers work reciprocally, i. e. they give off electric energy under the influence of mechanical forces.

Barium titanate has proved to be the material of choice. The construction of a probe is adjusted to its use: it can be damped so much that the ultrasonic pulses practically consist of a single oscillation.

The probes used have frequencies of 1, 2, 4 and 6 MHz and diameters of 10, 15 and 24 mm (Fig. 20 a to c). In the United States, England and Japan probes with frequencies of 2.25 or 2.5 MHz are preferred.

The construction principle of a probe is shown in Fig. 22. Fig. 4 indicates the dimensions of the near field of such a probe.

The best frequency for routine echo-encephalography is 2 MHz. In the brain, a wave length of 0.75 mm is short enough to provide adequately exact data about the site of the various interfaces from which the echoes come. Furthermore the area of amplification is large enough to pick up all reflections coming from the inside of the adult skull. *A probe with a frequency of 2 MHz and a diameter of 15 mm may be regarded as the universal probe* because it is able to solve nearly every echo-encephalographic problem. This probe produces a rectified ultrasonic beam of 80 mm in the direction of propagation, so that the structure in the centre of the brain can be easily localized. Furthermore the manipulation of the light probe is very simple. A suitable application point in the temporal area where the surface of the 15 mm probe can make complete contact, may be found in nearly every patient while the 24 mm probe presents coupling problems.

For special problems of investigation, e. g. the demonstration of epidural hematomas on the side of the hemorrhage or infantile hydrocephalus, a probe with a frequency of 4 MHz and a diameter of 10 mm has proved to be very useful, because the depth of ultrasound penetration required in such cases is only slight. Since ultrasound of a frequency of 4 MHz is attenuated much more in bone (see Fig. 12) this condition is fulfilled. The inside of the skull as far as the origin of the midline echo can be evaluated very well, because the initial echo is much shorter here than with the 2 MHz probe and a distortion-free reproduction of the echoes on the oscillographic screen is obtained from the outset. To achieve this with low frequency probes, special measures must be employed with the amplifier such as an alteration in time-voltage control or echo suppression. Sometimes it is impossible to obtain a clear midline echo in older patients when the normal probe is used. These difficulties can often be circumvented by using a 1 MHz frequency probe with a diameter of 10 mm which penetrates bone much better.

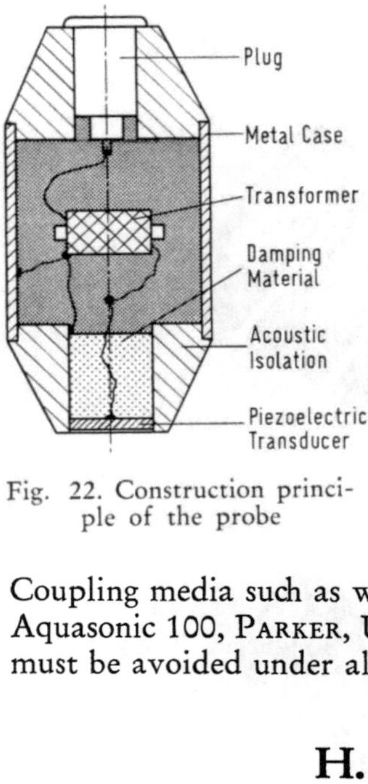

Fig. 22. Construction principle of the probe

For the perfect reception of ultrasonic reflections, it is necessary to have good contact between probe and scalp. Coupling media such as water, paraffin, vaseline, polyethylene glycol or various jellies (e. g. Aquasonic 100, PARKER, USA) are best for this. Air bubbles between the probe and the scalp must be avoided under all circumstances.

H. The Technique of Investigation

Our own investigations were carried out with a Siemens-Krautkrämer echo-encephalograph (Fig. 18)[1]. This is a modified ultrasound flaw detector, type USM 1, whis is used for non-destructive testing in industry. Other models such as USIP 9 and USIP 10 which were employed by other investigators (DREESE and NETSKY; JEPPSSON; KRAMER; LEKSELL; LITHANDER etc.) differ from this apparatus only by the narrower frequency range.

Footnote[1] see page 29.

1. Calibration of the Echo-Encephalograph

Most ultrasonic diagnostic equipments have *phantoms* included which correspond to a certain standard thickness of tissue. In the case of our apparatus, this phantom equals a tissue thickness of 2.5 cm. The standard sound velocity is the mean value of sound velocities in different media sounded during echo-encephalography (approx. 1550 m/s at body temperature). After setting the test range at 18 or 20 cm of tissue, the zero point is adjusted exactly. To do this, the apparatus must warm up for a few minutes, because the zero point is always displaced somewhat towards the left in the first few minutes. Once the zero point is set, as marked by the first rise of the initial pulse, it is made to coincide with the beginning of the centimeter scale and fixed there (Fig. 31). Subsequently the probe is applied to the phantom and a coupling medium is employed (Fig. 18). Now one sees numerous reflections on the screen of the cathode ray tube (Fig. 23 top) which are completely unintelligible at first. By switching on the sensitivity time control, (step 4 to 5) the interfering echoes can be screened. Now several echoes appear at regular intervals; their amplitude becomes progressively smaller towards the right. These reflections are multiple echoes from the back wall of the phantom. The third multiple echo — i. e. the fourth reflection — is now adjusted by turning the scale regulator on the centimeter scale in such a way that the ascent lies exactly at the 10 cm mark (see Fig. 23 bottom). The adjustment of the apparatus is now fixed after several months of use a calibration control is recommended.

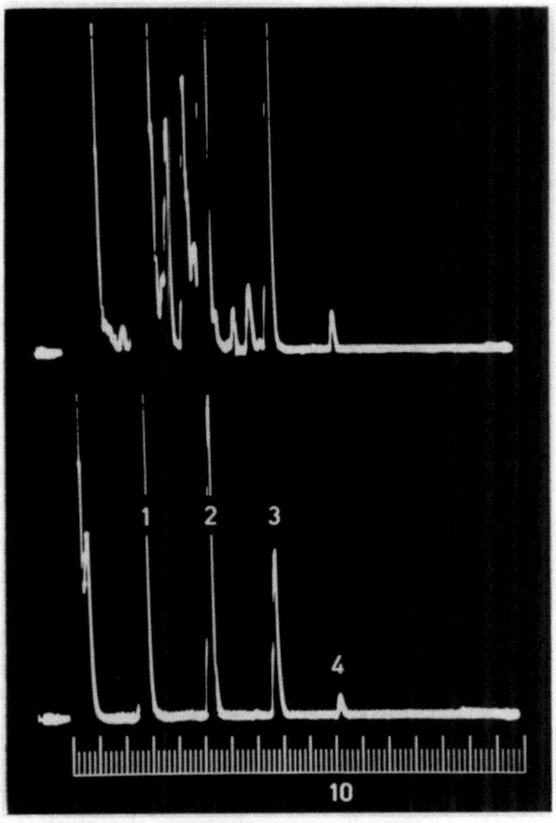

Fig. 23. Calibration of the A-scan trace. When the probe is placed on the phantom which equals 2.5 cm brain tissue, an unintelligible trace appears at first (top). The important reflections are demonstrated only after the sensitivity time control is switched on (bottom). Echo 1 develops at the back wall of the phantom, the rise of the third multiple echo (reflection 4) must be set at 10 cm

We have chosen to introduce the centimeter scale in order to avoid time-consuming calculations, so that the width of the ventricle, the thickness of the brain mantle, the diameter of the hematomas can be read directly off the scale in millimeters. KRAMER's objection (1964) that a calibration in centimeters or millimeters is inexact may be ignored. The maximum error which is due to the fact that ultrasound travels through bone as well as through brain tissue and CSF which hardly differ from

[1] The following firms are the chief manufacturers of ultrasonic diagnostic equipments: Alvar Elektronic, Montreuil-Paris, France; Kretz, Zip, Austria; Hewlett Packard Co., Palo Alto, Cal., USA; Hoffrel Instruments Inc., South Norwalk, Conn. USA; Japan Radio Co. Ltd., Tokyo, Japan; Kelvin & Hughes, Glasgow, Scotland; Metrix Inc., Denver, Colorado, USA; Physionics Ing. Inc., Longmont, Colorado, USA; Siemens AG, Wernerwerk für Medizinische Technik, Erlangen, Germany; Smith-Kline Instrument Co., Philadelphia, Pa., USA; Sonomedic Corporation, Westwood, New Jersey, USA; Ultrasonoscope Co. (London) Ltd., London, England

each other in sound velocity, is 1.0 to 3.0 mm over a distance of 15 cm, i. e. 1 to 2%. This error is of no significance for the diagnostic procedure of echo-encephalography. The American Kretz equipments blend in centimeter or millimeter scales electronically with the cathode ray (Fig. 194).

2. Technique of Investigation and Operation of the Apparatus

The first step in the investigation is *positioning the patient.* Patients, who are able to walk, are best seated in a chair with a head rest, facing the physician (Fig. 25). Other authors prefer a position behind the patient. If the investigator is right-handed, recumbent patients are best examined from the right side of the patient and facing him (Fig. 24). The special

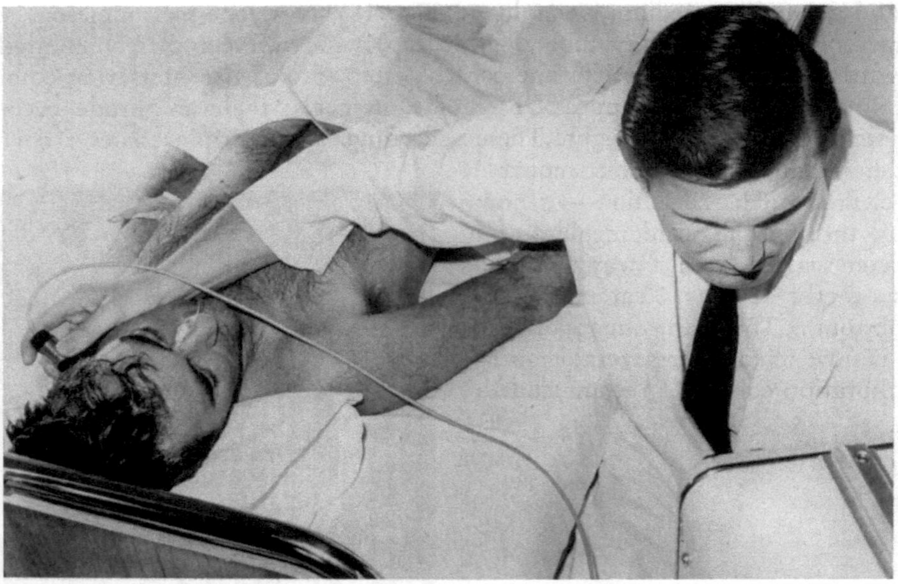

Fig. 24. Examination of a patient recumbent in a hospital bed

advantage of this position of the patient is that the physician always holds the probe in the right hand so that he can rest the little finger and the ball of the little finger on the patient's head near the zygomatic arch or behind the ear. Thus the probe can be held almost perfectly still (Fig. 25). This is particularly important with restless patients (small children, restless head-injured patients) since the hand can then follow the head effortlessly. The left hand is free to attend to the apparatus and the camera. The use of two probes, applied to both temples simultaneously, presents only difficulties and has no advantages in our opinion.

After the patient is positioned correctly, the *theoretical midline* of the patient's head must be determined. This can be done by way of the through-transmission procedure. To do this, the apparatus must be switched over and a second probe used. After applying the coupling medium to the probes, these are applied to the unshaved scalp somewhat above the ear, facing each other through the skull (Fig. 26). At a point corresponding to half the diameter of the head, a single reflections appears, because the ultrasonic beam covers the distance from one probe to the other only once (T-echo in Fig. 29). The second and simpler method and the one we prefer, is to measure the bitemporal diameter of the head by means of callipers (Fig. 27); half this value is used as the theoretical midline. Then one probe is applied to the temporal region and by means of the coupling medium we achieve adequate contact with the unshaved head.

On the oscilloscope a high reflection may be seen at the left edge of the screen without applying the probe to a surface. This is the previously mentioned transmitter pulse. The variations in electrical voltage which stimulate the crystal are responsible for this reflection.

The transmitter pulse is followed by a smooth zero line. Reflected quantities of ultrasonic energy are received.

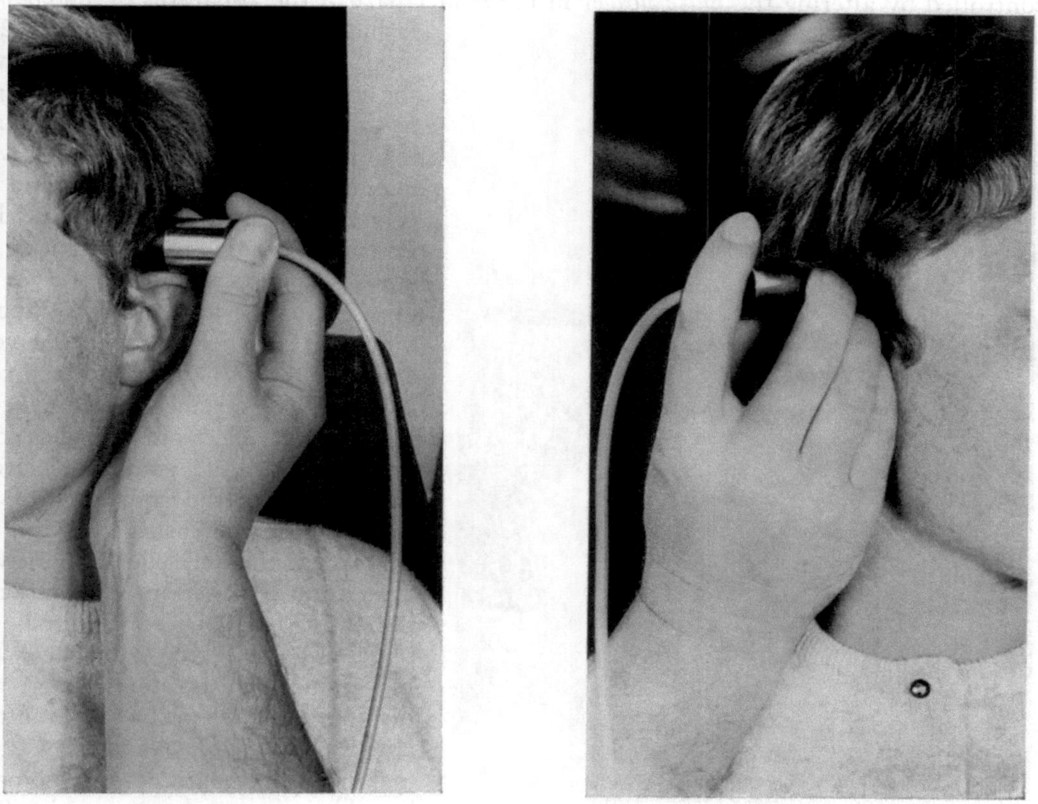

a b

Fig. 25. By resting the little finger and the ball of the little finger on the patient's head, it is possible to hold the probe very steady. a) Examination from the left b) Examination from the right

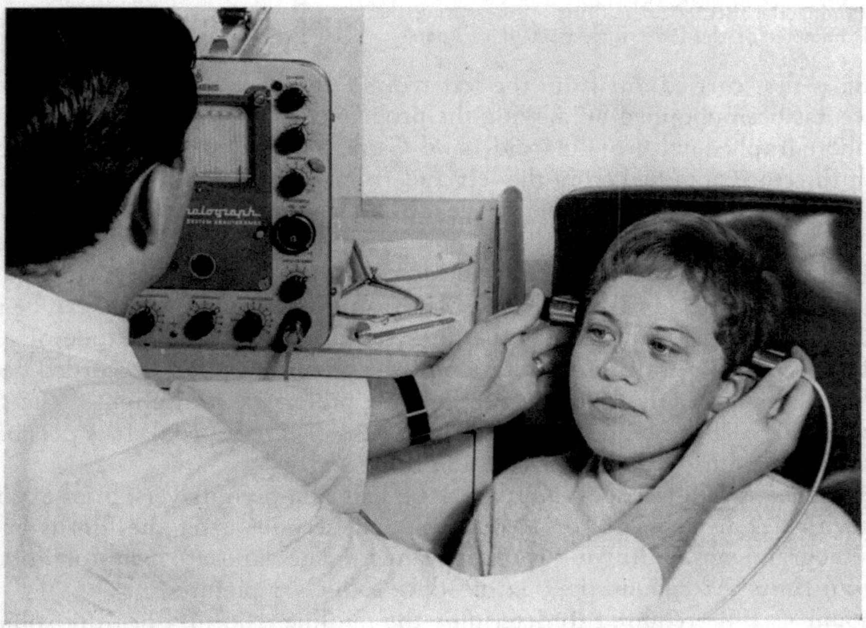

Fig. 26. Determining the theoretical midline by using the through-transmission technique

After the probe is applied, different reflections which will be discussed later, appear on the screen of the cathode ray tube. The height of the amplitude of the individual echoes can be controlled by altering the *pulse power* in the sender part of the apparatus in a step-wise fashion by a step attenuator (Fig. 28 a to d). A second method of altering the amplitude is to *regulate the amplifier* in the receiver part. At the beginning of the investigation we work with maximal sender performance and receiver amplification, which is then lessened according to the height of the echo. The next step is to *switch on the sensitivity time control* (abbreviated STC; synonyms are: time voltage gain — TVG, time voltage control — TVC, distance amplitude control — DAC, electronic distance compensation — EDC, gain time control) through which the amplitude of the reflections in the entire echogram appear approximately equal. Sensitivity time control functions by markedly decreasing the reflections from the structures close to the probe while reflections from structures at a greater distance from the probe are less attenuated (Fig. 28 e). Another way of making the echogram more comprehensible is to use *echo suppression*. This permits the elimination of smaller disturbing echoes ("cutting the grass"). Fig. 28 e/f shows two echograms from the same patient, one with and one without echo suppression.

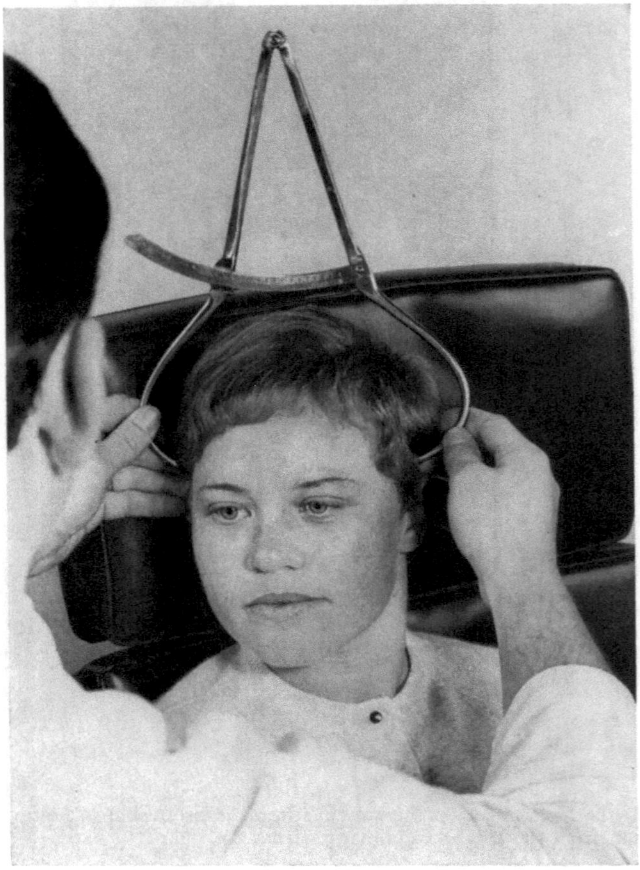

Fig. 27. Determining the theoretical midline by measuring the bitemporal diameter of the head with a set of callipers

Sounding is first carried out from the left temple, then from the right. Once a comprehensive trace has been obtained by moving the probe and regulating the cathode ray picture, it can be photographed with a Polaroid-Land-Camera (Fig. 30). It has proved useful to project both the trace obtained from the left and from the right on a single double exposure photograph. This can be done either by elevating the cathode ray for the first exposure (Fig. 29 b) or by projecting the cathode ray in the centre of the cathode ray tube for the first picture and then tilting the cathode ray upside down for the second picture (Fig. 29 c and 31). The latter method has the advantage that displacements of the midline echo can be recognized more easily. On the other hand, it has the disadvantage that the amplitude of a reflection (which is often important in the interpretation of the picture) is poorly recorded because only half of the screen is available for its projection. We prefer the projection of the two curves one above the other. The base line is elevated by means of a foot pedal. It is possible to blend in a centimetre scale during the documentation if desired.

The *film* used was a Polascope-Film type 410. This film permits very brief exposure times as its sensitivity lies at 41° DIN or 10,000 ASA. 10 seconds after the film is exposed, the developed echogram can be drawn out of the camera. The exposure time for Polascope film type 47 is four times as long and does not produce such clear pictures.

The amount of time required to determine the midline echo in a healthy subject is only 1 to 2 minutes; if there are pathological findings present on the other hand, the investigation takes longer (4 to 5 minutes including the taking of the picture, see JEFFERSON, 1962). This time expenditure is certainly justified even in acute cases since the findings obtained with echo-

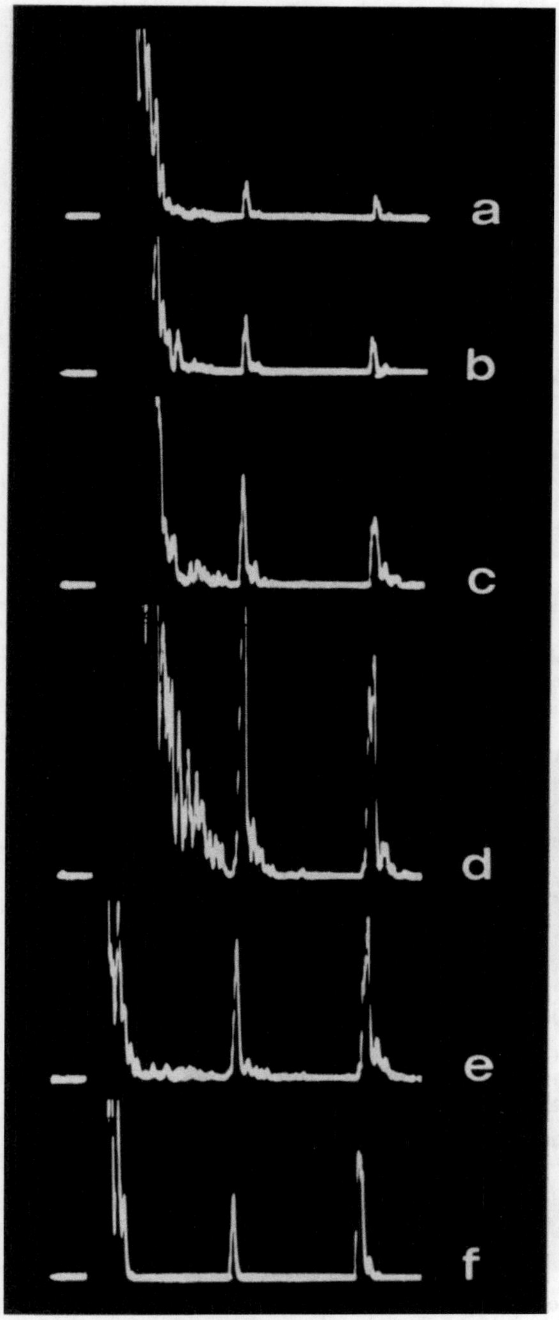

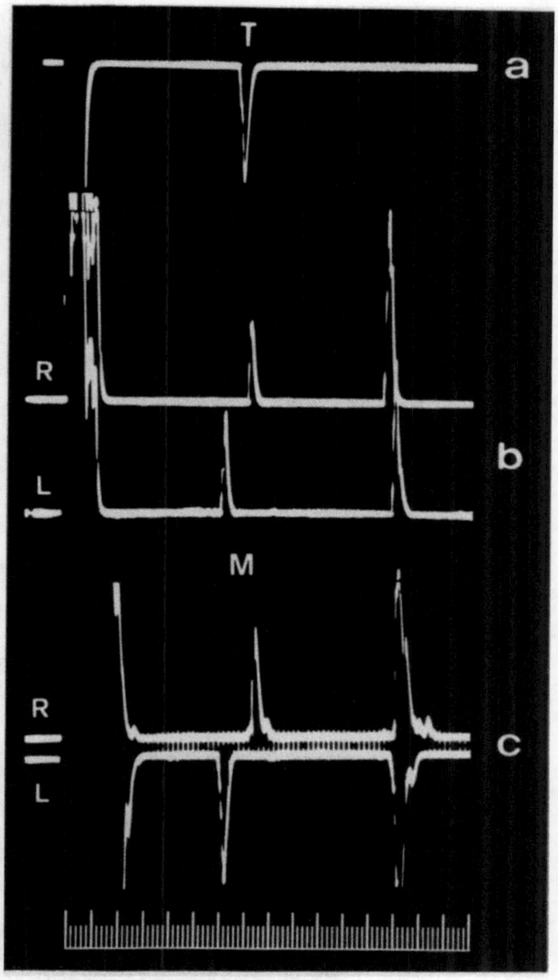

Fig. 29. Photographic documentation of the oscilloscopic trace. a) transmitter pulse (= theoretical midline echo). b) Echo-encephalograms on sounding from the right (R) and from the left (L) one above the other by *elevating* the cathode ray during a double exposure. c) Echo-encephalograms on sounding from the right (R) and from the left (L), *tilting* the cathode ray picture downwards during a double exposure

Fig. 28. Influence of pulse power, sensitivity time control and echo suppression on the cathode ray picture (=echogram). a) to d) Stepwise increase of the pulse intensity: all echoes become larger. At a high level of pulse power the initial echo extends as far as the midline echo. e) Switching on the sensitivity time control results in a damping of the structures close to the probe. f) The smaller echoes (noise) can be eliminated by switching on the echo suppression ("cutting the grass")

encephalography not infrequently indicate immediate trepanation without subsequent dye studies (see section on epidural hematomas).

More detailed echo-encephalographic investigations where the width of the third ventricle is determined, the inferior temporal horn is detected or tumor reflections are located, naturally require more time on occasion (10 to 20 minutes, sometimes even longer).

Fig. 30. Echo-encephalograph with mounted photographic equipment (Oscillophot Steinheil with polaroid film holder)

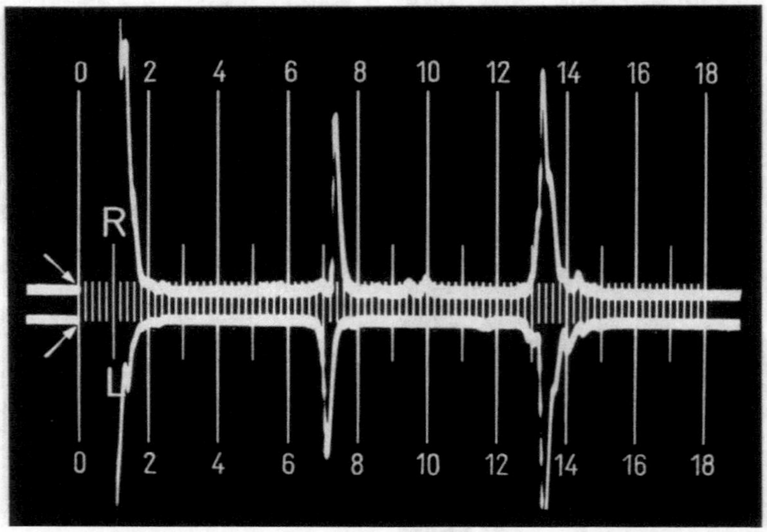

Fig. 31. Normal echo-encephalogram. Adjusting the zero point: the first interruption of the cathode ray must coincide with the beginning of the centimeter scale

The echo-encephalographic examination is not possible after pneumoencephalography because the interface between the brain and the air injected into the ventricles results in total reflection of the ultrasonic beam. This produces such a high deflection of the cathode ray on the screen that it is impossible to evaluate the trace. The investigation can be properly repeated only after all the air has been reabsorbed from the ventricles because even slight quantities of air in the scanning area cause high echoes.

Ford and Ambrose (1963) used this fact in order to verify the passage of air into the third ventricle during pneumoencephalography. Fig. 32 shows an echo-encephalogram before and during pneumoencephalography where air is entering the third ventricle. Tanaka has reported similar findings with posttraumatic pneumocephalus.

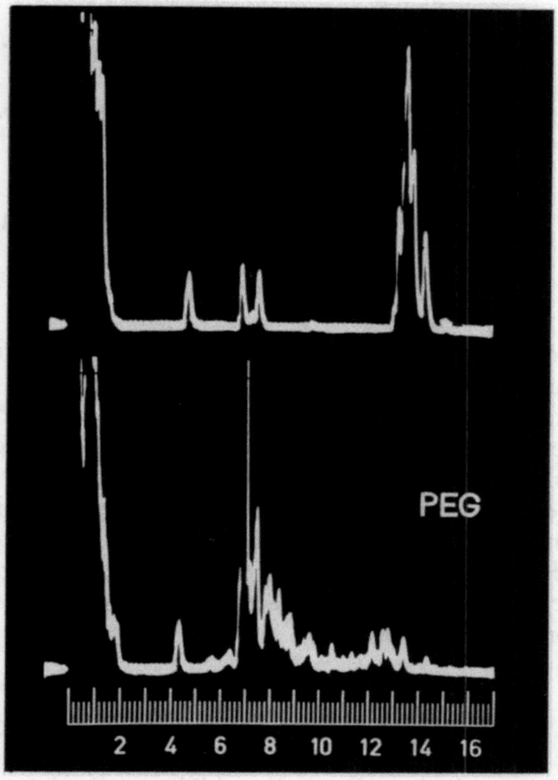

Fig. 32. Echo-encephalographic examination before and during pneumoencephalography. Top: double echo from the walls of the third ventricle. Bottom: when air enters the third ventricle, ultrasound is totally reflected at the interface ventricle wall-air (PEG)

Part II. General Echo-Encephalography

The Reflections on the Echo-Encephalogram and their Origins

A. The Normal Echo-Encephalogram

1. Initial Echo (Initial Complex)

Temporal sounding of the skull reveals a whole series of echoes which will be individually discussed in the following section. After the probe has been applied to the skull, the transmitter pulse is obliterated by multiple reflections originating from the scalp, muscle, fascia and bone directly under the probe. This entire complex is described as the "initial echo" (Fig. 33). The length of the initial echo depends on sender power, thickness of the scalp, muscle and bone on the side of the probe and on possible pathologic changes in the scalp (e. g. subgaleal hematoma). Furthermore the initial echo may be reduced by using the *sensitivity time control* so that an evaluation of the reflections appearing just behind the initial echo is possible in selected cases. The initial echo is always on the left side of the screen regardless of the position of the probe.

2. End Echo (Bottom Echo, Far Side Echo)

Whenever the probe is applied to the skull, a high amplitude reflection appears on the right edge of the screen (Fig. 33). This echo, or more precisely this echo complex which consists of several reflections, develops when the ultrasound beam hits the opposite side of the skull and is reflected by it. Smyth (1964) investigated the end echo in more detail. He states that the end echo complex is generally composed of 3 or 4 single reflections. According to Smyth the first rise occurs at the interface of brain and meninges. This is followed by two reflections from the bony skull (inner and outer table of the temporal bone) and finally by a reflection which originates from the external skin/air interface. In our clinical experience it has not always been possible to subdivide the end echo because in most cases all these echoes melt together to form a single broad-base echo (compare the following echograms). One thing is certain however, namely that *the first rise of the end echo is due to the reflection of ultrasound by the dura mater* (see also Oberschulte-Beckmann and Otto, 1968). Since the dura has a thickness of 0.3 to 0.6 mm in the temporal area according to measurements of our own, the reflection which develops here cannot be registered independently of the echo from bone. If a pulse wave length of 1 μs in brain tissue and the usual size of screen are used, it is necessary for 2 interfaces to be at least 0.75 mm apart in order to produce 2 distinct reflections. In evaluating echograms, the first rise of the end echo is regarded as the most important. This reflection always appears considerably ahead of the point corresponding to the skull diameter because the ultrasonic beam must traverse the entire distance from the dura to the skin/air interface yet. This distance which Feuerlein and Dilling (1965) call "head wall thickness" is 8 to 15 mm (men 10 to 15 mm, women 8 to 12 mm). A graphic demonstration of this thickness in infants and children is presented in Fig. 34. Thus one cannot simply add 8 to 9 mm to the distance initial echo—end echo and then take half the total distance in order to determine the theoretical midline echo, as FEUERLEIN and DILLING (1965) suggested. However, the real "egress" echo which indicates when the ultrasonic beam penetrates the external surface of the scalp, can be clearly differentiated from all other reflections of the end echo complex by using a simple device suggested by DE VLIEGER (1964). If the scalp opposite the probe is moved back and forth with a finger, fluctuations in the amplitude of the actual egress echo occur while the other reflections of the end echo remain unchanged.

Normally *the end echo should appear at the same place on the screen regardless of the side sounded.* If this is not the case, the following possibilities must be considered:

1. a different application point was used for the probe;
2. a different direction of sounding was used or
3. there are variations in the thickness of the scalp and bone.

The first two sources of errors can be readily checked on and eliminated. If the positions of the end echo are still not congruent, then the thickness of the scalp and bone on the two sides must differ. This may be due to an *asymmetrical skull,* the bony skull on one side being clearly thicker (very rare). More frequently this finding is obtained if the scalp as well as the musculature are thickned e. g. in the presence of an unilateral *subgaleal hematoma* following head trauma. The recognition of such an extracranial mass is of importance in the evaluation of the echo-encephalogram particularly with head injuries because the former can simulate a displacement. A detailed discussion with pictorial examples will follow on pp. 78 and 79. A trepanation defect in the temporal area may produce the same effect.

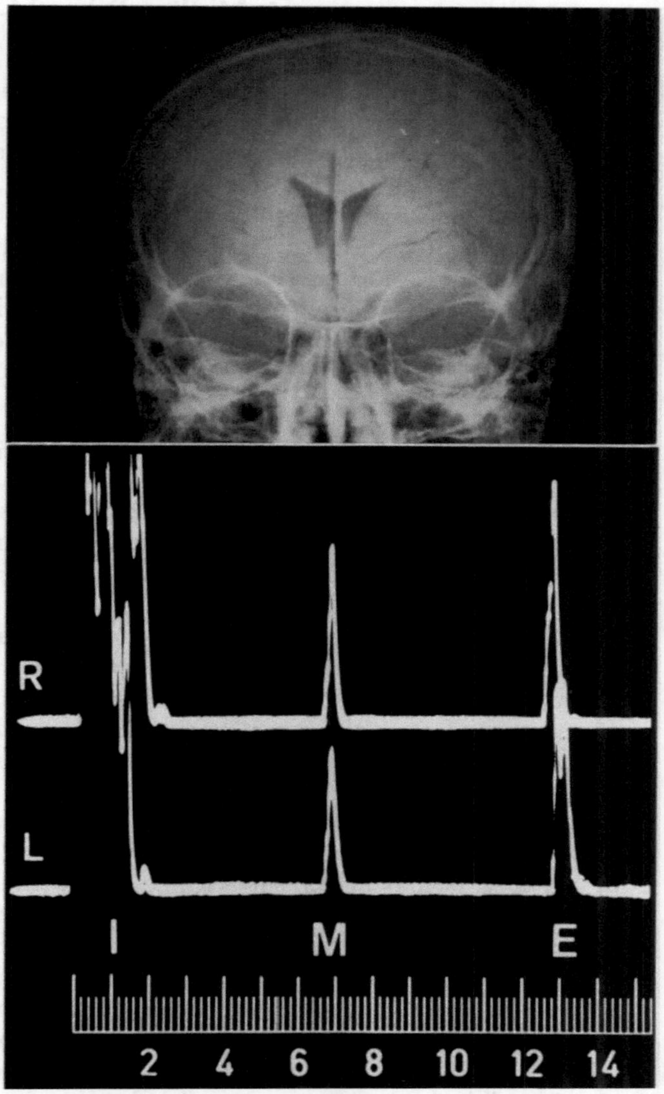

Fig. 33. Normal echo-encephalogram and pneumoencephalogram of the same patient. Lower trace sounding from the left temporal application point, upper trace from the right temporal application point. I=initial echo, M=midline echo, E=end echo

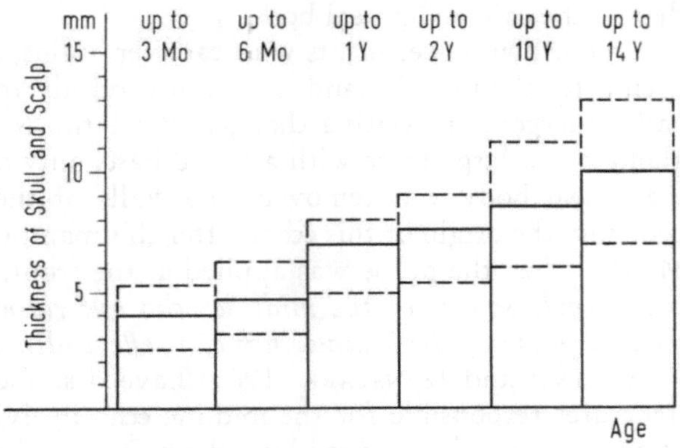

Fig. 34. Thickness of the scalp and skull up to the age of 14. Ultrasound measurements in 549 subjects

3. Midline Echo (M-Echo)

The decisive criterion for the interpretation of the echo-encephalogram is the so-called "midline echo", or simply M-echo. This reflection appears with great regularity near the centre of the echogram. Normally this echo can be found at the same place on the screen regardless of whether one is sounding from the left or from the right, (Fig. 33) i. e. the distance initial echo—midline echo is normally the same on sounding from either side. The same is true for the distance midline echo—end echo; however, when compared to the former distance it is shorter by the thickness of scalp and bone (Fig. 36).

Opinions are still widely divergent on the origin of the midline echo, although no significant new viewpoints have entered the debate since 1961. Every structure located in the median plane of the brain has been designated as the source of the midline echo.

a) Pineal Body (Leksell, 1958; Jeppsson, 1960, 1961)

By way of clinical investigations and experiments with cadaver brains, Leksell (1958) arrived at the conclusion that the main source of the midline echo is the calcified pineal body regardless whether it is radiologically visible or not. He sees the proof for this statement in the fact that the echo of the pineal body is the only reflection which can be registered from any application point. In cadaver experiments he found that the midline echo disappeared after the pineal body was removed. These investigations were repeated by Gordon (1959) as well as by Ter Braak et al. (1961). In contrast to Leksell's experiments the midline echo was still present even after the pineal body was removed. In 1961 Jeppsson published a very detailed experimental investigation on this same question in his monograph on echo-encephalography and arrived at the same result as Leksell. He emphasized above all the circular shape of the pineal body as proof that this is the only structure in the median plane of the brain which is in a position to reflect ultrasound equally in all directions. Our own investigations led to a partial agreement with Jeppsson and Leksell, and served to explain the disagreements between the different results. While Jeppsson immersed the cadaver brains into a 10% invertose solution and rinsed the brains thoroughly with this solution, our own experiments were performed with brains in physiological saline solution. Preliminary experiments with 10% invertose showed that this solution which has a specific weight of 1037 kg/m^3 at 22° C and a sound velocity of 1532 m/s, possesses an acoustic impedance of $1.5888 \cdot 10^6$ Ns/m^3. That is almost exactly the same value as in fresh brain tissue ($1.585 \cdot 10^6$ Ns/m^3). Thus it was not possible to obtain unequivocal reflections from the ventricular walls in the experimental method described by Jeppsson, because the differences in acoustic impedance which normally exist between brain tissue and CSF are practically effaced completely if the ventricles are filled with invertose solution. The single high echo in the midline was therefore that of the calcified pineal body.

In our own experiments with cadaver brains, an unequivocal midline echo could be found even after the pineal gland had been carefully removed. However, the shape of the M-echo had undergone a decisive change. While the M-echo originating from the calcified pineal gland was a large trace with a broad base, only a low amplitude double echo remained after the pineal body was removed; the walls of the posterior part of the third ventricle are probably the origin of this echo. After this manipulation it was no longer possible to record an M-echo when the probe was applied at the front. *This result alone already proves that there is no single source of the midline echo but rather that there are different structures in the median plane of the brain which can reflect ultrasound.*

Brylski and Izenstark (1965) have described an interesting method for localizing the structures responsible for the midline echo in living subjects. With the help of a transducer attached to a rigid arc-template, the points on the skull where the ultrasonic beam enters the skull and the opposite side of the cranial vault as defined by the arc-template were marked with metallic pins (electroencephalographic electrodes). The pins were placed on the skin at the instant at which the M-echo attained the highest amplitude. This procedure was carried out from both sides. Then x-rays of the skull were made in various views. In a number of patients there was a calcified pineal body; in several others without such a calcification, pneumoencephalography was carried out after the marking. In combination with the radiologically visible markers the two trajectories always met in the area of the calcified or non-calcified pineal body or in the suprapineal recess. Similar results were obtained by FISCHER (1966). LITHANDER (1961) denied the significance of the pineal body as the source of the midline echo and argumented that the round shape of the calcified pineal body presents far poorer reflection conditions than plane surfaces such as the walls of the third ventricle for example. However, it must be noted that the difference in acoustic impedance between brain tissue and calcium is about 50 times higher than between brain tissue and CSF. Since the pineal body is composed of a conglomerate of small calcifications, the ultrasonic beam always hits small interfaces, standing perpendicular to the direction of sounding thereby reflecting ultrasound back to the probe.

Our own observation that a calcified plexus (Fig. 43) and calcifications within a brain tumor produce extraordinarily high echoes, seem to favor the assumption that the calcified pineal body is a very important source of the midline echo.

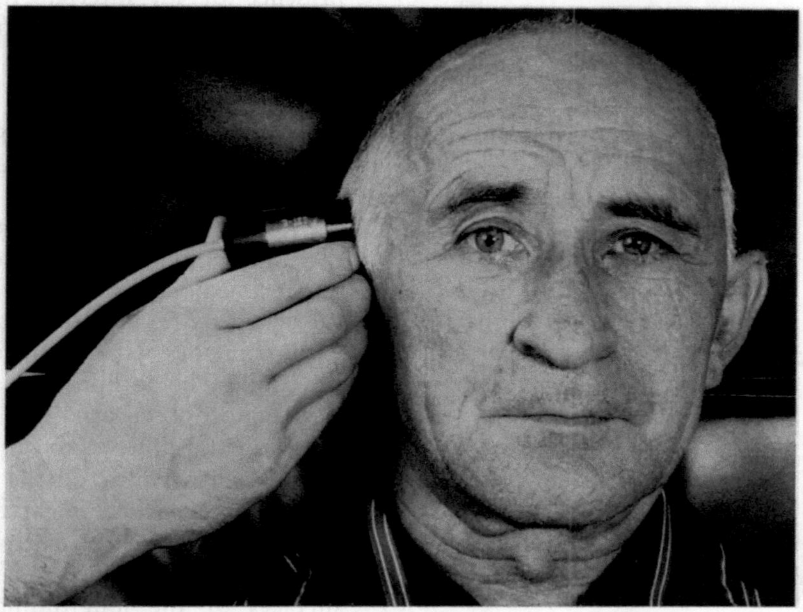

Fig. 35. Sounding of the pineal body: the probe is applied somewhat above and behind the fleshy attachment of the ear and is tipped slightly upwards

All authors, who regard the calcified pineal body as the source of the midline echo agree that a radiologically visible calcification is *not* necessary for the development of a significant midline echo. On the other hand the pineal body can become a source of the M-echo only if a certain amount of calcification has occurred. According to JEPPSSON this process begins towards the end of the first decade at the earliest. Therefore there must logically be other sources of the midline echo in children and young people which may be of importance in adults as well (see JEPPSSON, 1961).

Regarding the *technique of investigation,* we see that the echo originating from the pineal gland is easiest to record, if the probe is applied somewhat above and behind the root of the ear and is directed slightly upwards (Fig. 35).

b) Septum Pellucidum (Gordon, 1958, 1959; Lithander, 1961; Taylor, Newell and Karvounis, 1961; Jefferson, 1962)

According to Gordon, the midline echo is produced by the septum pellucidum. A number of other authors are of the same opinion (Lithander, 1961, Taylor, Newell and Karvounis, 1961; Jefferson, 1962). However, all these authors mention other structures as the source of the midline echo as well.

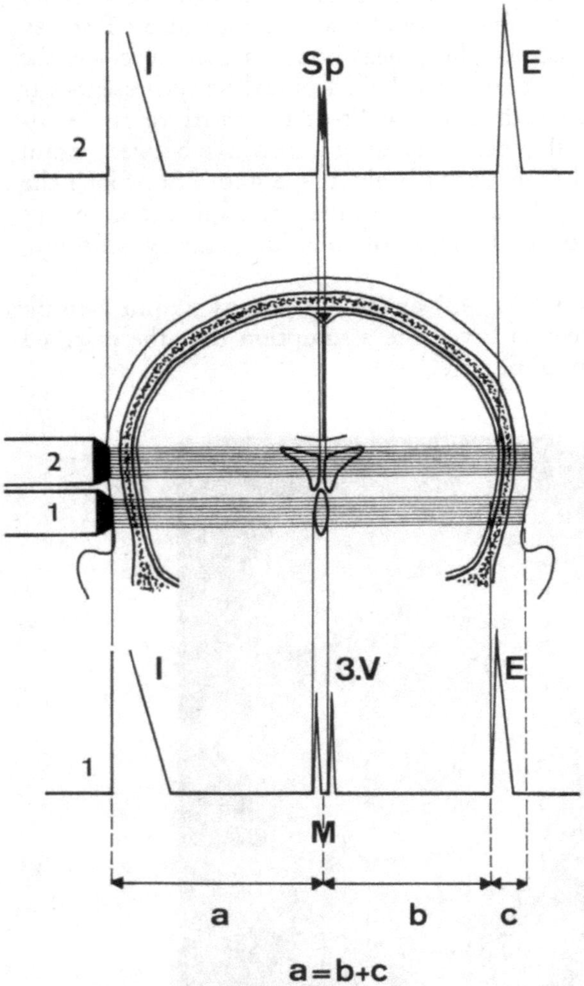

Fig. 36. Diagrammatic demonstration of the development of reflections from the walls of the third ventricle (1) and septum pellucidum (2) when the ventricular system is *not* displaced. The distance initial echo/midline echo (a) is longer than the distance midline echo/end echo (b) by the distance equalling the thickness of the scalp and skull (c)

If the common application point of the probe directly above the fleshy attachment of the ear is used, the septum pellucidum does not lie in the scanning area (Fig. 36). Only if the probe is placed 2 to 3 cm higher and somewhat ahead of the ear, can the walls of the septum pellucidum reflect ultrasound (Fig. 36). (Schiefer, Kazner and Brückner, 1963; Kazner and Schiefer, 1966). In the presence of an internal hydrocephalus, this fact attains some significance as discussed below.

c) Interhemispheric Fissure (de Vlieger and Ridder, 1959)

According to De Vlieger and Ridder, the longitudinal fissure is the main source of the M-echo. This fissure which separates the two cerebral hemispheres and contains CSF, coincides with the median plane of the brain. The authors base their view on a series of observations. They say that since the midline echo can be recorded from various points on the skull, a relatively large surface must be the source of the M-echo. Frequently the midline echo appears double because the two surfaces of the longitudinal fissure are somewhat separated from each other. They regard the fact that pulsations of the midline echo or of both midline echoes are frequently observed, as an important argument in their favor. This pulsation is synchronous with the heart beat and both reflections pulsate 180° out of phase. This is thought to be due to an increase in brain volume during the approaching pulse wave and a surface alteration in the sense of an amplified concavity or convexity.

In all these considerations one fact is disregarded however, namely that the cerebral falx fills the interhemispheric fissure nearly completely and that *this rigid membrane follows mass displacements only to a limited degree.* The main shifts occur in the area below the falx (third ventricle, pineal body!). Furthermore the interhemispheric fissure is situated so high up that it cannot be hit vertically from the usual application point. But this vertical impact of the beam on a plane or largely smooth surface is a prerequisite for the development of a usable echo. Only if the application point is located farther towards the frontal or occipital region, does it seem theoretically possible to record a midline echo from the interhemispheric fissure (see also JEPPSSON, 1961).

d) Cerebral Falx (JEPPSSON, 1961; LITHANDER, 1961)

The cerebral falx also produces a midline echo according to experiments on cadavers (JEPPSSON, 1961; LITHANDER, 1961). For the clinical investigation however, this echo is of little importance because it lies outside the range of the usual bitemporal scanning area (GORDON, 1959; DE VLIEGER and RIDDER, 1959). In adults it is rarely possible to detect echoes from this membrane from an application point lying perpendicular to the cerebral falx. The thick skull bone absorbs ultrasound almost completely in the upper part of the temporal and parietal bone so that no useful reflections can be recorded. In infants however, a very high echo may be obtained from this structure as a rule (see Fig. 37). Especially in hydrocephalic babies this reflection often shows peculiar fibrillations.

In his experiments on the origin of the M-echo, JEPPSSON examined the importance of the cerebral falx in great detail. He writes, "From the ultrasound-physics point of view, the dura is most interesting; especially the falx cerebri, a strong, fibrous process of dura mater in the median plane. The separate analysis of the part played by the falx as an echo source clearly proves it to reflect ultrasound strongly". JEPPSSON (1961) examined brains with an intact falx in a water tank by echo-encephalography. Here a plane was found whose dimensions corresponded to those of the falx and in whose domain an echo of high amplitude appeared. After removing the falx, this echo disappeared or it was present only as a reflection of a few millimeters in height originating from the interhemispheric fissure. According to JEPPSSON's investigations (1961) the "falx in its central-posterior portion (is) 15—20 mm distant from the pineal body. The region where the falx merges into the tentorium gives rise to echoes of a lower amplitude, with a tendency to become split at the top, which may look similar to the pineal body M-echo".

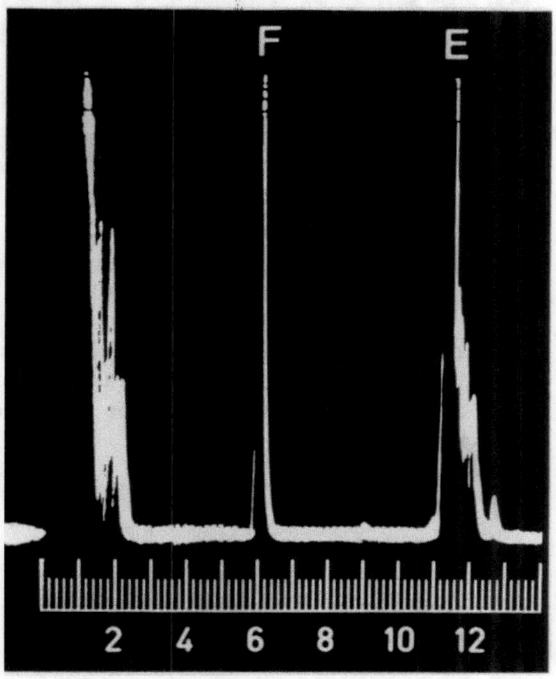

Fig. 37. A midline echo (F) from the falx in a child. The reflection can be registered only if the probe is applied high enough in the temporo-parietal region

These doubtless very significant observations can be confirmed by clinical examination of children again and again. If the probe is applied about 3 cm above the normal application point directly above the ear, the cerebral falx is picked up by the ultrasonic beam and appears on the screen as an echo whose height exceeds all other echoes (Fig. 37) (KAZNER, KUNZE, SCHIEFER, 1965). This is in good agreement with the observations of other investigators (EDLER and HERTZ, 1955; WILD and REID, 1953) who show that fibrous surfaces such as the endocardium or muscle fascia cause unexpectedly marked ultrasonic reflections.

e) Third Ventricle (GORDON, 1959; DE VLIEGER and RIDDER, 1959; LITHANDER, 1961)

Aside from the septum pellucidum, the lateral walls of the third ventricle are of the greatest importance in the development of the midline echo according to GORDON (1959). This opinion is shared by LITHANDER (1961), JEFFERSON (1962), TAYLOR, NEWELL and

KARVOUNIS (1961). DE VLIEGER and RIDDER (1959) as well as GORDON (1959) saw a double echo in the centre of the echo-encephalogram on the screen; they assumed this originated from the lateral walls of the third ventricle. In 1961 LITHANDER first reported on ventricular dilatations in infantile hydrocephalus diagnosed by such double echoes and reflections from the lateral ventricles using ultrasound. In healthy controls the width of the third ventricle was also determined. On the basis of physical calculations JEPPSSON (1961) believed that it was impossible to obtain reflections from the ventricular walls in most adults because ultrasound was largely absorbed by the bony skull. This would not apply to children and to adults with extremely thin skulls.

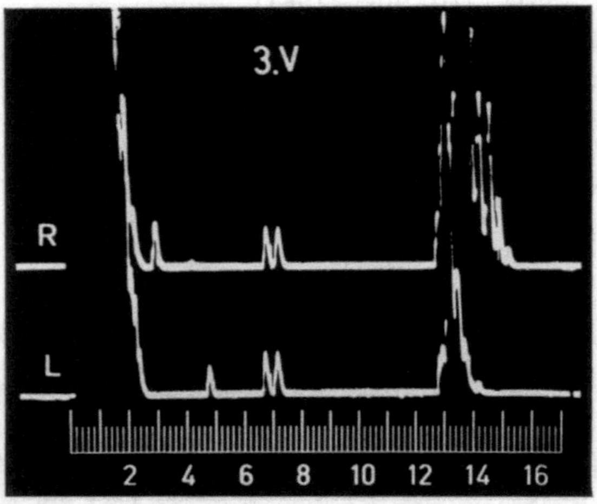

Fig. 38. Echo-encephalogram with a double midline echo originating from the lateral walls of the third ventricle. The width of this ventricle can be determined in this way. It is 4 mm in this case. Pat. S. W., 7 years of age. Echo-EG No. 626/63

SCHIEFER, KAZNER and BRÜCKNER (1963) were able to measure the width of the third ventricle by means of a characteristic double echo in a large series of healthy subjects of all ages. In 129 subjects the diameter of the third ventricle lay between 3 and 6 mm. The typical echo-encephalographic picture of a third ventricle is seen in Fig. 38. In the meantime we have measured the width of the third ventricle in all age groups. The values found during this investigation are presented in Fig. 39. From it the conclusion can be drawn that a third ventricle exceeding 7 mm in an adult under the age of 60 must be regarded as pathological. In children the upper limit is much lower yet.

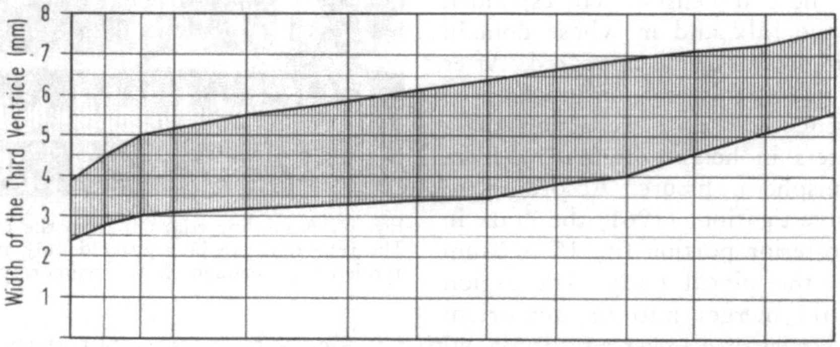

Fig. 39. Echo-encephalographic measurements of the width of the third ventricle in 280 healthy subjects (144 children, 136 adults). The values obtained lie in the hatched area. In the first years of life, the width of the third ventricle increases rapidly. Before the age of 60, a value of more than 7.0 mm must be regarded as pathological

Similar results were obtained by FEUERLEIN and DILLING (1967) on echo-encephalographic examination of 300 healthy subjects in different age groups. Only in a number of patients over 70 did the authors observe a considerable increase in the width of the third ventricle (up to 11 mm). However, there was probably a clinically not yet manifest cerebrovascular disease in progress which was accompanied by a dilatation of the ventricular system.

Even the most recent investigations in this field show similar results (HUBER and PATIRI, 1967; HUBER, BETZ and KLEINÖDER, 1968; KRÜGER, ZUMPE and VELTIN, 1967, 1968). Of course we found third ventricles whose width exceeded 7.0 and 7.5 mm in numerous subjects over the age of 60 as well. However further investigations of these subjects, who appeared to be healthy, showed that they suffered from headaches, hypertension, vertigo and decreased mental activity. Thus these patients probably had cerebrovascular disease, which explains the abnormal ventricular width. For this reason these patients were excluded from the series of normal subjects.

In the literature opinions tend to differ widely on the normal width of the third ventricle in the pneumoencephalogram. SCHIERSMANN, FROWEIN and HARRER, SCHMIEDER and GÖLLNITZ regard 5.0 mm as the upper limit for the width of the third ventricle. DAVIDOFF and DYKE calculated a diameter from 2.0 to 8.0 mm as the normal range of variation in 53 cases. In 32 normal pneumoencephalograms where the x-rays had been corrected, NÜRNBERGER and SCHALTENBRAND found values between 3.0 and 6.0 mm depending on the diameter of the skull. HUBER regarded values up to 5.9 mm as normal, diameters of 6.0 to 7.9 mm as slightly enlarged, from 8.0 to 9.9 mm as moderately dilated, from 10.0 to 11.9 mm as considerably dilated and values exceeding 12.0 mm as grossly dilated. Values higher than 8.0 mm are regarded as definitely pathological, an opinion shared by DAVIDOFF and DYKE as well as by ENGESET and SKRAATSTAD. Table 1 shows the limits reported by the individual authors.

Table 1. *Normal width of the third ventricle in the pneumoencephalogram according to the literature*

Author	Year	Normal Width of the Third Ventricle
LARSBY and LINDGREN	1940	4.5
SCHIERSMANN	1942	up to 5.0
DAVIDOFF and DYKE	1943	2.0 up to 8.0
FROWEIN and HARRER	1948	up to 5.0
SCHMIEDER	1948	up to 5.0
GÖLLNITZ	1951	up to 5.0
SCHIFFER	1951	up to 5.0
HUBER	1953	up to 5.9
NÜRNBERGER and SCHALTENBRAND	1955	3.0 up to 6.0
ROBERTSON	1957	3.0 up to 9.0
ENGESET and SKRAASTAD	1964	up to 8.0
TAVERAS and WOOD	1964	3.0 up to 7.0

In an earlier publication (KAZNER and SCHIEFER, 1966), we compared the echo-encephalographic and pneumoencephalographic values of the width of the third ventricle and found complete agreement between the two methods of measurement providing the enlargement caused by the divergence of x-rays was drawn into consideration (see also p. 117).

The measurement of the third ventricle in the echo-encephalogram has found wide application in the meantime (DREESE and NETSKY, 1963; AMBROSE, 1964; CALATAYUD-MALDONADO, GELETNEKY and LORENZ, 1965; LAPAYOWKER and CHRISTEN, 1965; FORD, 1966; GROSSMAN, 1966 and many others).

The Japanese research group under TANAKA believes that the midline echo is chiefly due to the third ventricle. Investigations by SIMON, MIKOL and HAZEMANN (1965) also support the hypothesis that the midline echo originates from the posterior part of the third ventricle.

The *technique of investigation* for the demonstration of the double echo from the lateral walls of the third ventricle is described in detail on page 57 f. (compare also Fig. 36 and 55).

On the basis of experimental investigations on cadaver brains LITHANDER (1962) came to the conclusion that ultrasound can also be reflected from the first part of the *aqueduct*. This opinion is shared by NADJMI (1968). When the common method of temporal sounding is used, this structure will certainly lie within the the ultrasonic cone so that it seems reasonable to record echoes from the aqueduct. However, such reflections have not attained diagnostic significance as yet. JEFFERSON and HILL (1966) expressed the belief "that the walls of the apex of the *interpeduncular fossa* may sometimes act as a source of echo-reflection. This might occur in individuals with very flat temporal bones, and perhaps also in infants and young children where the base of the brain lies relatively higher than in adults." NADJMI (1968) was able to show that echoes of high amplitude could be obtained from the basal cisterns as well.

The enumeration of the various structures which can serve as possible sources of the midline echo and the survey of clinical and experimental investigations of the midline-echo-question allow us to draw the conclusion that no one structure alone can be regarded as the source of this echo. Without doubt the *most important M-echo sources are the posterior part of the third ventricle and the calcified pineal gland.* In many cases reflections can be obtained from all the different midline structures depending on the direction of sounding. This opinion is generally accepted today. However, the relationships are altered considerably if the cerebral midline structures are displaced by a supratentorial space-occupying lesion. The reflection conditions existing then and the results which can be expected will be discussed in the section on pathological echo-encephalograms.

4. The Lateral Reflections

Besides the initial, midline and end echo complexes, a number of other reflections called lateral echoes appears in the echo-encephalogram. Since they can give rise to errors on the one hand, but often make a more detailed evaluation of the echo-encephalogram possible on the other hand, we attach great importance to the understanding of the lateral reflections.

a) Echoes from the Lateral Ventricles

When studying the reflections which originate from the lateral ventricles, a differentiation must be made between echoes from the cella media region and reflections from the temporal horn. Ter Braak, Crezée, Grandia and de Vlieger described a method for identifying lateral ventricle echoes in 1961. The author recorded an echogram during pneumoencephalography whereby the direction of the ultrasonic beam was indicated on the x-ray picture. The air-filled ventricle naturally produced much higher reflections on the screen since the interface brain/air resulted in almost total reflection. Echo-encephalograms performed after air had been removed showed distinct reflections of the cathode ray if the probe was pointed in the same direction where there had been total reflection previously. The individual echoes could be attributed to certain parts of the ventricles in this manner.

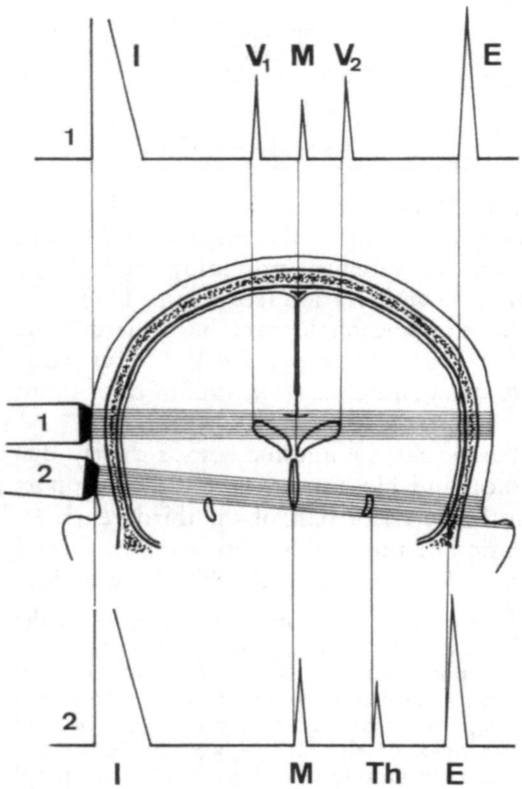

Fig. 40. Diagrammatic presentation of the development of reflections from the walls of both lateral ventricles (1) and the temporal horn (2)

Echoes from the *cella media region of the lateral ventricles* — calculated for tissue — normally lie at about 1.0 to 1.5 cm beside the midline echo. The technique of investigation necessary here is schematically represented in Fig. 40. Occasionally, especially in children, one is successful in obtaining a reflection from the left and from the right lateral ventricle at the same time (compare Fig. 41). Ambrose (1964) calculated a ventricular index comparable to the Schiersmann ventricular index in pneumoencephalography by using the site of the lateral ventricle echoes. Echo-encephalographic measurements of the lateral ventricles were carried out above all by Sjögren (1965, 1967, 1968), Ford and

McRae (1966), Uematsu (1966) and Ford (1968). The calculation of the ventricular index is particularly useful in cases of infantile hydrocephalus; this will be discussed in more detail later (s. also Umbach and Kley, 1965).

Reflections from the *temporal horn* appear on the screen when the probe is applied one finger's breadth above the root of the ear and the ultrasonic beam is directed somewhat downwards and often slightly posteriorly (see Fig. 40). Normally the temporal horn echo lies half-way between the midline and end echo (see Fig. 42). This lateral echo is mainly an ultrasonic reflection from the *wall of the inferior horn opposite the probe.* Using the position of this echo, Schiefer, Kazner and Kunze (1965) calculated an *echo-encephalographic brain mantle index* (BMI), which indicates the relationship between midline echo—end echo and temporal horn echo—end echo. Fig. 60 shows the calculation of this index diagrammatically. The normal value ranges from 2.0 to 2.2. Values above 2.4 indicate distinct dilatation of the lateral ventricles (see also p. 61). The width of the inferior horn usually cannot be measured with sufficient reliability by echo-encephalography. A reliable measurement in the presence of a ventricular dilatation can be obtained here only if the echo of the medial wall of the inferior horn also becomes visible.

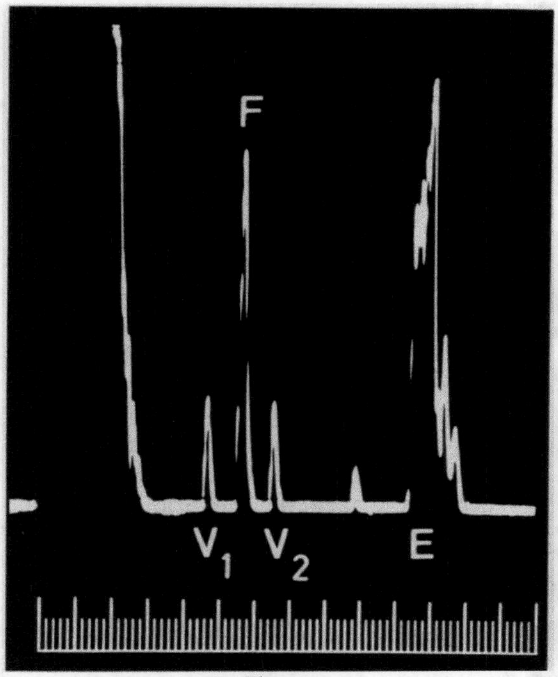

Fig. 41

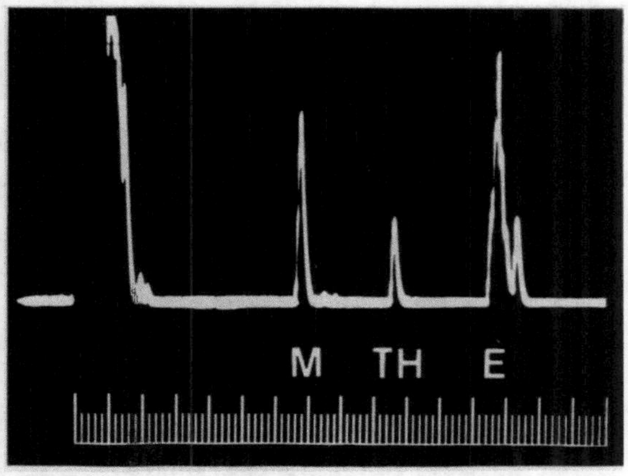

Fig. 42

Fig. 41. Recording echoes from the ventricular walls of a healthy infant. On applying the probe high up in the temporo-frontal area, ventricular wall echoes (V_1 and V_2) from the body region appear on both sides of the falx echoes (F). A. H., 1 year of age. Echo-EG No. 2889/67

Fig. 42. Normal echogram with reflection of the external wall of the temporal horn (TH) on the side opposite to the probe. Normally this reflection lies halfway between the midline and end echo

According to our observations the registrability of the lateral ventricle echoes is generally limited to children and adults with relatively thin skulls since the slight amounts of ultrasound reflected at the lateral ventricle walls, disappear through the absorption of bone in the presence of a thick temporal bone. However, this is not true if there is ventricular dilatation; the enlargement of the reflecting surfaces results in better conditions for the development of echoes. According to Jefferson and Hill (1966) a distinct reflection may also be obtained from the *ipsilateral temporal horn* if it lies in the path of the ultrasound beam (see also Fig. 58). Very often this reflection becomes a part of the initial complex however.

b) Reflections from the Choroid Plexus of the Lateral Ventricles

From an application point somewhat above and behind the ear where the pineal body can be picked up readily, in many cases marked echoes with a broad base may be found, originating from the choroid plexus of the lateral ventricles, according to our investigations.

This is usually a complex of several closely adjacent reflections which attain an extraordinarily great height if the plexus is calcified (Fig. 43 a to c). JEPPSSON 1961 drew attention to the fact that a calcified choroid plexus may produce a high echo which can be mistaken for the midline echo. The fact that the *calcified* plexus of the lateral ventricle produces unusually marked ultrasonic reflections may be regarded a further proof for LEKSELL's midline echo theory. The diagnostic importance of this echo is only slight however.

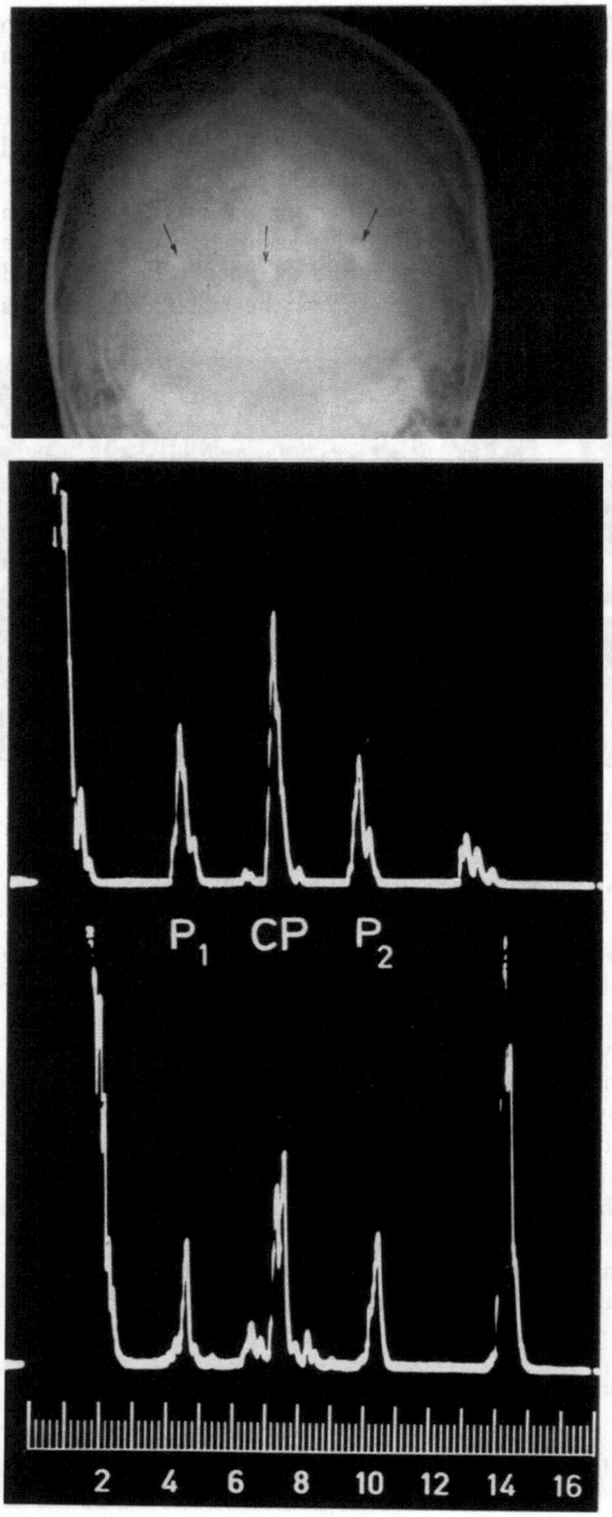

c) Reflections from the Sylvian Fissure Region

On horizontal sounding from a temporal application point reflections from the Sylvian fissure can be detected with great regularity (LITHANDER, 1961; TAYLOR et al. 1961; JEFFERSON, 1962; SCHIEFER et al., 1963). These are usually several echoes lying close together about 2 to 2.5 cm ahead of the end echo (Fig. 44). Often a pulsation of one or more of these reflections can be recognized. The cause of this pulsation is the middle cerebral artery or the branches of the middle cerebral artery in the Sylvian fissure (see also p. 48 and 76).

d) Reflections from the Subarachnoid Spaces

Not uncommonly, smaller echoes originating at the interfaces of the subarachnoid spaces are found ahead of the end echo. Especially in infants and children such reflections can be seen frequently (see DISTEL and KASPER, 1968). They appear to an even greater extent in older people with atrophic brain lesions (Fig. 45).

Fig. 43. Reflections from the choroid plexus of the lateral ventricles. a) On the plain film of the skull, two areas of choroid plexus calcification appear on either side of the pineal body. b) Echo-encephalogram of the same patient: beside the echo of the pineal body (CP), high reflections with a broad base originating from the calcified plexus are seen (P_1 and P_2). Because ultrasound is strongly absorbed by the calcium, only a small end echo remains. Pat. G. P., 57 years of age. Echo-EG No. 2598/66. c) Ultrasonic display of a 20-year-old man with reflections from the choroid plexus of the lateral ventricles without radiologically visible areas of calcification. Pat. W. N., Echo-EG No. 2025/65

e) Other Lateral Reflections

Aside from the lateral reflections described above, all of which develop at the interface between brain tissue and CSF or from vessels and calcifications, ultrasound is also reflected from the interface between white matter and gray matter (thalamus, internal and external capsule)

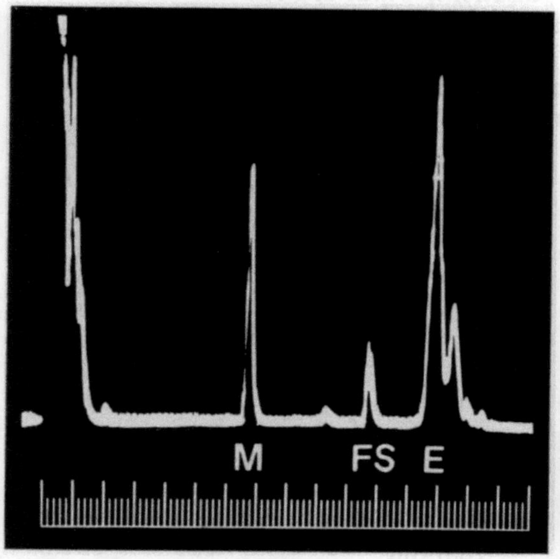

Fig. 44

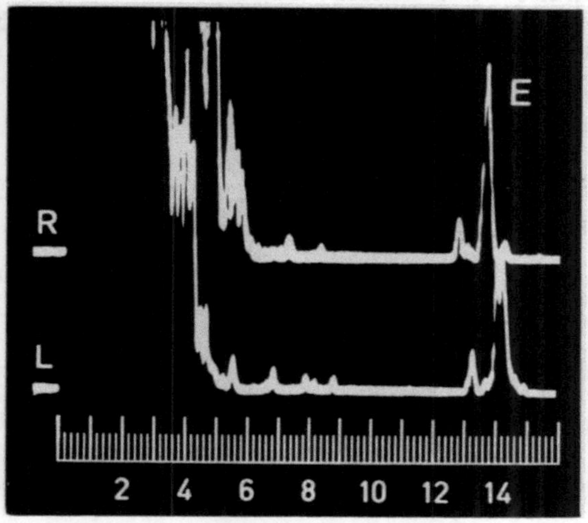

Fig. 45

Fig. 44. Normal echogram with reflections from the Sylvian fissure (FS)

Fig. 45. Reflections from the subarachnoid spaces appear as small peaks ahead of the end echo in an older man

under optimal reflection conditions (LITHANDER 1961). The experiments were carried out on formalin hardened cadaver brains, however. This alters the conditions of reflection. Therefore we share JEPPSSON's opinion that such reflections do not reach a diagnostically useful amplitude in vivo. DREESE and NETSKY (1964) also showed reflections between white and gray matter, but these too, were not obtained under circumstances comparable to clinical conditions. The authors sounded considerably deformed brain sections which were about 1 cm thick and preserved in a plastic container filled with 10% formalin. The echograms obtained in this manner were superimposed on photographs of the brain specimens. The authors believed that individual reflections could be grouped with certain cerebral structures in this way.

In experimental investigations on formalin hardened cadaver brains GROSSMAN (1966) found reflections from the area of the internal capsule where it adjoins the thalamus and the lentiform nucleus. He assumes that these reflections can be detected in live subjects as well.

A further investigation on this topic has been published by EMERY (1967). He sounded 40 brains of children with normal or hydrocephalic ventricles who had just died. Echoes from the ventricular walls of newborns could not be detected. He believes this is due to the slight difference between the acoustic impedances of brain tissue and CSF in this age group. On the other hand echoes seemed to develop at the junction between myelinated and non-myelinated brain tissue. The most consistent echoes came from tracts of myelinated nerves running almost at right-angles to the ultrasonic beam. However, it is definitely necessary to undertake further investigations of this problem before anything definite can be said. Without doubt, numerous small echoes develop when ultrasound passes through normal brain tissue, which can be especially well registered in children because of their thin skull bones. SJÖGREN (1965) describes these reflections as "*parenchymal echoes*". These echoes may be of decisive importance in differentiating brain parenchyma from fluid-filled cavities.

5. Echo-Pulsations

LEKSELL, in his first paper on echo-encephalography (1955/56) first pointed out pulsation phenomena synchronous with the heartbeat. DE VLIEGER and RIDDER (1959) tried to explain these pulsations by changes in the cerebral volume during the systolic injection of blood into the cerebral hemispheres. In our opinion these pulsations, which are 180° out of phase, are due to a change in the curvature of the walls of the third ventricle during the impact of the pulse wave whereby the reflection conditions are altered and the echo peaks become correspondingly larger or smaller (see also MCKINNEY, 1964). Fig. 46 is to show a theory to explain this phenomenon, which grew out of DE VLIEGER's interpretative attempts.

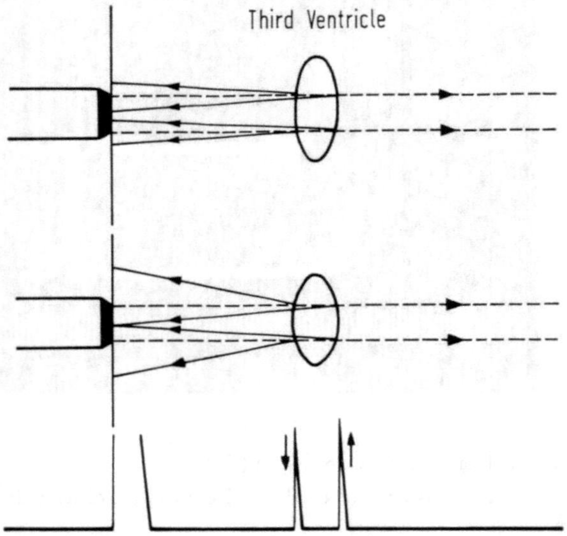

Fig. 46

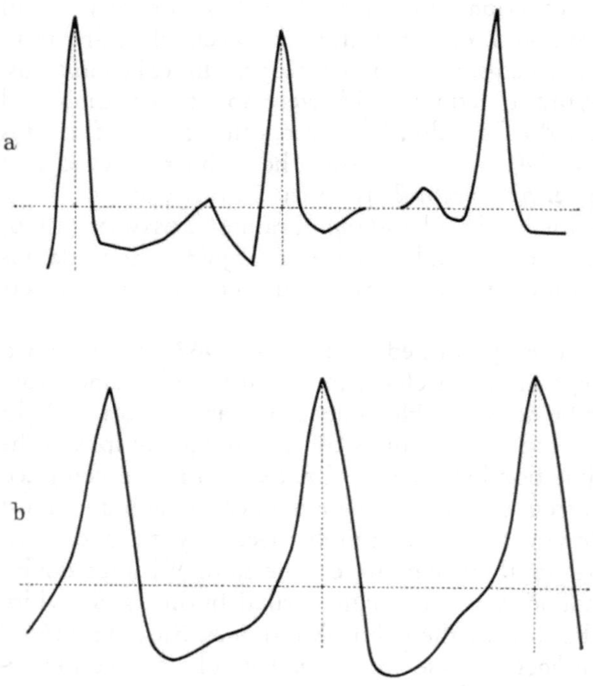

Fig. 47

JEPPSSON (1964 and 1968) as well as TER BRAAK and DE VLIEGER (1965) described an echo-encephalographic technique for the *measurement of intracranial pressure* which makes use of the pulsations just described. According to these authors deductions about the intracranial pressure relationships (see Fig. 47) can be made from the rise times of continously recorded pulsation curves. Perhaps it will be possible to develop a method for measuring the absolute pressure on the intact skull from this.

While the above-mentioned pulsations probably come from the walls of the ventricles, individual intracerebral vessels can give rise to strongly pulsating echoes as well. When the common sounding technique is used, pulsating echoes are most easily obtained from the middle cerebral artery i. e. the vessels of the Sylvian fissure. Such pulsations are more clearly seen in children than in adults. According to FREUND (1965) they can be recorded symmetrically in healthy subjects and in patients with an intact arterial cerebral circulation. In cases of unilateral carotid occlusion, the homolateral pulsations are said to be absent or markedly decreased in comparison with the healthy side. Attempts to register

Fig. 46. Diagram of the development of pulsations of ventricular wall echoes. The curvature of the walls of the third ventricle is altered by the influence of the pulse wave. For this reason ultrasound is reflected back to the probe in different amounts from the two sides which results in echo pulsations 180° out of phase

Fig. 47. Continuous recording of echo pulsations in a patient with increased intracranial pressure a) before, b) after the infusion of urea. When the pressure is lowered, the rise time is prolonged (from JEPPSSON, ST.: Acta chir. scand. **128**, 218, 1964)

pulsations in the echo-encephalogram continously and to develop a clinical method for the diagnosis of carotid artery thromboses and vascular occlusions as well as stenoses of larger vessels are not as yet complete (FREUND, 1965, 1966; FREUND, KAPP and KENDEL, 1967).

BUSCHMANN (1964) described "carotid echography" for the diagnosis of carotid thromboses for which we often have to consult the ophthalmologist because of the ocular symptoms and the need for ophthalmodynamometry. An ultrasonic pulse apparatus with the requisite miniature probes (10 MHz) was constructed especially for ophthalmological use and was found to be suitable. The wall of the carotid artery adjacent to the probe and the one opposite the probe produce echoes whose pulsations can be demonstrated on the screen and registered continuously. Thereby variations in the lumen, in the thickness of the wall and in the pulsation as well as pathological deposits within the lumen of the vessel can be recognized.

B. The Pathological Echo-Encephalogram

Disorders of the brain can produce a series of diagnostically significant changes in the echo-encephalogram:

1. a displacement of the midline echo
2. the appearance of abnormal ventricular echoes
3. the appearance of unusual reflections in cases of brain tumors ("tumor echoes"), hematomas ("hematoma echoes"), cysts and pathological calcifications
4. the appearance of abnormal echo pulsations.

1. Demonstration of a Supratentorial Mass Shift by a Displacement of the Midline Echo

The midline echo shift is the most important pathological echo-encephalographic sign indicating a lateral displacement of the cerebral midline.

Most authors speak of a pathological midline echo if there is a difference of at least 4 mm between the two measurements (calculated for brain tissue) from the left and from the right. The genuine displacement of the M-echo from the theoretical midline is then 2.0 mm since the displaced midline echo is first registered 2.0 mm to the left of the calculated midline and 2.0 mm to the right the second time (see diagram in Fig. 48). Thus the actual displacement of the midline structures is half the difference between the two measurements, providing the end echo is recorded at the same position on the screen on sounding both from the right and from the left. We arrived at this borderline value on the basis of personal investigation of 3600 subjects. Only 9 cases were found where the displacement of the M-echo exceeded 1.5 mm in the absence of a space-occupying lesion. A number of these patients is still under observation today (compare also p. 89). Among 149 healthy adults the greatest deviation from the theoretical midline of the brain was 1.5 mm (5 cases), the average however, was only 0.28 mm (Table 6). According to JEPPSSON (1961) "when an M-echo shift of more than 2 mm is recorded in adults, an intracranial expansivity is present with a probability of 99%".

On the other hand, several publications reported a borderline value of 3.0 mm (LITHANDER, 1961; KRAMER, 1968). PLANIOL et al. (1964) even reported a midline echo shift of 4 mm as being significant, while values between 2 and 4 mm were regarded as questionable. However, such a high displacement of the M-echo can no longer be considered to be normal under any circumstances. Today PLANIOL (1968) is of the opinion that an M-echo displacement around 2 mm must be regarded as pathological in cases of suspected brain tumors.

Very important in ultrasonic investigations which reveal only small midline echo displacements (2.0 to 4.0 mm) is the *determination of the width of the third ventricle, at the same time*, because otherwise such shifts can be simulated even in healthy subjects (see p. 80).

As already mentioned, in most cases the diameter of the third ventricle can be determined echo-encephalographically. Exceptions to this rule are the more pronounced M-echo shift and lesions which lead to *oblique* positioning of the third ventricle so that the ultrasound reflected from the walls of the third ventricle no longer reaches the probe. By applying the probe in the temporal area somewhat above and in front of the ear, ultrasound can be pointed directly at the lateral walls of the third ventricle so that a double echo subsequently appears on the screen of the cathode ray tube. As mentioned above, the distance between the two reflections corresponds to the diameter of the third ventricle. Not infrequently there is still another echo visible between the two reflections which often has a double peak caused by the posterior part of the septum pellucidum (if the ventricular

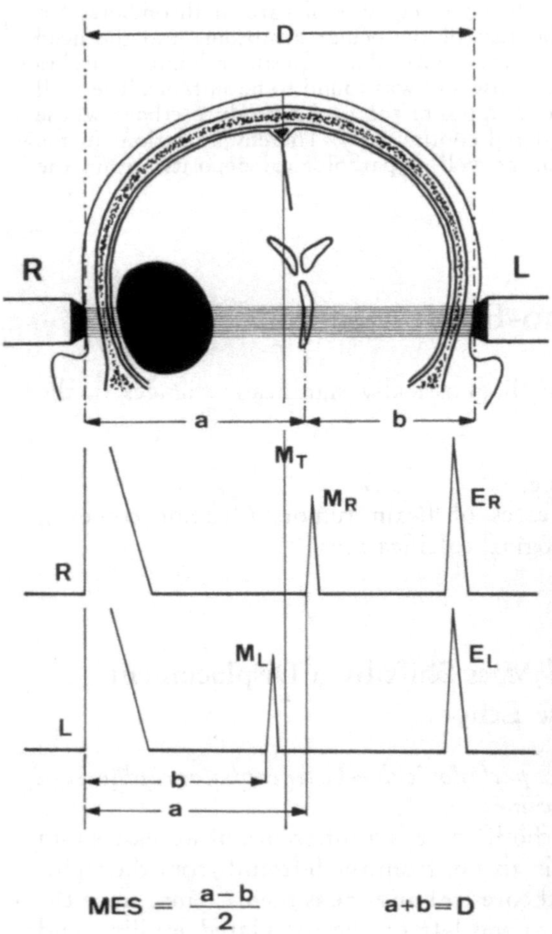

Fig. 48

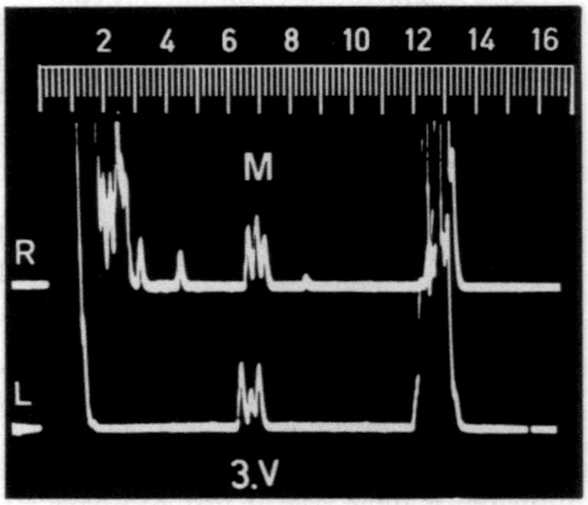

Fig. 49

Fig. 48. Diagrammatic presentation for calculating the amount of midline echo shift (MV). D=bitemporal diameter of the head, M_T=theoretical midline echo, M_R=midline echo from the right, M_L=midline echo from the left. The midline echo shift (MES) equals half the difference between the two midline echo measurements

Fig. 49. Echo-encephalogram of a healthy (as far as the brain is concerned) 13-year-old boy demonstrating the third ventricle and the midline echo. Upper trace: the midline echo (M) which comes from the septum pellucidum rises above the two reflections from the walls of the third ventricle which is 5 mm wide. Lower trace: the septum pellucidum echo appears between the echoes of the third ventricle (3. V). This reflection sometimes has a small double peak. Pat. F. N., Echo-EG No. 2160/66

system is midline and the application point of the probe is sufficiently high and displaced somewhat towards the front!) (Fig. 49). This reflection or the echo from the pineal gland obtained more posteriorly then represents the actual midline echo. Therefore one should always try to obtain all 3 or all 4 echoes in borderline cases in order to avoid errors in measurement. If the same wall of the third ventricle is recorded once from the left and once from the right, an apparent displacement of the midline echo is obtained which equals half the diameter of the third ventricle. Schiefer, Kazner and Brückner were the first to draw attention to this error in 1963. In most cases this results in a midline echo shift of 2 to 3 mm because the average width of the third ventricle in a healthy subject is 4 to 6 mm. In a hydrocephalic patient such an error is much greater (see also p. 80 and Fig. 82).

The degree of midline echo shift is largely dependent on the location of a space-occupying lesion. The echo-encephalographic measurements can best be compared with pneumoencephalographic results or with sagittal plain films of the skull, if a calcified pineal body is visible here. As already mentioned, the pineal gland and the posterior part of the third ventricle generally play the most important roles in the development of the midline echo in echo-encephalography. As expected, space-occupying lesions in the frontal, parasagittal and mediobasal region as well as in the oral brain stem produce the least lateral displacement of the third ventricle and the pineal gland (Fig. 88 and 91). The most prominent midline echo shifts are found with temporal space-occupying lesions, followed by occipital processes whose great displacement may be explained by the fact that the occipital lobe's only avenue of escape is anteriorly since it is enclosed between the falx and the tentorium (see Tönnis and Schiefer, 1953). This opinion contradicts the results of other authors who report that occipital tumors frequently do not cause a displacement of the midline structures (Planiol et al., 1964, 1967, 1968; Akermann and Guiot, 1965; Bostem et al., 1968). In all cases of occipital space-occupying lesions our measurements were very carefully compared to the neuroradiological findings and complete agreement was found between the two methods (see also Fig. 87 and 88).

When the usual method of sounding from the temporal area is applied, the septum pellucidum is out of the question as a source of the midline echo in cases of *supratentorial midline shifts* in our opinion. The following two cases will support this point.

Case 1: Helmut B., 29 years of age. Echo-EG No. 2399/66. Left fronto-lateral astrocytoma. In the echo-encephalogram a midline echo shift of about 2 mm is found. The pneumoencephalogram on the other hand shows a septum pellucidum shift of 7 to 8 mm from left to right. Repeated control echograms also failed to reveal more extensive midline echo shift thus excluding the possibility of an error in measurement (Fig. 50).

Case 2: Heide Z., 2½ years of age. Echo-EG No. 1927/65. Right temporal astrocytoma. 5 to 6 mm M-echo shift to the left in the echo-encephalogram. The pneumencephalogram shows a slanted and displaced third ventricle, the septum pellucidum, however, is exactly midline (Fig. 51).

According to Ford (1968) the choice of a higher and more anteriorly situated application point allows the detection of reflections from the septum pellucidum, which are of diagnostic importance especially in case of frontal tumors. With such tumors the echo of the septum pellucidum is displaced even if the echo of the third ventricle or the pineal gland remains midline. This echo is often flanked by the reflections from the lateral ventricles. In 1961 already Taylor, Newell and Karvounis reported a midline echo shift in the case of a frontal tumor while the pineal body radiographically was midline. We ourselves made a similar observation (Schiefer et al., 1963). In these cases echoes must have been recorded from the shifted septum pellucidum.

In correlating the angiogram and the ultrasonic measurements, the *internal cerebral vein* must be used as a point of reference. Particularly in cases of frontal space-occupying lesions there may be considerable discrepancy between the behavior of the midline echo and the shift of the anterior cerebral artery (Fig. 52). On the other hand, a shift of the anterior cerebral artery is largely missing in cases of temporal or occipital space-occupying lesions although the internal cerebral vein is considerably displaced (Fig. 53 and 87). By comparing echo-encephalograms and angiograms showing displacements Jeppsson (1961) and Lithander (1961) were able to show that the agreement is greatest between the midline echo shift and the position of the internal veins.

The most detailed investigation on the *correlation of the midline echo with the median structures of the brain on x-ray* was carried out by Brinker, King and Taveras (1956) on 287 patients, the results of which appear in Table 2: the last difference was found between the displaced midline echo and the calcified pineal body (1.00 mm), the posterior part of the internal cerebral vein (1.09 mm) and the posterior part of the third ventricle (1.20 mm).

On the other hand, the differences between the position of the M-echo and the anterior cerebral artery, the pericallosal artery and the anterior part of the third ventricle were quite considerable in some cases. These investigations present further proof for the hypothesis that the midline echo is a reflection from the calcified pineal body and the posterior part of the third ventricle.

Table 2. *Correlation of midline echo and midline structures* (Brinker, King, and Taveras, 1965)

Differences between the Midline Echo and	Mean (mm)
Pineal gland, posterior third ventricle, and posterior internal cerebral vein	1.15
Pineal gland	1.00
Posterior internal cerebral vein	1.09
Posterior third ventricle	1.20
Anterior third ventricle	2.55
Anterior internal cerebral vein	2.65
Anterior cerebral artery	3.43
Pericallosal artery	3.83
Septum pellucidum	2.00

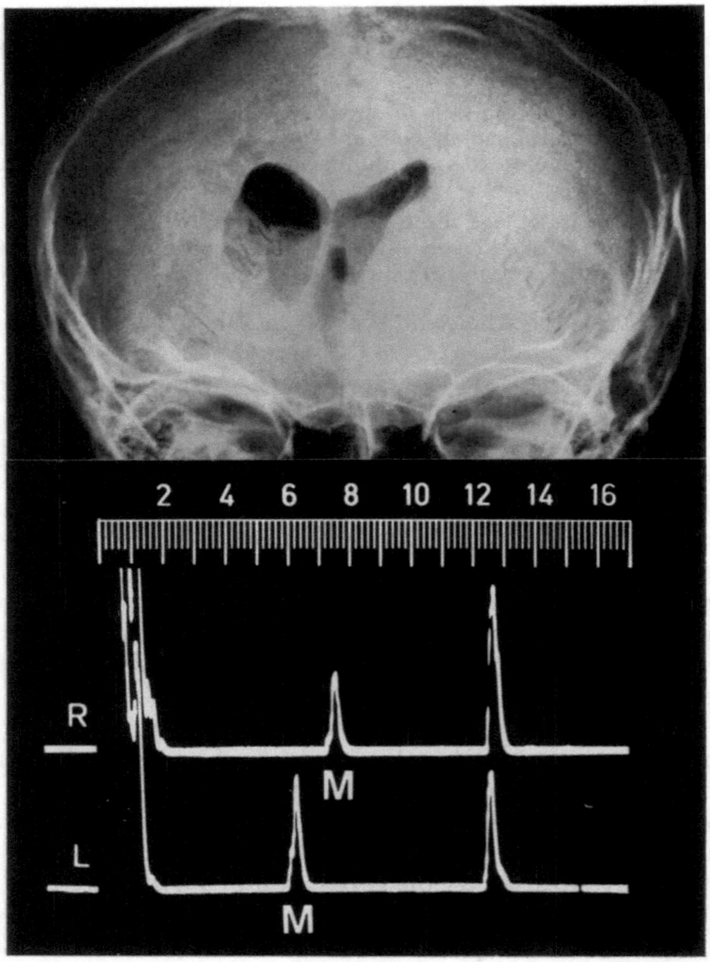

Fig. 50. Dissociation between the displacement of the septum pellucidum and the midline echo. Bottom: echo-encephalogram of a 29-year-old man with a 2 mm displacement of the midline echo from left to right due to a left fronto-lateral astrocytoma. Top: in contrast the pneumoencephalogram of the same patient reveals a considerable displacement of the septum pellucidum while the third ventricle shows considerably less shift. Pat. H. B., Echo-EG No. 2399/66

Identification of the Midline Echo

In most cases the recognition of the midline echo does not pose any problems since it is the most prominent and most constant reflection in the central area of the echo-encephalogram. Additional criteria for the identification of the Me-echo were described by LEKSELL (1955/56): the broad base, the rise and fall of the echo synchronous with the pulse and the fact that the midline echo often shows multiple peaks. According to our own experience, the pulsation of the midline echo is often absent especially in patients with displaced midline structures so that this sign contributes very little to the identification of the midline echo. With the help of the other criteria mentioned here, however, the correct identification of the midline echo by the experienced investigator is achieved in 98% of cases. But we want to point out here too, that the type of diagnostic ultrasound equipment influences the results also (compare also BOSTEM et al., 1968; JEFFERSON and HILL, 1968). With some equipments it is very difficult to separate the actual midline echo from other reflections in the median area because many of the echoes have a very great amplitude. The scepticism of some authors about results which

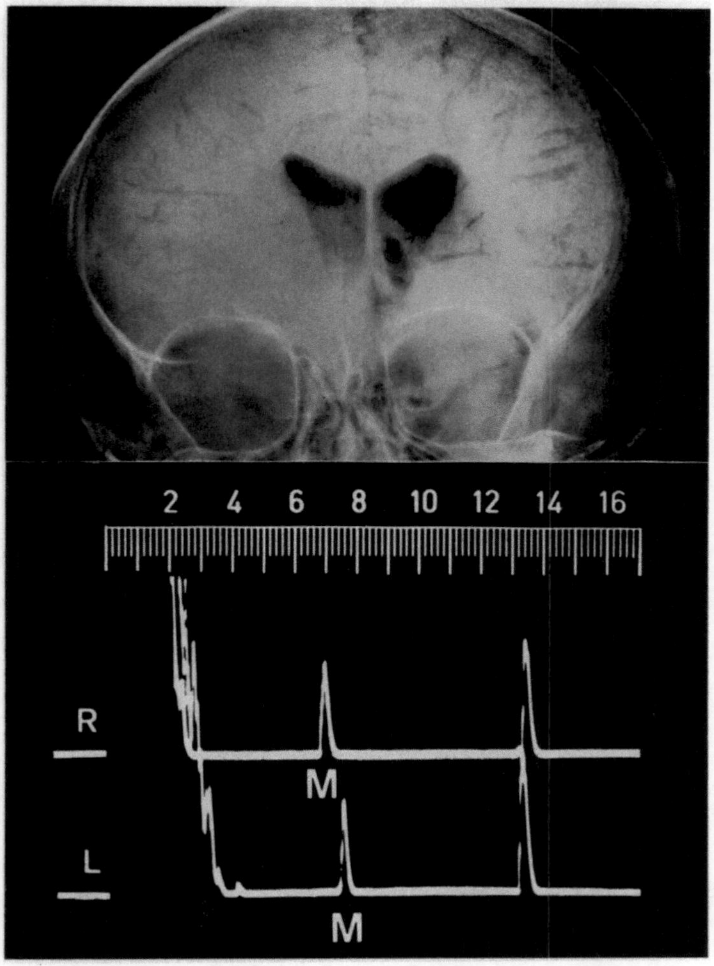

Fig. 51. Dissociation between septum pellucidum displacement and midline echo shift. Bottom: Displacement of the midline echo by 5 to 6 mm from right to left in a $2^1/_2$-year-old girl with an astrocytoma located in the right temporal lobe. Top: the pneumoencephalogram of the same childshows a normal septum pellucidum, while the third ventricle is clearly displaced. Pat. H. Z., Echo-EG No. 1927/65

a

b

c

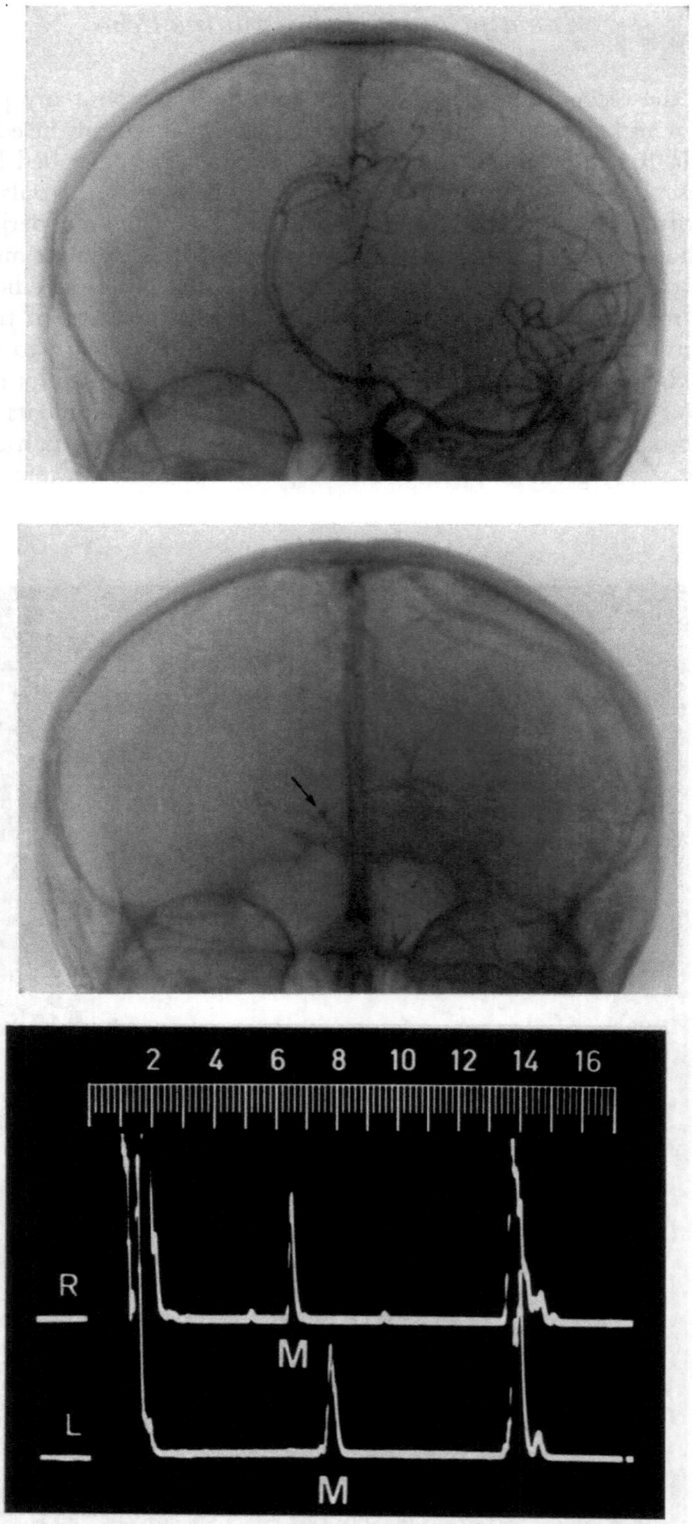

Fig. 52. Comparison between angiogram and echo-encephalogram of a frontal tumor. a) arched displacement of the anterior cerebral artery 2 cm to the right, b) on the other hand, the internal veins are shifted towards the right by only 6 mm (arrow); c) the echo-encephalogram also indicates a 6 mm displacement towards the right so that it can be correlated only with the phlebogram. Pat. J. P., 61 years of age. Echo-EG No. 1961/65

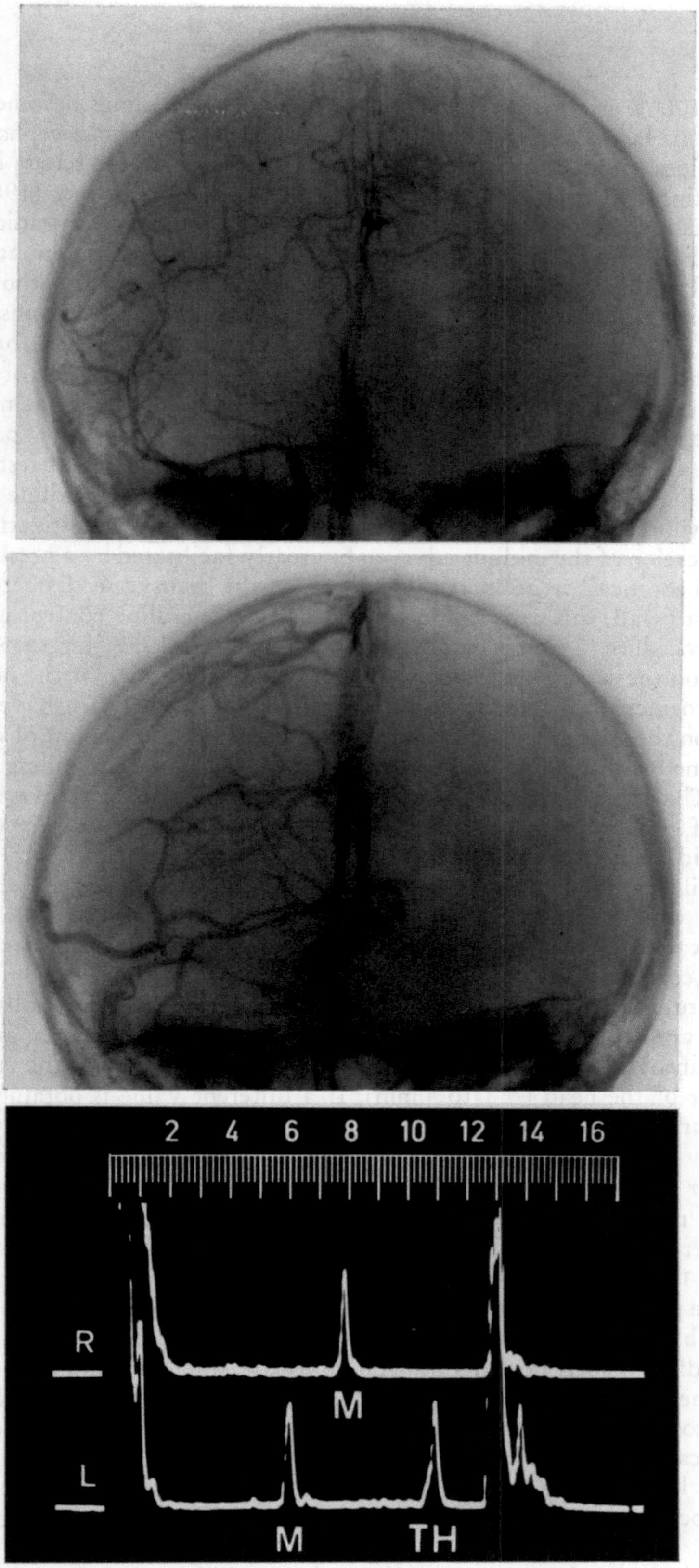

Fig. 53. Comparison between angiogram and echo-encephalogram of a tumor located deep in the right temporal lobe. a) No displacement of the anterior cerebral artery, b) internal veins are shifted 10 mm towards the left, c) in agrement with the phlebogram, the echo-encephalogram shows a midline echo shift of nearly 10 mm Pat. H. B., 29 years of age. Echo-EG No. 1726/65

show a nearly 100% agreement with the neuroradiological findings becomes understandable in the light of this fact. Every physician who is familiar with echo-encephalography, knows that it is possible to produce a high amplitude echo anwhere on the screen by employing the appropriate tilting and rotating movements of the probe; McKinney (1964) demonstrated this very impressively in a film shown at the VII. Symposium neuroradiologicum in New York. The skill in echo-encephalography lies in the ability to differentiate such reflections from the actual midline echo. This is only possible if the investigator knows exactly which area he is scanning with the ultrasonic beam at any given moment. Care must always be taken to see that the sound axis is at right angles to the median plane of the brain because even slight deviations from the perpendicular may lead to the appearance of echoes from undesirable structures and thus to errors in measurement (e. g. echo from the midline falx in the case of a space-occupying lesion, see Jeppsson, 1961). The application point of the probe must always be checked as well. It is best to investigate with the hand resting firmly on the head (see also Fig. 25). Mentally the physician must project the midline structures on the temporal area which is being scanned with the probe during the investigation.

The identification of the midline echo can be greatly facilitated by a preceding examination in which the theoretical or geometrical midline of the transverse diameter of the skull is calculated. First of all, this can be done by the so-called midline control using the through-transmission procedure (2 probes) (de Vlieger and Ridder, 1959; Jeppsson, 1960). The echo which appears on the screen here, shows the mathematical midline of the head providing the calibration is correct. In a normal case the midline echo would then have to appear at the same position on the screen. If the midline echo is displaced, the amount of deviation distance from the midline control is the same regardless of whether one is sounding from the right or from the left. The other method is to determine the diameter of the skull with callipers at the exact points of probe application (see Fig. 27). By taking half of this value we find the "theoretical midline". In practice the midline echo usually appears somewhat in front of this line (in children about 1 mm, in adults 1 to 2.5 mm), because the velocity of sound in bone is higher while the apparatus is calibrated to the velocity of sound in brain tissue. Several American equipments (e. g. Hoffrel) with fixed scales which can be blended in electronically, correct this distortion a priori. In this way the ultrasonic distances which are of special interest to us, namely those of brain tissue or CSF especially in children with hydrocephalus are depicted as being somewhat larger than they actually are. We regard this effect as undesirable. When the two midline echo values are added, the sum must be equal to the diameter of the head (±1 to 4 mm). If a different value is obtained the M-echo has not been measured correctly.

Hardly any other method of investigation is so dependent *on the personal experience of the investigator* in obtaining good results as is echo-encephalography. Therefore the beginner should start by recording as many echo-encephalograms as possible including those of healthy subjects in order to become familiar with the interpretation and evaluation of the oscillogram. White (1964, 1965, 1966) has pointed out this disadvantage repeatedly and has tried to standardize the procedure so that it can be performed by a technician. In our opinion this will barely be possible since echo-encephalography has progressed so far that the mere determination of the midline echo is no longer adequate today. Interest in the echo-encephalographic findings obtained in cases of ventricular dilatations, hematomas and tumors continues to rise. In order to arrive at satisfactory results here, the investigator must be informed on the entire clinical case through a knowledge of the history and the clinical findings, and possess adequate knowledge on the pathological-anatomical relationships in intracranial disorders such as space-occupying and atrophic lesions. This can hardly be expected of paramedical personnel.

In case difficulties arise in the recording of the midline echo, the following possibilities which are an extension of the points mentioned by Grossman (1966) must be considered:

1. Incorrect setting of the apparatus (e. g. insufficient pulse power, switch from through-transmission to reflection method forgotten).

2. Wrong application point of the probe.
3. Poor contact between the probe and scalp due to insufficient amounts of coupling medium or due to hair lying between probe and scalp.
4. Unusually thick skull bone, possibly with pneumatization extending into the temporal bone particularly in older people. Try 1 MHz probe.
5. Subgaleal hematoma, possibly combined with a comminuted fracture of the temporal bone and a contusion of the underlying brain. Investigation is often possible only from one side then.
6. Presence of a space-occupying lesion near the third ventricle, the quadrigeminal region, the basal ganglia, the pons or clivus.
7. Presence of a space-occupying lesion in the temporal lobe directly under the probe. The M-echo is then usually easy to identify from the opposite side.
8. Presence of an internal hydrocephalus especially in infants and children. The echogram is remarkable because of exceptionally numerous and high reflections.
9. Massive hematoma into the ventricular system resulting in a destruction of the normal anatomic structures.

2. Demonstration of Ventricular Dilatation

a) Recording an Abnormal Double Echo from the Walls of the Third Ventricle

In healthy subjects the width of the third ventricle gradually increases during life and reaches a maximum of 7 to 7.5 mm at an advanced age. This corresponds to a pneumoencephalographic finding of 9 to 10 mm, due to the enlargement caused by the divergence of x-rays. In younger patients (under 40 years) a third ventricular width of 7.0 mm must be regarded as pathological (SCHIEFER and KAZNER, 1967; FEUERLEIN and DILLING, 1967; HUBER and PATIRI, 1968) (see Fig. 39).

If a *double echo* from the third ventricle with a diameter of more than 7.0 mm is found in the echo-encephalogram (Fig. 54) there is probably a pathological intracranial lesion present (a disturbance in the flow or resorption of CSF, brain atrophy).

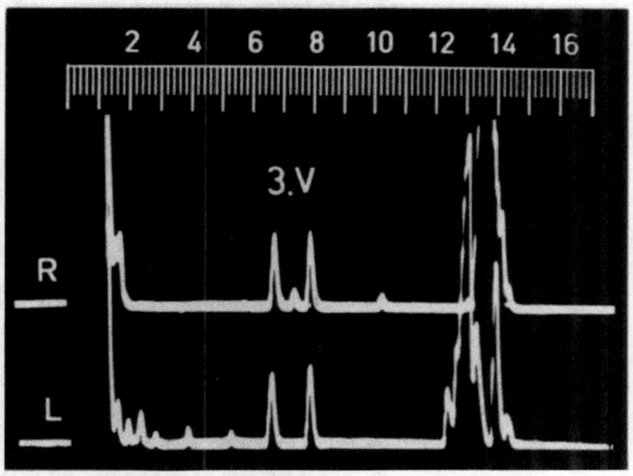

Fig. 54. Abnormal double echo from the walls of a dilated third ventricle (3. V). The distance between the two reflections equals the width of this ventricle and is 13.0 mm in this case. Pat. F. R., 10 years of age. Echo-EG No. 1440/64. The patient had had tuberculous meningitis

The investigation technique for demonstrating the normal or abnormal double echo from the walls of the third ventricle is shown diagrammatically in Fig. 55. By directing the ultrasonic beam from an application point directly above the root of the ear either perpendicularly or tipped slightly upwards on the third ventricle, one obtains two distinctly separated reflections in the central area of the echo-encephalogram if the third ventricle is dilated. After some practice the two echoes from the lateral walls of the third ventricle can be made visible on the screen without great difficulty in most cases in a short time. The distance between the two reflections corresponds to the diameter of the third ventricle (Fig. 56). Occasionally one sees a pulsation 180° out of phase which is produced by changes in the form of the lateral ventricular walls during the arrival of the pulse wave (see also p. 48). At this probe application point the diameter of the posterior part of the third ventricle is measured. The development of high reflections is favored by the fact that the temporal bone and parts of the walls of the third ventricle are largely parallel (FISCHER, 1966). If the probe is moved so that the ultrasonic beam slowly leaves the vicinity of the third ventricle and is directed at the pineal body, one

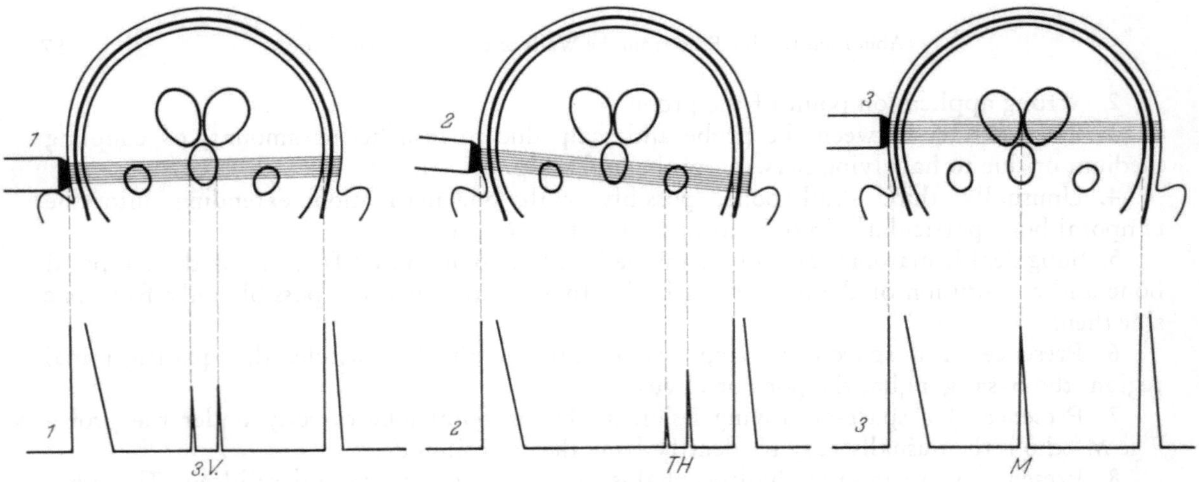

Fig. 55. Sounding technique in the presence of ventricular dilatation. 1. Demonstration of the double echo originating from the walls of the third ventricle (3. V). 2. Demonstration of reflections from the walls of the inferior horn (TH) opposite the probe. 3. Midline echo from the septum pellucidum. To obtain this echo, the probe must be placed somewhat higher and more anteriorly than usual

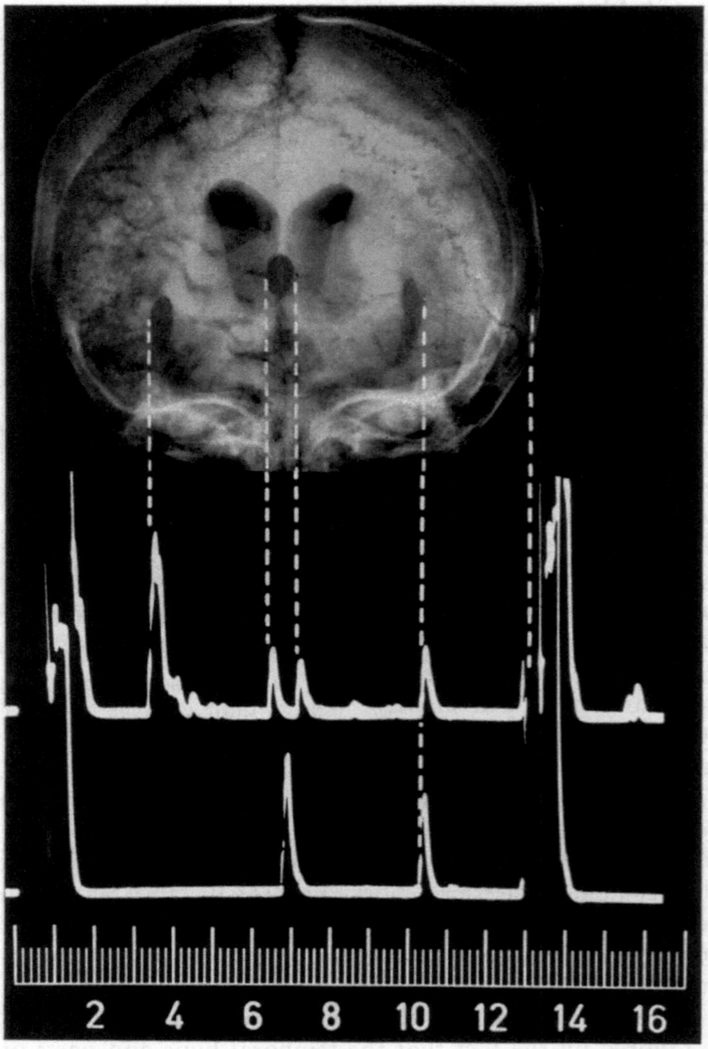

Fig. 56. Echo-encephalogram and pneumoencephalogram of a 2-year-old child with slight ventricular dilatation following encephalitis. Complete agreement between the two methods. Pat. U. W. Echo-EG No. 2091/66

recognizes how a single reflection with a broad base, originating from the pineal body, evolves out of the double echo from the third ventricle and that the double echo finally disappears (see Fig. 57).

Even in cases of ventricular dilatation, it is not always possible to obtain the double echo of the third ventricle from the application point directly above the root of the ear. Especially in cases of cerebellar tumors that displace the third ventricle anteriorly and upwards by pressing on the pons a more frontally and higher situated application point for the probe must be chosen (see Fig. 58). Such a technique of investigation is also necessary for tumors in the posterior part of the third ventricle.

Particularly the beginner has *difficulties in identifying the two reflections from the lateral walls of the third ventricle* and in demonstrating them on the screen without difficulty. The important thing is to consider the possibility of a ventricular dilatation. The first indication in this direction is an unusually large number of high echoes. If a general symmetry is found during the determination of the midline echo, an attempt should be made to pick up reflections from the lateral walls of the third ventricle. When the probe is slowly moved back and forth and the direction of sounding altered, it is nearly always possible to demonstrate the needle-sharp echoes from the lateral walls of the third ventricle. Through minimal alteration in the direction of the probe, a clear echo of the third ventricle may usually be obtained. Not infrequently the reflections reach the height of the entire screen even in adults. The amplitude of the echoes from the third ventricle depends on the thickness of the scalp and skull on the one hand and on the width of the third ventricle on the other. In the case of a dilated third ventricle, the reflection conditions for ultrasound improve because the reflecting surface of the third ventricle is enlarged and the radius curvature is altered in this portion of the ventricle.

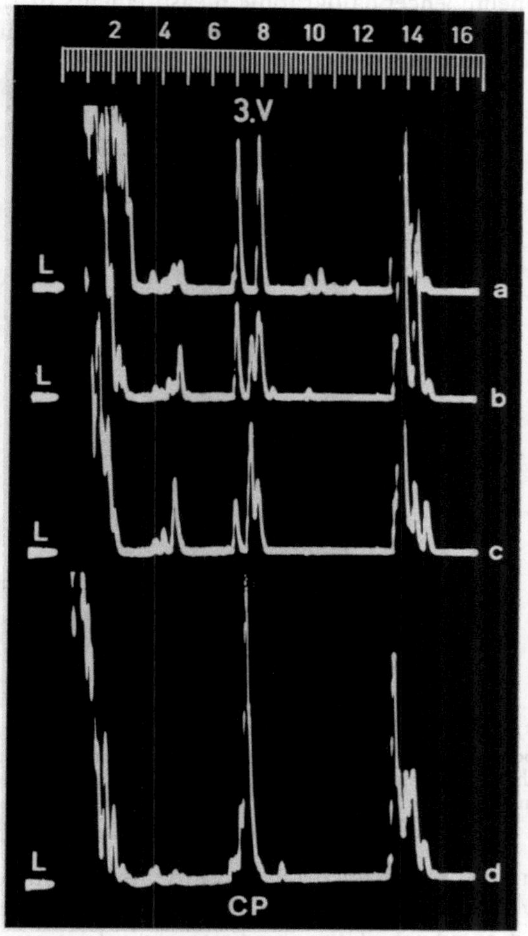

Fig. 57. Development of the M-echo from the calcified pineal body out of the double echo from the third ventricle in a 50-year-old man with a cerebellar tumor. a) Double echo from the lateral walls of the third ventricle. b) By directing the ultrasonic beam slightly posteriorly, the amplitude of the double echo decreases and an additional echo appears between the two reflections from the third ventricle. c) The echo from the pineal body already rises above the echoes from the third ventricle. d) Further angulation of the probe results in an increase in the amplitude of the echo originating from the pineal body (CP), but the double echo from the third ventricle can no longer be obtained because the ultrasonic beam is not scanning this structure any more. (Pat. M. P. Echo-EG No. 3598/68)

The echo-encephalographic demonstration of the ventricle is most successful between 7.0 and 20.0 mm. If the ventricle is narrower, errors in measurement may occur; the same is true for extremely wide ventricles, especially in infants and children.

The question to what degree there is actually an agreement *between the ultrasonic measurement of the third ventricle and the pneumoencephalogram,* was examined in detail by us. In 187 patients who showed a double echo in the center of the echo-encephalogram which had originated from the lateral walls of the third ventricle in our opinion, comparisons were made with pneumoencephalograms or ventriculograms (SCHIEFER and KAZNER, 1967). In 183 cases, there was very good correlation between the echo-encephalographic and pneumoencephalographic measurements of the third ventricle, considering the usual enlargement by x-rays. The differences in the measurements obtained with the two procedures were

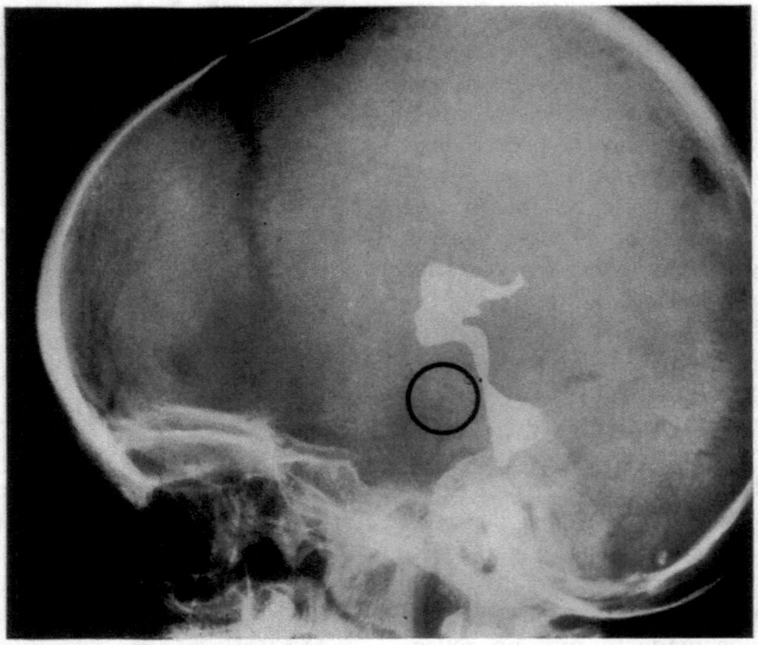

Fig. 58. Pantopaque ventriculogram of a 3-year-old boy with a cerebellar astrocytoma. The third ventricle has been displaced so far anteriorly and upwards that it is no longer possible to pick up the double echo of the third ventricle from the usual application point of the probe (black circle). Therefore the probe must be placed more anteriorly and higher up. There is no constant probe application point for the echoes from the third ventricle. Pat. F. Sch. Echo-EG No. 2532/68

in the range of 1.5 mm. Only in 4 patients was the echo-encephalographic width of the ventricle not confirmed. Here reflections from the walls of the lateral ventricles were regarded as echoes of the third ventricle. It is remarkable that the diagnosis of two of these cases was fresh subarachnoid hemorrhage. With this clinical picture, one often sees very many reflections on the screen, so that misinterpretations can occur in our experience. Aside from a few exceptions, pneumoencephalography was performed 1 to 2 days after the ultrasonic measurement. Therefore echo-encephalography can be regarded as a reliable method for the determination of the width of the third ventricle. With a correlation of 97.8%, a degree of reliability is found which is almost as high as that of a mass shift with a midline echo displacement.

The great accuracy of the echo-encephalographic determination of the width of the third ventricle is confirmed by numerous authors (FORD and MCRAE, 1966; PIA and GELETNEKY, 1968; ULBRICHT and DE SEIXAS, 1968 among others). JEFFERSON and HILL (1966) agree that it is possible regularly to identify an enlarged third ventricle, but they are not convinced that there is always an exact relationship between the radiological and ultrasonic width of the third ventricle. Only WHITE (1967) is of the opinion that it is impossible to detect any structure other than the midline echo with certainty.

Comparative investigations of several autopsy specimens of patients examined by using echo-encephalography shortly before death, showed that the measurements obtained with ultrasound could be verified completely (Fig. 59); however, we do not regard this comparison as being very reliable, since the removal of the CSF during autopsy and fixation leads to significant alterations about the third ventricle. This is particularly true for more marked ventricular dilatations.

In summary it may be stated that the diagnostic application of echo-encephalography is by no means limited to the detection of a supratentorial mass shift. It is this very ability to measure the exact width of the third ventricle routinely within a few minutes, especially dilatations of the ventricles, that has established the position of this procedure in clinical diagnosis and transformed the ultrasonic investigation of the skull into *echo-encephalography*.

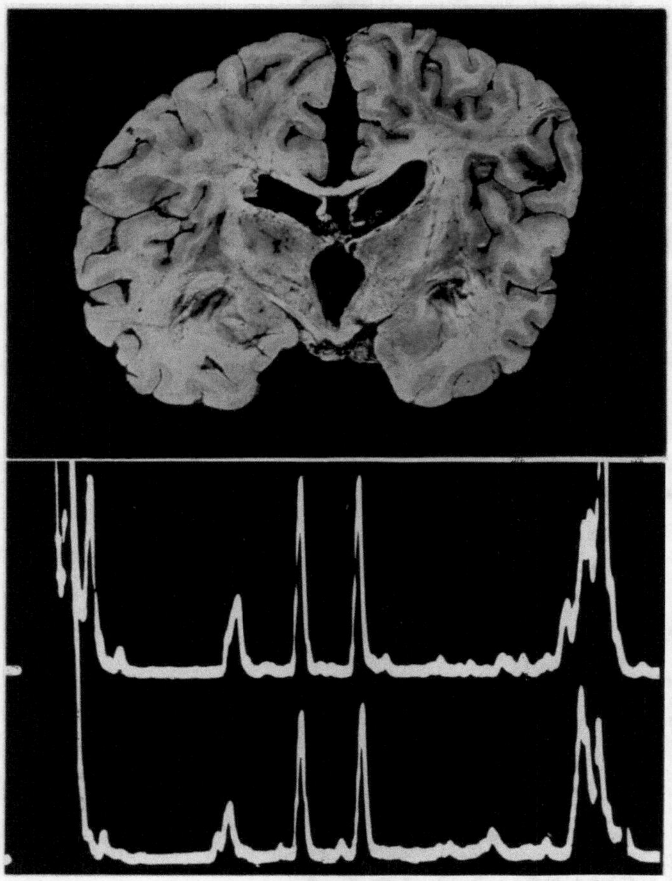

Fig. 59

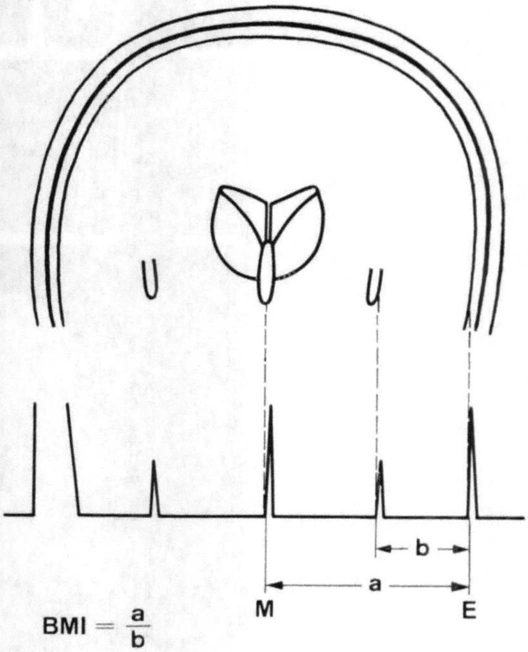

Fig. 60

Fig. 59. Determining the width of the third ventricle. Agreement between the echo-encephalogram and autopsy finding. The high reflection preceding the double echo of the third ventricle comes from the medial wall of the inferior horn on the side of the probe. Pat. A. L., 44 years of age. Echo-EG No. 1143/20/65

Fig. 60. Diagram for the calculation of the echo-encephalographic brain mantle index (BMI, SCHIEFER, KAZNER, KUNZE, 1965)

b) Registration of Abnormally Located Lateral Ventricle Echoes — Echo-Encephalographic Brain Mantle Index (BMI)

Besides the double echo of the third ventricle in cases of ventricular dilatation, the reflections originating from the temporal horn of the cerebral hemisphere opposite to the probe are of special interest. The application point for the probe differs very little from the position used for measuring the third ventricle. However, the ultrasonic beam must be directed downwards because the temporal horn cannot be picked up completely on horizontal sounding in many cases. The direction of the ultrasonic beam is shown in the diagrammatic sketch of the investigation technique in Fig. 55. Although the temporal horn of healthy subjects causes only a low-amplitude reflection which is sometimes absent entirely, the conditions of reflection improve as ventricular width increases. The concave curvature of the external wall of the opposite temporal horn enables us to register reflections from this structure on the echo-encephalogram much more easily (hollow mirror effect). For the same reason, one can often pick up a high reflection from the medial wall of the homolateral temporal horn

(Fig. 59). The inner wall of the contralateral temporal horn, on the other hand, rarely produces such a high echo as, for example, the case in Fig. 100 because the ultrasonic rays are reflected in divergent directions. We have described an *echo-encephalographic brain mantle*

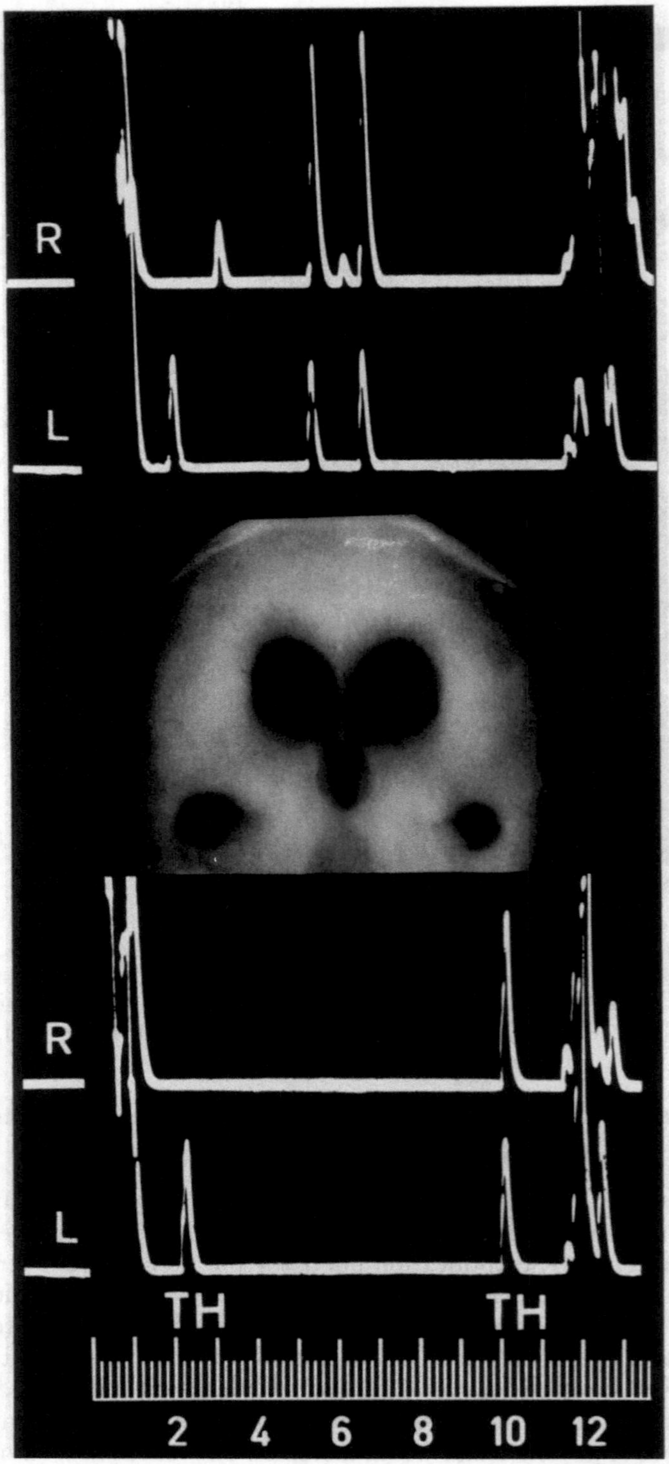

Fig. 61. Agreement between the echo-encephalographic findings and ventriculotomography. Top: double echo from the walls of the third ventricle 13.0 mm apart. Bottom: Echo from the lateral wall of the temporal horn. By using the site of this reflection, it is possible to calculate a BMI of 3.2, indicating marked ventricular dilatation. Pat. I. L., 9 years of age. Echo-EG No. 2438/66

index (BMI) in 1965 using the *position of the lateral wall of the temporal horn.* The calculation of this index is seen in the diagram in Fig. 60. The index shows the relation between the distances M-echo—end echo and echo from the lateral wall of the temporal horn—end echo. Normally this index is 2.0 to 2.2, in other words the thickness of the brain mantle in the temporal horn region is about half the thickness of the diameter of the hemisphere. If the lateral ventricles become more dilated, the temporal horn echo moves in the direction of the end echo and the brain mantle index increases (Fig. 107). The following example will serve as an illustration:

Distance midline echo—end echo from both sides	61.0 mm
Distance TH-Echo—end echo from both sides	20.0 mm

$$\text{BMI} = \frac{61}{20} = 3.05$$

Such a figure points to a marked dilatation of the lateral ventricles such as is usually found in cases of cerebellar tumors (see also Fig. 119). But the BMI can also be determined separately for each of the temporal horns so that unilateral ventricular dilatation can be easily picked up. Example:

Distance midline echo—end echo from both sides	60.0 mm
Distance TH echo—end echo from the right	25.0 mm
Distance TH echo—end echo from the left	28.0 mm

From this a BMI of 2.15 can be calculated for the right lateral ventricle and a BMI of 2.4 for the left ventricle.

Thus we are dealing with a slight dilatation of the left temporal horn here. Such findings can be observed occasionally in patients with encephalomalacia.

Using the echo-encephalographic width of the third ventricle, it is possible to diagnose a hydrocephalus on the basis of echo-encephalography alone with great accuracy. The brain mantle index can also be used to detect a dilatation of the lateral ventricles even if the width of the third ventricle cannot be determined. This is of special importance with tumors filling part or all of the third ventricle where the determination of the midline echo is often impossible. However, the degree of ventricular dilatation can be measured without great difficulty in cases of glioma of the oral brain stem and craniopharyngioma by determining the position of the temporal horn echo (see p. 104).

The *correlation of the results of echo-encephalographic investigation* and pneumoencephalographic studies in respect to the measurement of the

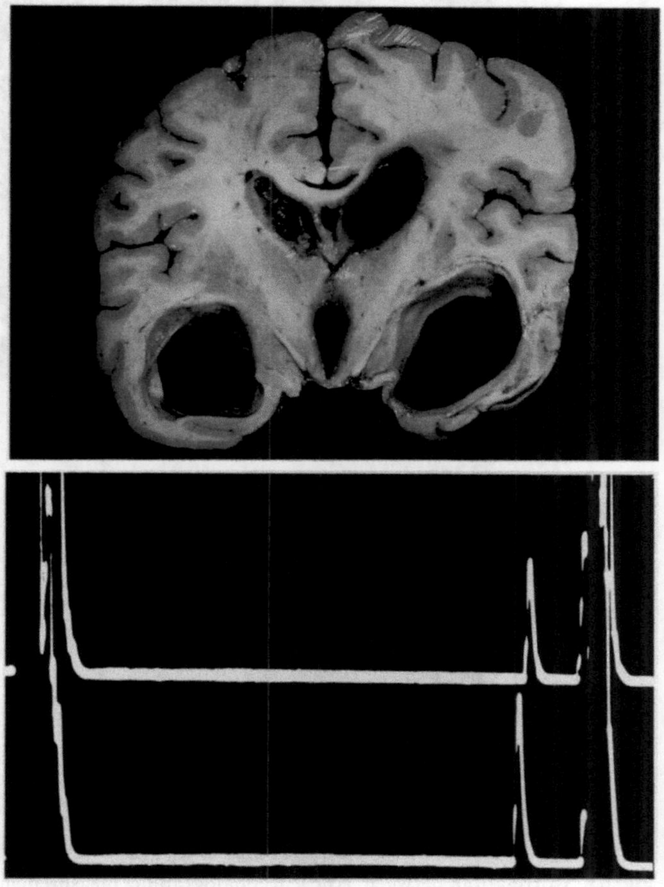

Fig. 62. Agreement between the echo-encephalographic measurement of the position of the external wall of the temporal horn and the autopsy finding. The smooth oscillogram is produced by switching on the sensitivity time control (step 6 to 7). The echograms were recorded one week before death. Pat. S. N., 29 years ofage. Echo-EG No. 1321/64

temporal horn position was examined on 114 pneumoencephalograms and ventriculotomograms (see Fig. 61). Taking the roentgenological enlargement into consideration, a general agreement between the results was found in 111 cases (97%). A misinterpretation occurred three times, i. e. a reflection other than the one from the temporal horn was assumed to originate from it. In all 3 patients there was no ventricular dilatation which had been erroneously diagnosed when the echo-encephalogram was interpreted. In several cases, autopsy findings also confirmed the ultrasonic measurements (see Fig. 62). The reliability of echo-encephalography in determining the position of the temporal horn in cases of ventricular dilatation with such a low error quota seems to justify the routine use of the method. However, the personal experience of the investigator is important here too, because the identification of the temporal horn echo may cause difficulties.

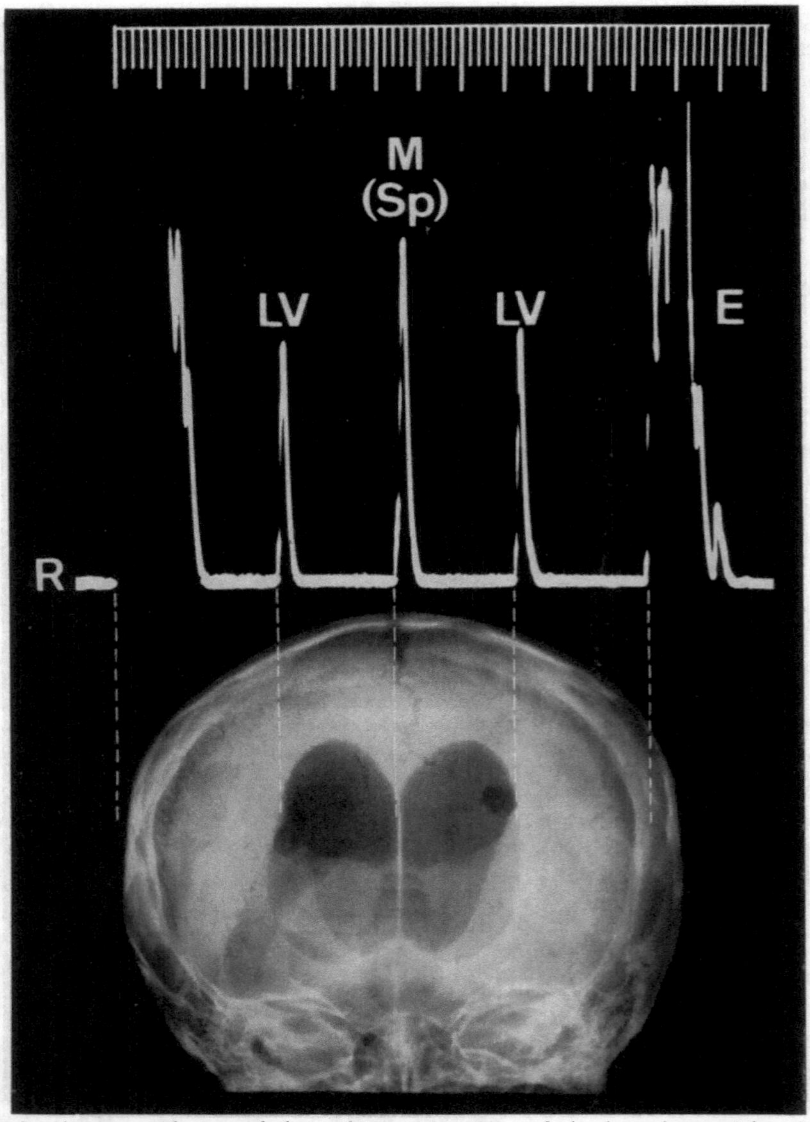

Fig. 63. Correlation between echo-encephalographic measurement of the lateral ventricle width and a later ventriculogram in a 10-year-old girl with an ependymoma of the fourth ventricle. The M-echo in this record probably originates from the septum pellucidum. (Pat. T. G. Echo-EG No. 3687/68)

Besides echoes from the third ventricle and from the temporal horn, reflections from the external wall of the lateral ventricle in the cella media region can be detected occasionally if the probe is applied 2 to 3 finger breadths above and in front of the root of the ear (LITHANDER, 1961; FORD and AMBROSE, 1963; UMBACH and KLEY, 1965; SJÖGREN, 1965, 1967, 1968; FORD and MCRAE, 1966; UEMATSU, 1966; WEST, 1967; PIA and GELETNEKY, 1968). In suitable cases, this reflection can be registered beside the midline echo on both sides. This is chiefly true for infants and small children with thin skull bones where this special investigation seems to be valuable. Fig. 40 illustrates the correct technique of investigation. FORD and MCRAE (1966) as well as UEMATSU (1966) were able to carry out such measurements succesfully on adults as well. The echogram in Fig. 63 was recorded in a 10-year-old girl with a cerebellar tumor. For the sake of comparison, the ventriculogram which was made later, is shown above it and illustrates the excellent agreement with the ultrasonic finding.

UMBACH and KLEY (1965) use the SCHIERSMANN ventricular index (the distance initial echo—end echo/width of both lateral ventricles) when evaluating the width of the ventricles in such echograms. SJÖGREN (1965) also used the width of the lateral ventricles to set up an index — the lateral ventricle index — which facilitates the interpretation of echo-encephalograms in children with hydrocephalus. The distances are the same as with the SCHIERSMANN ventricular index, but SJÖGREN relates the width of the lateral ventricles to the diameter of the entire skull. Incidentally the author calls the measurement of ventricles in children echoventriculography (EVG). The identification of the echoes from the lateral ventricles which flank the midline echo is made easier by the fact that there is an *echo-poor* or *echo-free zone* between those echoes on the echogram which correspond to fluid-filled spaces (LITHANDER, 1961; SJÖGREN, 1965 to 1968).

The special methods of investigation used in cases of infantile hydrocephalus will be discussed in the clinical section.

3. Direct Demonstration of Tumors, Hematomas and Cysts

a) Tumor Reflections, Cyst Wall Echoes

The pulse echo procedure was utilized for the direct demonstration of tumors at a relatively early date. The technique of investigation described in the first papers, was possible only on the operatively exposed brain (FRENCH, WILD, NEAL, REID, 1950, 1952 and KIKUCHI et al., 1957). It is used primarily in Japan (TANAKA; MITSUNO et al., 1966) and

Table 3. *Localization of brain tumors by means of tumor reflections* (TANAKA, ITO, ISHIKAWA, YUKISHITA, and OHARA, 1966)

Localization of Tumors		Ultrasonoechography through the Intact Skull No. of Examined Cases	Correct Diagnosis
Supratentorial	Cerebral hemispheres	72	70 (97%)
	Chiasmal region	26	22 (85%)
	Rostral brain stem	14	8 (57%)
Infratentorial	Caudal brain stem	5	0
	Cerebellum	15	13 (87%)
	Fourth ventricle	1	0
	Cerebello-pontine angle	17	4 (24%)
Total		150	117 (78%)

China today (CH'EN and P'AN, 1964), in recent years it has also found application in the USA (WALKER and UEMATSU, 1966; DYCK, BARROWS and KURZE, 1967, 1968) and in Europe (MÜLLER and LÉVY, 1968) and will not be described in more detail here.

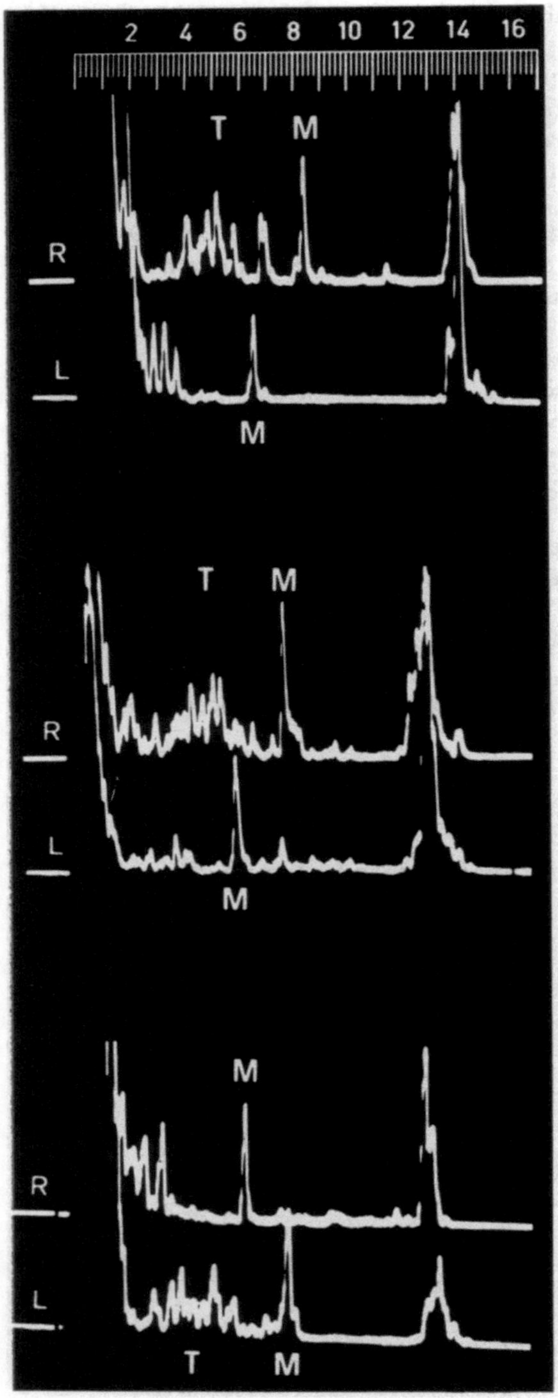

Fig. 64. Tumor echo complexes (T) of tumors located in the temporal lobe. A differential diagnosis of the type of tumor by echo-encephalography is usually not possible. However, tumor echo complexes permit a more exact localization and orientation about the presence of acoustically effective interfaces in the tumor. a) Right temporal glioblastoma. Pat. J. H., 53 years of age. Echo-EG No. 2507/66. b) Sphenoid wing meningioma on the right. Pat. Ch. R., 22 years of age. Echo-EG No. 2558/66. c) Astrocytoma in the left temporal lobe. Pat. A. W., 39 years of age. Echo-EG No. 1524/65

On the other hand, there is great interest in every possibility of diagnosing brain tumors directly through the intact skull by means of special reflections on the echo-encephalogram and not only indirectly by demonstrating a midline displacement or ventricular dilatation.

Especially Tanaka et al. have been reporting for years that it is possible to localize brain tumors even through the skull bone and not infrequently even to determine their histology. Table 3 taken from a publication by Tanaka et al. (1966) shows that 117 of 150 (78%) tumors in both the infratentorial and supratentorial space were correctly localized by means of tumor reflections. In tumors of the cerebral hemispheres, the percentage of cases where tumor echoes are found, is said to be 97%. Tanaka (1966, 1968) generally differentiates between two types of tumor echoes: *type-M echoes* (continuous spikes with distinct peaks at the tumor border) and *type-A echoes* (irregular undulation of the baseline without definite echoes at the border of the tumor). Type-M echoes were detected in cases of meningioma, metastatic tumors, craniopharyngioma, acoustic neurinoma and pituitary adenoma. Type-A echoes appeared in cases of glioblastoma, astrocytoma, oligodendroglioma and medulloblastoma. It is noteworthy here to mention that sounding was carried out not only from the temporal area but also from various other application points on the skull. A transorbital or transoral technique with a special probe was employed in cases of tumors of the caudal brain stem and acoustic neurinomas. Unfortunately these papers make no mention of the amount of time necessary for this method of investigation and to what degree this eliminates the use of dye studies.

Schiefer, Kazner and Kunze (1965) investigated the frequency of tumor reflections with supratentorial neoplasms in a large number of patients and tested their usefulness. However, sounding was carried out from the temporal area only because first of all changes in the application point of the probe result in new and unknown reflections even in healthy subjects; the second reason is that the thickness of the adult skull is too great in many parts of the head to permit the registration of usable echoes.

In 214 cases of supratentorial tumors, tumor echoes or cyst wall reflections were found only 48 times (22.4%). Consideration must be given to the fact that only about half the tumors lay in the ultrasonic cone using a temporal application point. But even so, the results do not sound very promising. Mention must also be made of the fact that the localization of the tumor was known to the authors before sounding was carried out in many cases (e. g. tumor calcification on the plain films of the skull). With increasing experience, on the other hand, tumor echoes were recognized with greater accuracy.

Our own results do not permit a more specific differentiation of the tumors based on the shape of the tumor echoes either. Fig. 64 for example, shows echograms with tumor reflections due to a glioblastoma, a meningioma and an astrocytoma. Therefore we cannot confirm the distinguishing characteristics mentioned by the Japanese ultrasound research group. A detailed presentation of our own results with tumors of the cerebral hemispheres and tumors of the oral brain stem will be given in the special section.

Differences in the acoustic impedance of the structures sounded are necessary for the development of an echo at the interface between tumor and brain tissue or within the tumor itself. Up to the present only a few scattered reports existed on sound velocity measurements in brain tissue and tumors. The values presented in the literature vary so widely and the authors often arrive at such contradictory results that is was necessary for us to make our own investigations.

For this purpose we carried out 604 *measurements of sound velocity* on 213 different tissue and fluid samples with an KRAUTKRÄMER interferometer: 43 brain specimens removed during operative lobectomy, 35 CSF samples, 34 glioblastomas, 27 meningiomas, 9 oligodendrogliomas, 8 astrocytomas, 5 ependymomas, 4 neurinomas, 1 spongioblastoma, 1 sarcoma of the meninges, 3 carcinoma metastases, 10 different tumor cyst fluids, 19 times the contents of different chronic subdural hematomas, 10 times coagulated blood from epidural, acute subdural and intracerebral hematomas as well as pus from 4 brain abscesses and subdural empyemas.

About 10 ml of material are required for the examination; measurements were always performed at 22° C since a temperature difference of 1° C is enough to produce a 2.5 m/s change in the sound velocity (BARTHEL, 1954). Furthermore the measurements were carried

Table 4. *Sound velocity and acoustic impedance in brain tissue and brain tumors* (own measurements with a KRAUTKRÄMER interferometer)

Medium	T °C	Frequency MHz	Sound Velocity m/s	Density kg/m³	Acoustic Impedance 10⁶ Ns/m³
Brain tissue	22	4	1532	1036	1.59
Cerebellar tissue	22	4	1537	1040	1.60
Ventricular CSF	22	4	1502	1004	1.51
Bone [a]	—	2.5	3380	1800	6.1
Calcium [b]	—	—	3810	2650	10.1
Contents of tumor cysts	22	4	1504—1524	1006—1028	1.51—1.57
Gliomas (incl. glioblastomas)	22	4	1525—1547	1030—1040	1.57—1.61
Meningiomas	22	4	1540—1550	1045—1060	1.61—1.64
Ependymomas	22	4	1537—1545	1040—1045	1.60—1.62
Spongioblastoma	22	4	1532	1040	1.59
Acoustic neurinomas	22	4	1522—1547	1032—1050	1.57—1.62
Carcinoma metastases	22	4	1535	1040	1.60
Arachnoidal sarcoma	22	4	1530	1035	1.58
Pus from abscess or empyema	22	4	1521—1523	1030—1035	1.57—1.58

[a] GÜTTNER, FIEDLER, and PÄTZOLD, 1952.
[b] CRAWFORD, 1955.

out as soon as possible after operative removal of the tumor or lobectomy in order to avoid errors due to the loss of water. The values obtained in this way are presented in Table 4. Using our interferometer method, it is possible to achieve a maximal accuracy of

0.1% in the measurement of sound velocity, since the length of the comparative distances in water is determined micrometrically and the velocity of sound in water is known exactly (KRAUTKRÄMER, 1961). The time of interference between test and control echo can be determined exactly. Each measurement was repeated a number of times in order to check it.

The sound velocity in fresh brain tissue, measured in 126 separate measurements, averaged 1532 m/s. The maximal deviation from the mean was 3 m/s. The mean value of the CSF measurements was 1502 m/s at 22° C (1499 to 1505 m/s). Here the sound velocity was obviously dependent on the protein content of the various CSF samples. The average sound velocity difference between brain tissue and CSF was 30 m/s. As already mentioned this difference suffices to permit the development of distinct ultrasonic reflections at the interfaces between the two media, especially since there is a considerable difference in the specific weights (1036 kg/m³ and 1004 kg/m³). The acoustic impedances for the two media can be calculated from this as follows: $1.59 \cdot 10^6$ Ns/m³ for brain tissue and $1.51 \cdot 10^6$ Ns/m³ for CSF at 22° C. JEPPSSON (1961) found practically the same value ($1.58 \cdot 10^6$ Ns/m³) for brain tissue at 24 to 25° C.

The velocity of sound in gliomas differs very little from that in normal brain tissue (contrast on the other hand TANAKA and ITO, 1966). A higher velocity was found only in calcified tumors. Meningiomas also evidenced a distinct difference in the sound velocity as compared to normal brain tissue. This is probably due to the high blood content of these tumors. The mean valve of sound velocity here was calculated to be 1545 m/s at 22° C. Contents of tumor cysts were also examined. The values varied considerably depending on the protein content and the specific weight of the fluid here (1504 to 1524 m/s, Fig. 65). The greatest contrast to brain tissue was found in older blood clots from intracerebral hematomas; here sound velocities of approximately 1600 m/s were obtained.

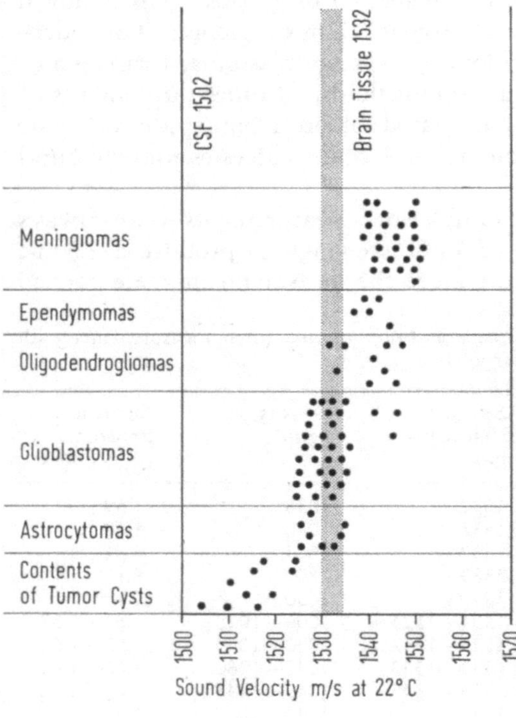

Fig. 65

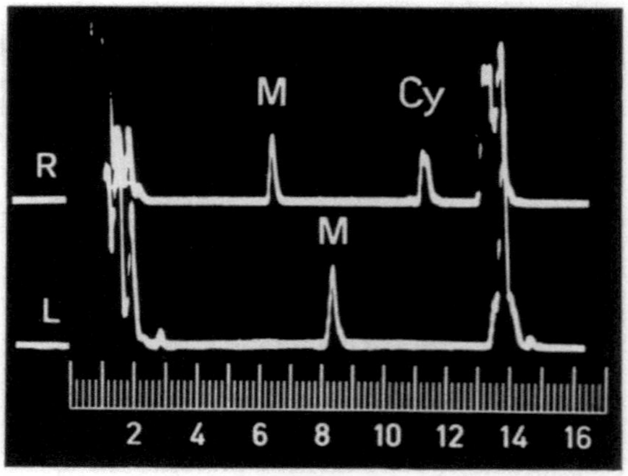

Fig. 66

Fig. 65. Sound velocity in brain tumors, normal brain tissue, CSF, blood and tumor cysts examined with a KRAUTKRÄMER interferometer. See text for further details

Fig. 66. Echo-encephalogram of a patient with a cystic astrocytoma in the left temporal lobe. Displacement of the M-echo by 10.0 mm towards the right. Recording from the right reveals an additional reflection (Cy) which could later be attributed to a tumor cyst. Pat. K. M., 49 years of age. Echo-EG No. 999/64

According to these sound velocity measurements, high amplitude *interface echoes* may be expected only with meningiomas hit vertically, tumor hemorrhages and cyst walls. This is actually the case. Fig. 66 shows the echo-encephalogram of a patient with a cystic astrocytoma in the left temporal lobe. Besides the displacement of the midline echo there is an additional echo preceding the end echo on sounding from the right. According to the operative finding, this must have been caused by the wall of a larger tumor cyst. If such reflections are found during the preoperative stage, they do not allow us to draw the conclusion that there is actually a cyst wall present here. *However, a large cyst should be suspected if there is an echo-free area in an echo-encephalogram* showing a midline echo shift. Fig. 67 is an example of this. This is the case of a 10-year-old boy with an unusually large tumor cyst

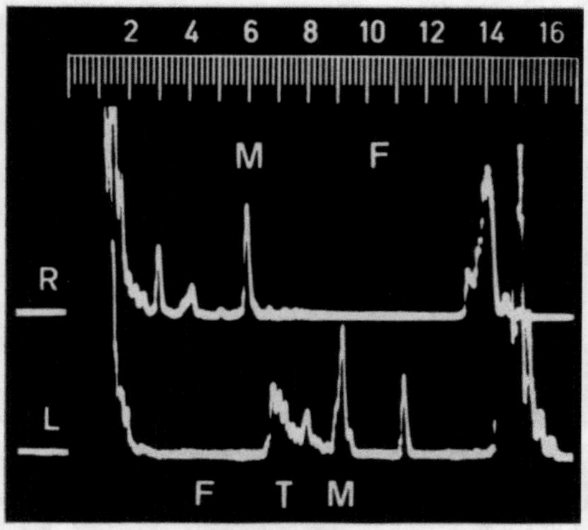

Fig. 67

Fig. 67. Echo-encephalogram of an 8-year-old boy with a cystic spongioblastoma in the left temporal lobe. Displacement of the midline echo 15.0 mm towards the right. The record of a wide echo-free zone (F) enabled us to make the diagnosis of a cyst by echo-encephalography already. Additional tumor echo complex (T) near the midline. Pat. M. W., Echo-EG No. 1211/64

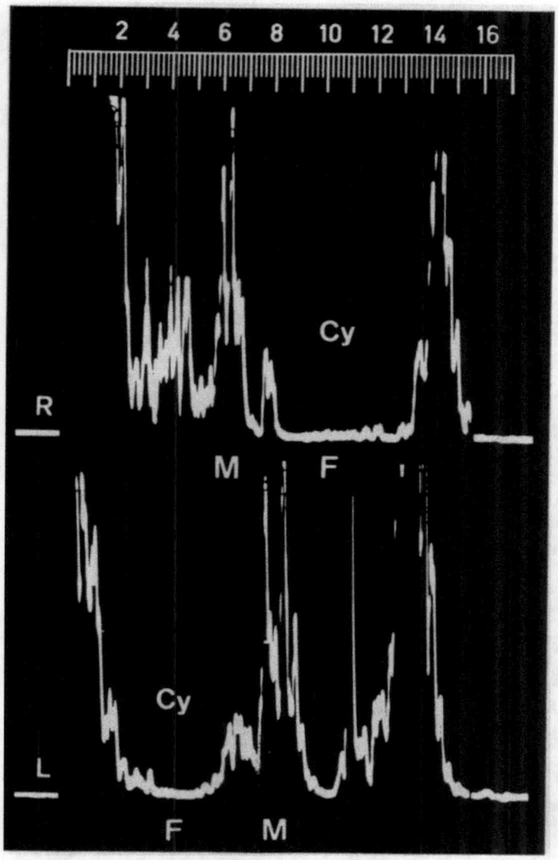

Fig. 68

Fig. 68. Echo-encephalogram of a 9-year-old girl with a cystic gangliocytoma in the left temporo-occipital region. The large echo-free zone F which indicates a large cyst (Cy) remains nearly echo-free even at maximal intensity and switched-off sensitivity time control, while all the other echoes become much higher (Pat. F. B., Echo-EG No. 3475/67)

(120 cm^3 contents) on whose medial wall a spongioblastoma was found. It is most characteristic for the echo-free zone to remain constant even at maximal sender power and receiver amplification (Fig. 68).

The *tumor echo complexes* which obviously originate in the tumor itself cannot be explained by the velocity differences between tumor and surrounding brain tissue. Therefore there must be structural differences within these tumors which lead to ultrasonic reflections. Radiological or microscopic calcifications were found in nearly all oligodendrogliomas which produced tumor echoes. There is not the least doubt that ultrasound is strongly reflected by calcium deposits (see calcified pineal gland as the source of the midline echo, Leksell, 1958; Jeppsson, 1960, 1961). Fig. 69 shows an example of a calcified ependymoma in a child, Fig. 70 the echogram of a calcified oligodendroglioma in a 30-year-old man and the x-ray of the skull in the same patient.

In our opinion the fact that tumor echo complexes are found relatively frequently in the very glioblastomas whose sound velocity differs only very little from that in normal brain tissue is due to the irregular structure of these tumors (including smaller cysts, pathological vessels, necrotic zones, hemorrhages). Illustrations of the echo-encephalograms in cases of glioblastoma are seen in Figs. 64 and 93 on p. 66 and 99.

While tumor echoes are not so important in cases of tumors of the cerebral hemispheres because there is usually a midline echo shift, indicative of a space-occupying lesion, present

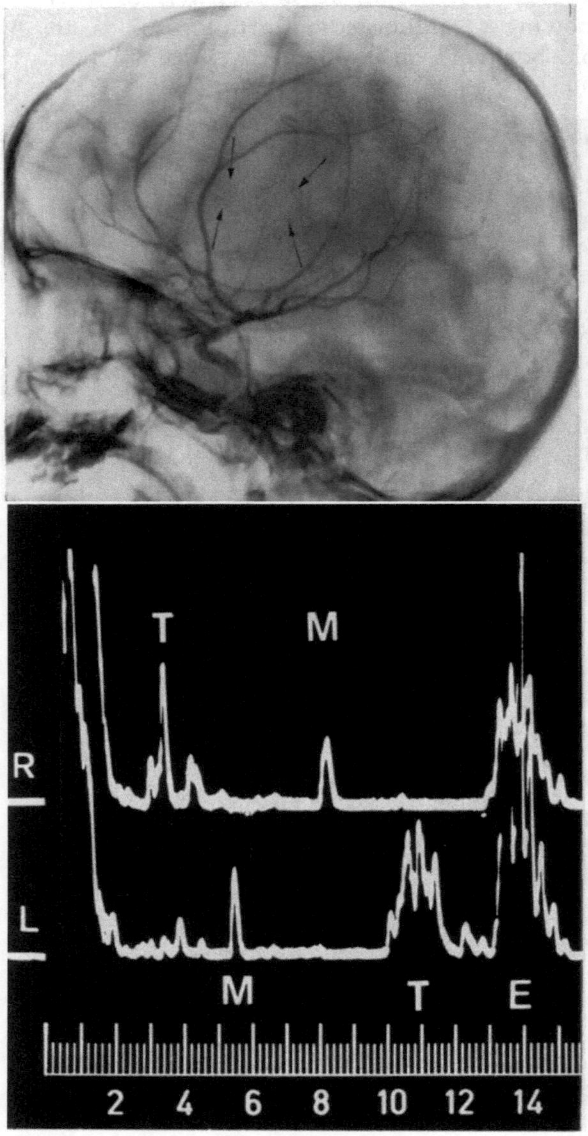

Fig. 69

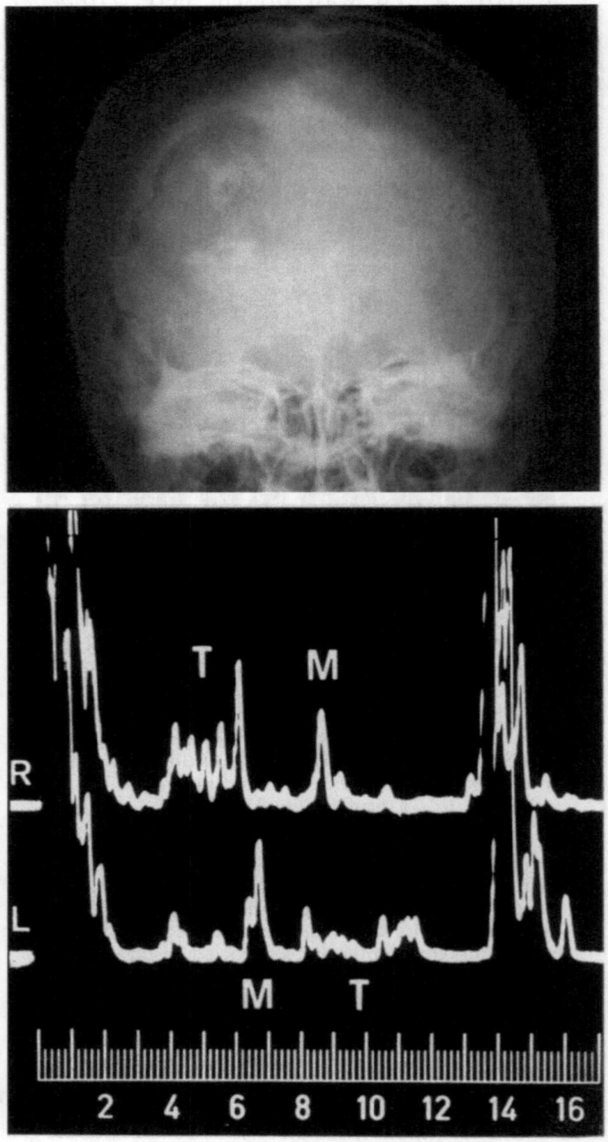

Fig. 70

Fig. 69. Echo-encephalogram of a 4-year-old girl with a calcified ependymoma above the Sylvian fissure on the right. M-echo shift 12.5 mm towards the left. In both traces additional reflections originating from the calcification may be seen (T). Pat. M. M. Echo-EG No. 2024/65

Fig. 70. Tumor reflections (T) in a patient with a calcified oligodendroglioma in the right occipito-temporal area. M-echo shift 10.0 mm. The skull film shows pronounced calcification (top). Pat. M. B., 30 years of age. Echo-EG No. 1301/64

as well, they acquire increasing importance in tumors of the third ventricle and the oral brain stem. The information provided by echo-encephalography is so great in some cases here that a nearly complete diagnosis can be made from the echo-encephalographic examination of the brain alone. Our own results will be presented in more detail in the section on tumors of the oral brain stem on p. 103 ff.

Occasionally reflections resembling those produced by tumor cysts can be found in brain abscesses lying within the scanning area (see Fig. 192 g). A specific diagnosis is not possible with this method, however. Echoes from the interfaces of parietal subdural empyemas were not obtained in a single case as yet, even if the ultrasonic beam was directed upwards as in the investigation technique for chronic subdural hematomas.

Summarizing the topic of tumor reflections it may be said that *one-dimensional echo-encephalography in its present form is limited in its ability to localize tumors exactly*. However, in individual cases, particularly in temporal glioblastomas, ependymomas and oligodendrogliomas as well as in tumors near the third ventricle, tumor echoes not infrequently provide additional information which increases the value of this investigation, especially in children and young adults with relatively thin skull bone. Perhaps further progress could be achieved with the two-dimensional echo method in certain cases if the localization were suitable. The first reports on the use of such a method have already been published (ADAPON et al., 1965, 1966; GROSSMAN, 1964, 1965, 1966; HOVIND et al., 1967; LOMBROSO and ERBA, 1968). Since there are no larger series as yet, a final evaluation of the procedure is not possible at this stage (see also appendix p. 193).

Even if tumor echoes are detected more frequently because of improvements in the investigation technique, choice of other application points and tomographic procedures as well as increasing experience, this method of investigation will never replace contrast medium studies. Rather it is a — sometimes very valuable — supplement to the pneumoencephalographic and angiographic findings. The demonstration of tumor echoes can also simplify the choice of the most suitable dye study on occasion (see Fig. 71).

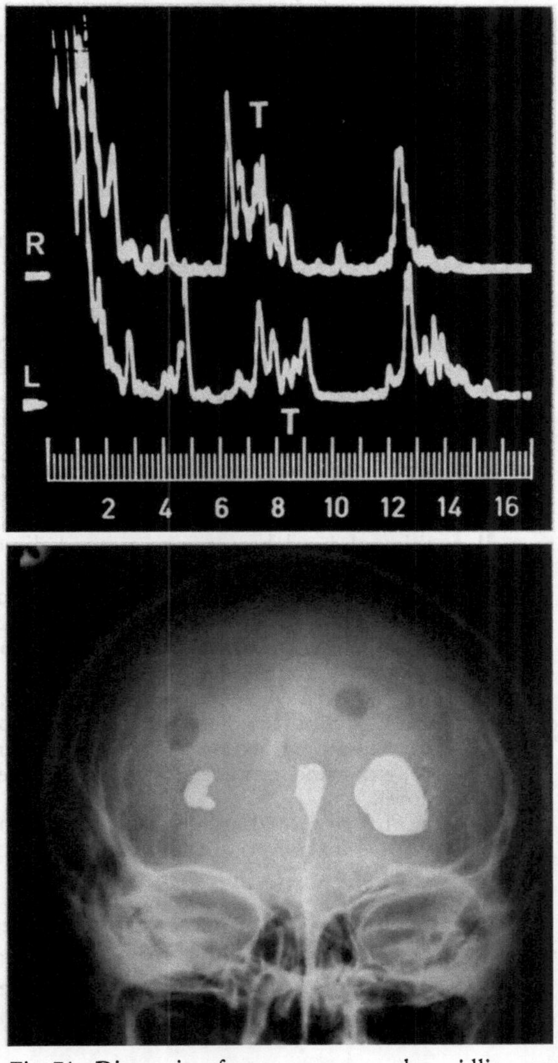

Fig. 71. Diagnosis of a tumor near the midline on account of a tumor echo complex (T). Subsequent pantopaque ventriculography revealed a narrowing of the third ventricle by a thalamus tumor situated below and to the right of the third ventricle. Pat. W. C., 27 years of age. Echo-EG No. 2589/66

b) Demonstration of Hematoma Interfaces

In 1959 TER BRAAK, GRANDIA and DE VLIEGER first reported on the observation that additional echoes appear before the end echo in cases of extradural hematomas. They described these as "*hematoma echoes*" and suspected that they originated at the interfaces between brain tissue and blood.

The authors observed the appearance of additional echoes not seen before during an injection of blood into a calf's brain; these echoes disappeared when the fluid was removed. In a formalin-hardened brain containing an intracerebral hematoma, distinct echoes were registered from the hemorrhage as well. Furthermore the same authors were also able to produce an artificial subdural "hygroma" at the autopsy of a patient without a cerebral disorder by injecting blood plasma. After injecting 80 ml, an additional reflection was seen ahead of the end echo complex; this disappeared after the fluid was removed.

In a summary of the echo-encephalographic results of 55 intracerebral hematomas, DE VLIEGER (1964) reported that 39 cases (71%) were diagnosed by ultrasound. Of 12 patients with epidural hematomas, only one showed no hematoma echo. Three acute subdural hemorrhages also revealed special reflections. The result with chronic subdural hematomas was far less favorable because of the parietal location (25 of 37). By the use of an angulated probe (45°) the results could be improved. According to his own reports, DE VLIEGER was able to differentiate the hematoma echo from other reflections preceding the end echo in most cases quite well.

In a later series of 83 extracerebral hematomas (DE VLIEGER, 1968), hematoma echoes were registered 54 times (65%). The poorest results were obtained in cases of frontal and occipital hematomas as well as with bilateral symmetrical hematomas.

DREESE and NETSKY (1963) reported on a patient with an acute subdural hematoma in whom it was possible to register additional reflections from the side of the hematoma.

AMBROSE (1964) also observed reflections with epidural and subdural hematomas which he traced to the displaced dura or to the interface between brain and blood.

SUGAR and UEMATSU (1964) also mentioned 2 cases of hematoma with abnormal echoes on the side of the bleeding. UEMATSU, SUGAR and WALKER (1966) were able to observe a hematoma echo in 5 of 9 cases of epidural hematoma; in 10 subdural hematomas, however, this observation was made only twice.

Since 1961, the Japanese ultrasound research group has repeatedly reported on extra echoes which can be used for the localization of hematomas with different forms of intracranial hemorrhage. In a series of 72 hematomas, a hematoma echo was recorded 55 times (76%). Frontal and occipital as well as thin subdural hematomas escaped direct demonstration in the echo-encephalogram (ABE, SUGAWARA, ISHI, ITO, KIKUCHI and TANAKA, 1966).

Our own experience with echo-encephalography in extracerebral hematomas was first reported in 1963. Here characteristic differences between the traces of the various forms of hemorrhages could be emphasized (Fig. 162) which enabled us not rarely to make the differential diagnosis (SCHIEFER, KAZNER, BRÜCKNER, 1963). A later survey included 81 extracerebral hematomas of whom 56 (69%) revealed hematoma echoes (KAZNER and SCHIEFER, 1966). A detailed presentation of the diagnostic possibilities of echo-encephalography with epidural hematomas was presented in 1965, where 36 such hematomas examined with ultrasound were described (KAZNER, KUNZE and SCHIEFER). In this paper, experiments were set forth which led to a clarification of the source of the hematoma echo (see below).

The possibility of recording special reflections in cases of intracranial hematomas was later confirmed by CALATAYUD-MALDONADO, GELETNEKY and LORENZ (1965) as well as by RADDA and SCHIMA (1965) and by FISCHER et al. (1967).

However, several authors regarded the hematoma echo as not significant or were unable to demonstrate such reflections. In his monograph on echo-encephalography (1961), JEPPSSON reported the results of investigations in 44 cases of extracerebral hemorrhage, 9 of which were epidural hematomas. In spite of his diligent efforts, he was unable to confirm the "extra echoes" on the side of the hematoma described by DE VLIEGER. On the contrary, he found similar additional echoes in completely healthy subjects; in his opinion these are due to the uneven inner surface of the skull with at times extremely marked digital impressions.

LITHANDER, who reported on 40 extracerebral hematomas in 1961, found additional echoes on the side of the hematoma in one third of them. However, she did not regard

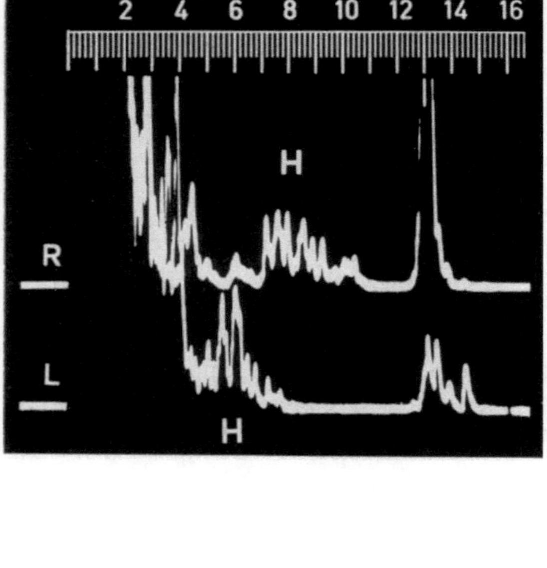

Fig. 72. Echo-encephalogram of a 28-year-old man with hypertensive hemorrhage into the left basal ganglia. Abnormal echo complex in both traces so that an exact localization of the lesion is possible (Pat. H. H. Echo-EG No. 3400/67)

these echoes as sufficient evidence for making the diagnosis of a hemorrhage because similar reflections can be observed in healthy subjects as well. In order to find the answer for ourselves, we performed experiments which should reveal what structures are in a position to produce distinct reflections with intracranial hematomas (KAZNER, KUNZE, SCHIEFER, 1965).

In cases of intracerebral hematoma, additional reflections may frequently be picked up if the hematoma is in the sounding area; these echo complexes resemble those found with brain tumors and may contribute to the diagnosis (Fig. 72) (TANAKA et al., 1961, 1968; KAZNER et al., 1965; SCHIEFER and KAZNER, 1967; PIA and GELETNEKY, 1968). The demonstration of special hematoma echo complexes in cases of hypertensive cerebral hemorrhage into the basal ganglia plays a more important role in the diagnosis than cerebral angiography, according to MITSUNO et al. (1966).

Experimental Investigations on the Question of Hematoma Echoes

As we know, it is necessary to have differences in the acoustic impedances of tissues through which ultrasonic energy passes in order to obtain a reflection. In addition, the interface between two media must be hit largely at right angles and the interface must be as even as possible. This is absolutely requisite for structures where the differences in acoustic impedance are slight.

As noted previously, the difference between brain and CSF is enough to bring about distinct echoes. With a Krautkrämer interferometer, we have carried out a number of sound velocity measurements in order to determine the differences between the acoustic impedances of the structures sounded in the presence of intracranial hematomas. The relevant results are summarized in Table 5. From this table it is obvious that the differences between

Table 5. *Sound velocity and acoustic impedance in brain tissue, CSF and blood* (own measurements with a Krautkrämer interferometer)

Medium	T °C	Frequency MHz	Sound Velocity m/s	Density kg/m³	Acoustic Impedance 10^6 Ns/m³
Water [a]	22	15	1488.1	997	1.48
Ventricular CSF	22	4	1502	1004	1.51
Brain tissue (fresh)	22	4	1532	1036	1.59
Blood (fresh, coagulated)	22	4	1570	1059	1.66
Blood (old clots)	22	4	1598—1607	1068	1.71—1.72
Citrated blood	22	4	1556	1050	1.63
Dura mater	22	4	1555	1108	1.72
Capsule of a chronic subdural hematoma	22	4	1525	1055	1.61
Contents of chronic subdural hematomas	22	4	1520—1530	1018—1045	1.58—1.59

[a] BARTHEL, 1954.

the acoustic impedances of brain tissue and coagulated blood or dura are at least as great as those between brain tissue and CSF. These differences lead us to expect that sufficient ultrasound energy will be reflected from the interfaces of the first named media.

Clinical experience has shown that an additional echo of great amplitude very frequently appears in epidural hematomas (see Fig. 73 a). Because the dura is forced away from the inner table of the skull in this case, we performed an experiment to establish the role of this fibrous membrane in the development of the hematoma echo.

Experimental procedure: In a plexiglas container with a base of 10 cm × 7 cm, we simulated the anatomical relationships of an extracerebral hematoma. Particular care was taken that the interface between brain, blood and dura was absolutely perpendicular to the direction of the incident ultrasonic beam. Under the conditions of an epidural hematoma, a very high sharp echo appeared at the interface between brain tissue, dura and coagulated blood (see Fig. 74 top). Subsequently the dura was removed. The high echo trace disappeared leaving only a small reflection whose height was a quarter that of the dura echo (see Fig. 74 bottom).

Thus it is clear that it is the *dura mater* which produces such a high hematoma echo in cases of epidural hematomas. The reflection originating at the interface between the surface of the brain and blood is much less prominent and has much less clinical significance. Our experience with patients corresponds to this view entirely (see Fig. 73 b).

a
b
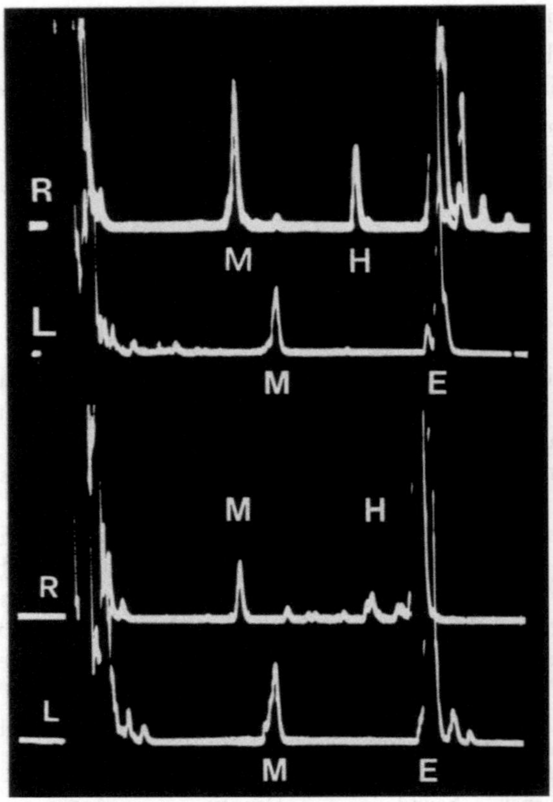

Fig. 73

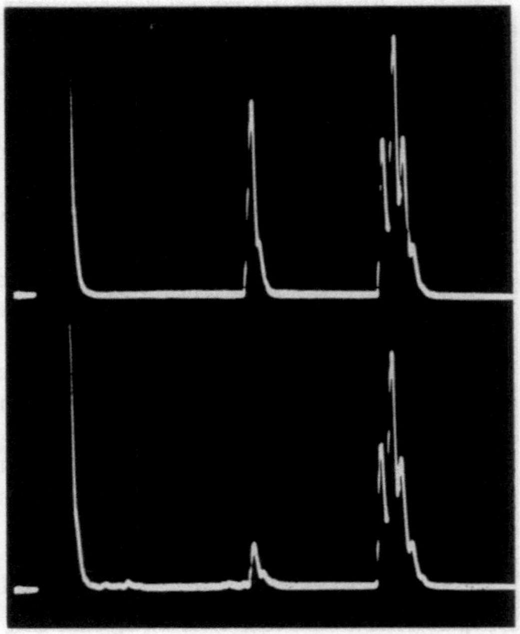

Fig. 74

Fig. 73. Echo-encephalographic demonstration of a space-occupying intracranial hemorrhage following head injury. a) Midline echo shift and high hematoma echo (H) of an epidural hematoma located over the parietal area. b) Midline echo shift and smaller hematoma echo (H) in an acute subdural hematoma. The thickness of the hematoma equals the distance between the hematoma echo (H) and the end echo (E). Pat. J. N., 55 years of age. Echo-EG No. 1309/64; pat. G. R., 61 years of age. Echo-EG No. 1008/64

Fig. 74. Experimental demonstration of the dura echo. Top: high amplitude echo at the interface of brain, dura and blood (middle reflection) in an artificial epidural hematoma within a plexiglas container. Bottom: after the dura is removed, the echo has a very low amplitude

The variations in the form and amplitude of the hematoma echo in the various types of intracranial hematoma and their importance in the differential diagnosis which we have worked out have recently been confirmed by Pia and Geletneky (1968). Ito, Tanaka and Abe (1968) arrived at the same results.

Similar conclusions are reached when fresh brain is examined under water. At the interface brain/water only a small echo is seen; if dura is placed on the surface of the brain opposite the probe however, a high "dura echo" can be recognized on the screen here as well (Fig. 75).

More recent investigations by Oberschulte-Beckmann and Otto (1967) with artificial epidural hematomas on cadaver skulls, confirm the importance of the dura mater for the development of hematoma echoes in this form of hemorrhage.

The particularly good conditions of reflection at the falx which is a fibrous process of the dura, were pointed out in the discussion of this structure under sources of the midline echo already (see p. 41).

Of decisive significance for the size of the echo produced by the forced-off dura is the angle at which this membrane is hit by ultrasound. A piece of fresh dura (7 cm × 7 cm) stretched on a frame was placed in a water-filled plexiglas container and sounded at different angles of incidence. As soon as the angle of incidence deviated as little as 5° from the perpendicular, the amplitude of the "dura echo" decreased by 75% to 80% (Fig. 76). For clinical use such an angle of incidence represents the limit for a diagnostically useful hematoma echo. Furthermore this explains the fact that temporo-basal epidural hematomas which are no doubt hit by ultrasound fail to produce unequivocal hematoma echoes.

Howry (1955) found that the amount of reflected ultrasound at an angle of incidence deviating 6° from the perpendicular, sinks to 1% of the initial value

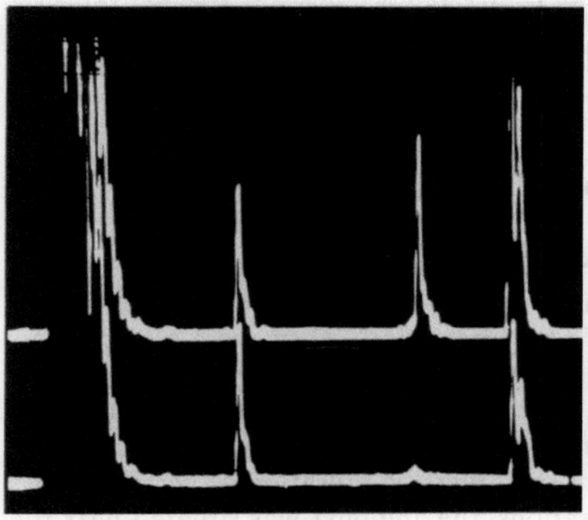

Fig. 75

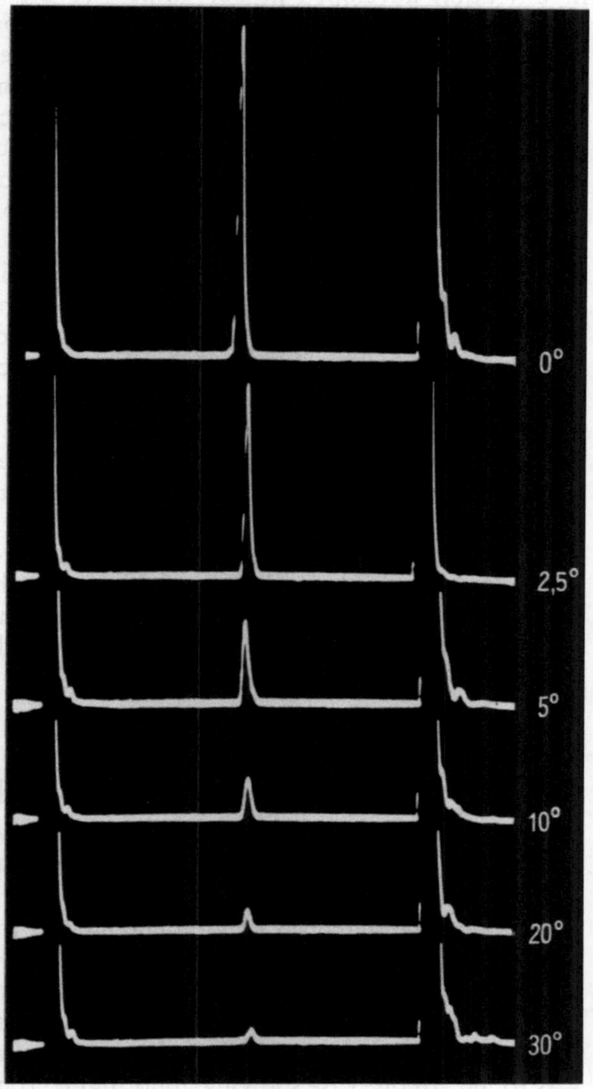

Fig. 76

Fig. 75. Ultrasound investigation of a cadaver brain in a water bath. Bottom: at the interface of brain/water only a minimal reflection develops. Top: the dura rests on the surface of the brain. Now we obtain a very high echo

Fig. 76. Sounding the dura in a water bath with the ultrasonic beam at different angles of incidence. Rapid decrease in the amplitude of the echo when the angle deviates from the perpendicular

while the amplitude on the screen is reduced to 1/10. Carried over to sounding the dura, a good agreement was found with the results of these examinations.

A detailed discussion of the special method we have developed for the localization of hematomas together with its clinical results and the ways in which the dura echo may be differentiated from other lateral reflections may be found on p. 132 ff.

Besides epidural and acute subdural hematomas where these interface reflections are usually found, such echoes also occur in chronic subdural hemorrhages and effusions. The difficulty in recording an echo from an interface consisting of a more or less thin membrane lies in the fact that nearly all such hematomas are located high up in the parietal area. Only by moving the probe to and fro until the ultrasonic beam is perfectly at right angles to the membrane interface, can an evaluable reflection be obtained (Fig. 157). The technique of investigation developed at our clinic was described in detail by Brückner and Kazner ("oblique sounding technique", 1963). A similar procedure was reported by de Vlieger (1964).

Experimental investigations showed that the inner layer of the hematoma capsule is of decisive importance in the development of the hematoma echo in the case of a chronic subdural hematoma. The registration of such reflections is more frequently successful in cases of unilateral than in bilateral hematomas or effusions.

At the interface produced by subdural accumulations of fluid in children, special echo peaks can also be picked up so that echo-encephalography can contribute to the diagnosis here.

Geletneky (1965) used the oblique sounding technique in infants and children with subdural effusions in order to pick up hematoma echoes. The echo-encephalogram in 16 of 19 children with this clinical syndrome revealed such additional echoes.

Jacobi, Kazner and Wollensak (1966) reported on the results of echo-encephalographic investigations in 31 children with subdural effusions or hematomas. In 22 cases, additional reflections originating from the inner hematoma membrane were observed ahead of the end echo on diagonal sounding. From the distance between the hematoma echo and the end echo, it is possible to make deductions about the thickness of the fluid accumulation.

Aside from diagonal sounding, Jacobi and Schuch (1966) tried out another technique of investigation in which the probe is applied directly above the hematoma so that hematoma echoes become visible in the near-field of the echo-encephalogram.

Today echo-encephalography occupies a firm place in the recognition and post-operative treatment of subdural effusions in infancy and childhood (Gerlach, Jensen, Kraus and Koos, 1967; Distel and Kasper, 1968).

4. Abnormal Echo-Pulsations

In 1955 Leskell first directed attention to the abnormal pulsating echoes which can be produced by arteriovenous aneurysms. Ford and Ambrose (1963) reported on 4 such cases where the diagnosis of an angioma was made by ultrasound. They were also successful in recording abnormal pulsating echoes from a large suprasellar aneurysm.

Ter Braak and de Vlieger (1965) also pointed out that abnormal dicrotic pulsation curves may be observed with angiomas on occasion. The authors examined 14 patients with arteriovenous aneurysms. In 9 of these, such an abnormal dicrotic pulsation of certain echoes was found. In one case showing abnormal echo pulsations, the angioma was situated near the midline structures. Ter Braak and de Vlieger have recorded these special pulsations continuously, similar to ultrasonic cardiography so that the difference to the normally present echo pulsations could be demonstrated graphically.

McKinney, Thurstone, Avant and Wallace (1965, 1966) also reported on amplified and abnormal echo pulsations which occurred in patients with hypertension, increased intracranial pressure, vascular anomalies and vascular headaches. As a point of interest, it was noted that the pulsations increased during an attack of headache or several hours before the onset of symptoms in such patients. In the symptom-free period however, there was no increase in the amplitude of the pulsations.

In 1966, Dreese, Hayes and Kempe demonstrated the clinical value of this method in the case of a patient with an aneurysm where circumscribed abnormal echo pulsations were recorded which could be made to coincide exactly when the angiogram and the echo-encephalogram were superimposed on each other.

Planiol and Pottier (1968) studied the recordings of cerebral echo pulsations in cases of arteriovenous angiomas in comparison to the rheo-encephalographic findings in great detail. However, it was not possible to establish an accurate diagnosis with this method.

Recently we were successful in diagnosing an aneurysm on the basis of abnormal echo pulsations for the first time in one of our own patients. When the ultrasound beam was directed towards the third ventricle, a number of strongly pulsating, fluttering echoes appeared in the central area of the echogram on the screen. Subsequently, carotid angiography revealed a pea-sized aneurysm of the anterior cerebral communicating artery. Abnormally high echo pulsations in the area between the M-echo and the end echo were also frequently observed in children with dilated subarachnoid spaces; however, no diagnostic conclusions could be drawn from this.

The registration of normal or pathological echo pulsations is in its infancy as yet. Whether it can ever become a routine clinical examination for cerebrovascular disorders, is uncertain so far (see also p. 48).

C. Misinterpretations of Echo-Encephalograms

The careful and conscientious interpretation of echo-encephalographic curves is the basis for a successful use of the echo technique in hospital and practice. More so than with almost any other complementary investigation, the result of the diagnostic effort depends on the skill and personal experience of the examiner.

Very often numerous single echoes appear which cannot all be attributed to the right intracranial structures. In the echo-encephalographic literature there is a large number of reports on misinterpretations and sources of error in the evaluation of echo-encephalograms see Jeppsson, 1961; Schiefer et al., 1963; Jefferson and Hill, 1966, 1968; Bingel, 1967; Pia and Geletneky, 1968 among others).

Beginners especially will be repeatedly confronted with echo-encephalographic findings which turn out to be erroneous on subsequent neuroradiologic investigations. But the physician, who has practiced the method for many years is also not spared an occasional error. This fact must be brought to mind again and again. The more differentiated the interpretation of the echo-encephalogram, the higher is the risk of an error in interpretation. Therefore the beginner should always start only with the midline echo and then proceed to employ the other diagnostic possibilities of echo-encephalography.

We have devoted a special chapter to the identification of the midline echo and mentioned several aids which will simplify this task. If an error in measurement has been made, however, the first duty is to find the cause of this misinterpretation. The study of photographic documentations does little to help, since it is impossible to remember exactly where the probe was applied and which sounding direction was used during the photograph. Unfortunately no method of standardization has been developed here yet. Each case demands a different setting of the probe and one can never record the midline echo from one specific point on the head alone. In a recent paper, Nadjmi (1968) has shown how much the structures responsible for the midline echo can vary in relation to the upper root of the ear under normal circumstances already. In cases of space-occupying intracranial lesions which do not lead to a shift of the midline structures, but rather to a displacement either anteriorly or posteriorly, these variations are naturally much greater. Therefore it is impossible to set up a standard application point for sounding the midline structures and parts of the ventricles as some authors have repeatedly suggested. Every echo-encephalographic investigation is practically a scan of a large part of the temporal area. For this reason we are continuing to use the term A-scan, instead of the word A-mode suggested by Grossman for the time-amplitude-method.

In every case of an echo-encephalographic error in measurement, the physician must repeat the investigation as soon as he knows the results of neuroradiological investigation. In most cases, the causes of this error can be uncovered quite rapidly then. But a few findings remain unexplained in spite of this. We will now examine chiefly misinterpretations of the midline echo or of midline echo shifts. Difficulties in differential diagnosis with other echo-encephalographic findings are discussed in detail in special sections (p. 66 f., 135 f., 171).

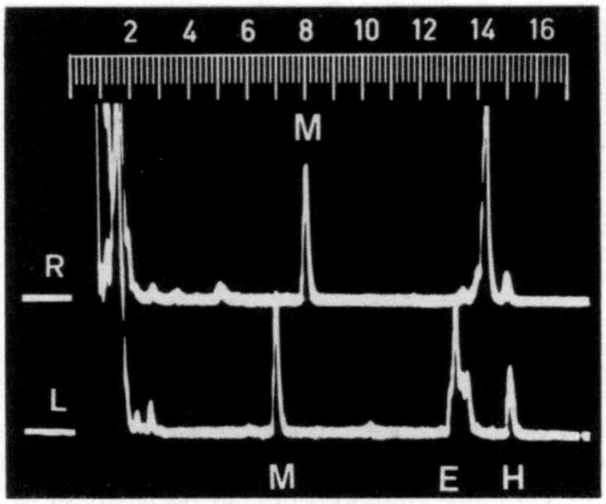

Fig. 77 a. Factitious midline echo shift in a patient with a right temporal subgaleal hematoma. The distance from the midline echo to the end echo is the same when sounding from either side; this fact distinguishes it from a genuine mass shift. On sounding from the left an additional reflection (H) can be seen which originates from the thickened scalp. Pat. D. R. Echo-EG No. 1782/65

If a displacement of the midline echo is found, the next step is to compare the distances midline echo—end echo. It is never enough just to measure the distance from the probe to the midline echo. For if one finds a displacement of the end echo in addition to a shifted midline echo, then there is obviously *no* intracranial mass displacement; rather there is a thickening of the skull (due to asymmetry of the skull) or of the galea on the side of the prolonged distance to the midline echo. This phenomenon can be observed quite frequently in cases of unilateral subgaleal hematoma or swelling of the scalp following head injuries. Not infrequently this leads to errors in echo-encephalographic measurement (Jeppsson, 1961; Chadduck and Crutchfield, 1964; Feuerlein, 1965; Schiefer et al., 1966). *In these cases the distance midline echo—end echo is equal* unless there is an additional space-occupying lesion present. If this fact is drawn

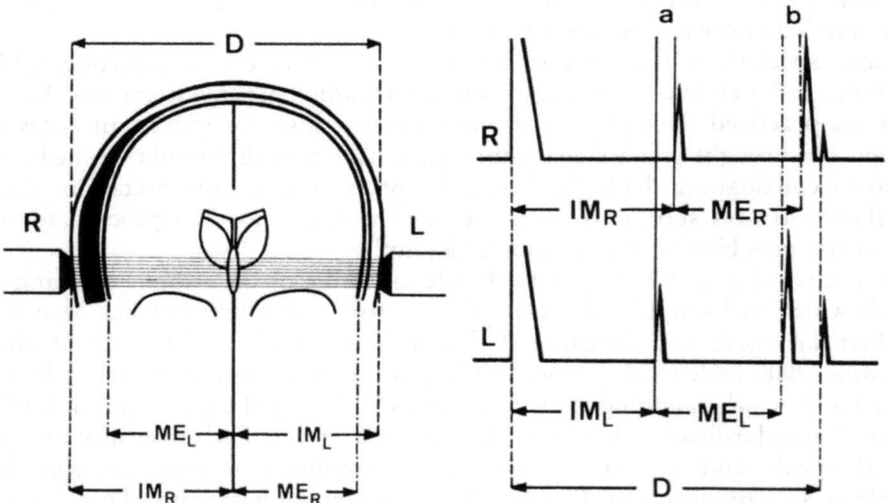

Fig. 77 b. Diagram of an echo-encephalogram in the presence of a subgaleal hematoma. The midline echo is apparently displaced. However, this shift is compensated by a similar shift of the end echo (a=b). The distance which is decisive for tissue displacements, namely the distance midline echo — end echo (ME$_R$ and ME$_L$) is of equal length. (IM=distance initial echo to midline echo, R=sounding from the right, L=from the left, D=bitemporal diameter of the skull)

into consideration, it is easy to distinguish this phenomenon from an actual intracranial space-occupying lesion. Fig. 77 a presents such an echo-encephalogram. In addition to the factitious displacement of the midline structures, a distinct reflection representing the actual egress from the skull can be seen behind the end echo when sounding is carried out from the healthy side (L). This echo develops at the interface scalp/air. We call this reflection the "external hematoma echo" in the case of a subgaleal hematoma (in the literature it is also called the skin/air interface). The interpretation of an echogram with an apparent midline echo shift, due to a subgaleal hematoma is demonstrated in Fig. 77 b.

In contrast the "egress echo" in healthy patients, i. e. in patients without an injury to the scalp usually cannot be distinguished from the end echo complex. Only by employing the device suggested by DE VLIEGER (1964) can the point where the ultrasound leaves the skull, namely the "egress echo", be identified even under normal circumstances (p. 37).

Occasionally it is impossible to record an unequivocal midline echo in cases of *subgaleal hematoma* because the ultrasonic beam travels back and forth between the crystal and the bony skull, thereby producing many multiple echoes (Fig. 78). In these cases a slight alteration in the application point of the probe is enough to pick up the midline echo again.

A rare occurrence is the case of a more or less obvious discrepancy of the end echo in the face of a midline or slightly shifted midline echo. Such a finding may be obtained if there is a *subgaleal hematoma on one side and an intracranial space-occupying lesion on the opposite side* (see Fig. 79). Here too, a comparison of the distances between the midline echo to end echo will elucidate the actual intracranial relations. Another misinterpretation may occur in the case of a *subgaleal hematoma together with a homolateral space-occupying hemorrhage.* Here the actual mass displacement is amplified by the subgaleal hematoma if only the distances probe-midline echo are measured (Fig. 80).

In addition to these apparent midline echo shifts or apparently normal echo-encephalographic findings which are *due to extracranial factors,* a further source of error lies in *mistaking the midline echo for one of the other reflections near the midline.* In our

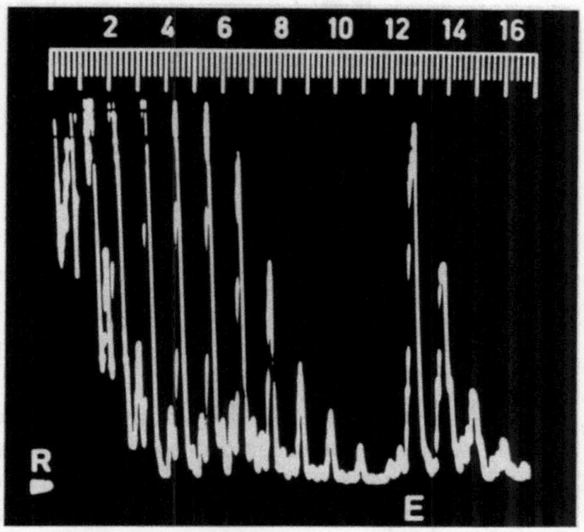

Fig. 78. Echo-encephalogram of a 7-year-old boy with a right temporal subgaleal hematoma. Sounding from the right reveals numerous multiple echoes caused by back and forth reflections of ultrasound between the probe and the external table of the skull. It is impossible to evaluate the position of the midline structures from this side. Pat. M. H. Echo-EG No. 3675/68

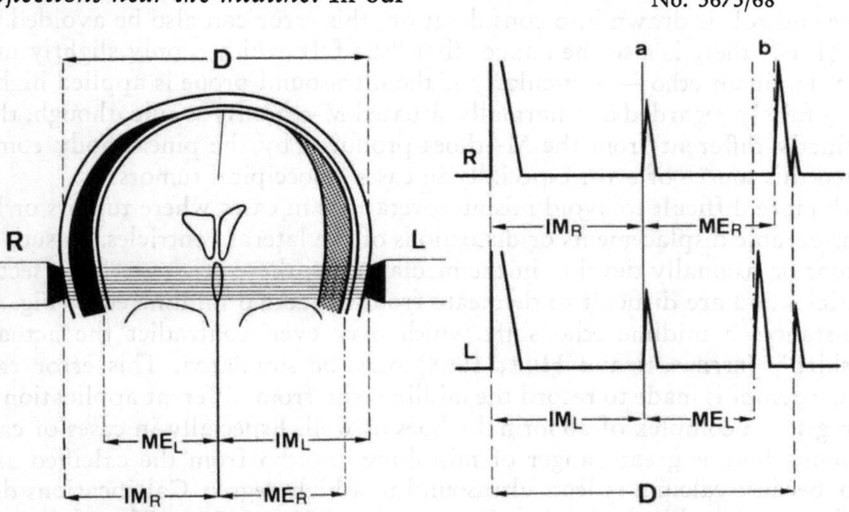

Fig. 79. Factitious midline echo-encephalogram in the presence of a subgaleal hematoma on one side and an intracranial hematoma on the opposite side (diagrammatic presentation). The actual displacement of the midline structures is equal to the difference between the end echoes (b/2)

experience the most frequent mistake is confusing the midline echo with an echo from the lateral wall of the third ventricle. If the third ventricle is dilated and we sound the same wall of the third ventricle from both sides, a considerable midline echo shift may be simulated

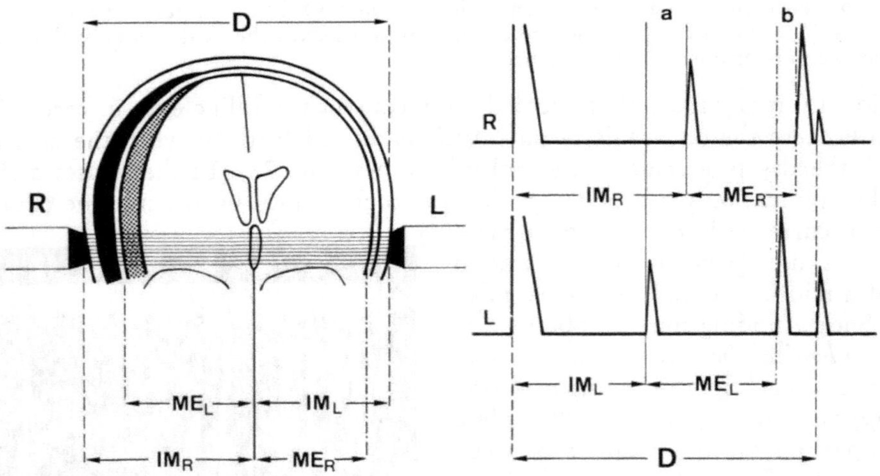

Fig. 80. Amplified midline echo shift in the presence of an intracranial hematoma and a subgaleal hematoma on the same side (diagrammatic representation). The actual shift of the midline strucutes may be calculated by taking the difference between the end echoes (b/2). The displacement of the midline echo is increased by the thickness of the subgaleal hematoma

(Fig. 81). The degree of shift corresponds to half the diameter of the third ventricle. In healthy subjects, the error due to this cause is naturally much smaller, but apparent midline echo shifts up to 3 mm may be found. In order to avoid such a misinterpretation, the width of the third ventricle should be determined in all patients, especially in those cases where the clinical signs suggest a space-occupying lesion in the posterior fossa. High reflections between the midline echo and the end echo point to a hydrocephalus and originate from the dilated lateral ventricles.

Even if one lateral wall of the third ventricle is mistaken for the midline echo from one side only and the actual midline echo is recorded from the opposite side, a slight midline displacement can still be simulated. If the theoretical midline echo or the through-transmission echo (midline control) is drawn into consideration, this error can also be avoided. According to Jeppsson (1961) there is also the danger that "the falx which is only slightly mobile, may give rise to a dominant echo — particularly if the ultrasound probe is applied high up on the ear vertical — falsely regarded as a normally situated M-echo. As as rule, though, the thin falx spike is distinctly different from the M-echoes produced by the pineal body complex". We have observed this source of error especially in cases of occipital tumors.

It is much more difficult to avoid misinterpretations in cases where tumors or hematomas produce considerable displacements or distortions of the lateral ventricles. In such cases, very high reflections occasionally develop in the median area; these originate from sections of the lateral ventricles and are difficult to delineate from the actual midline echo (Fig. 82). Under these circumstances a midline echo shift which may even contradict the actual relations ("reversed shift", Jefferson and Hill, 1968) may be simulated. This error can only be avoided if an attempt is made to record the midline echo from different application points and perhaps to register a complex of abnormal echoes as well. Especially in cases of calcifications within a tumor, there is great danger of mistaking an echo from the calcified area for the midline echo, because calcium reflects ultrasound to a high degree. Calcifications due to other causes must be mentioned here also (see Jefferson and Hill, 1966). Therefore x-rays of the skull in two planes should be performed in every patient, so that these can be compared with the echo-encephalographic findings and such errors avoided.

The interpretation of the echo-encephalogram is especially difficult for the examining physician *if the space-occupying lesion has affected the midline structures themselves.* In these cases it is often impossible to identify a single reflection rising above the other reflections

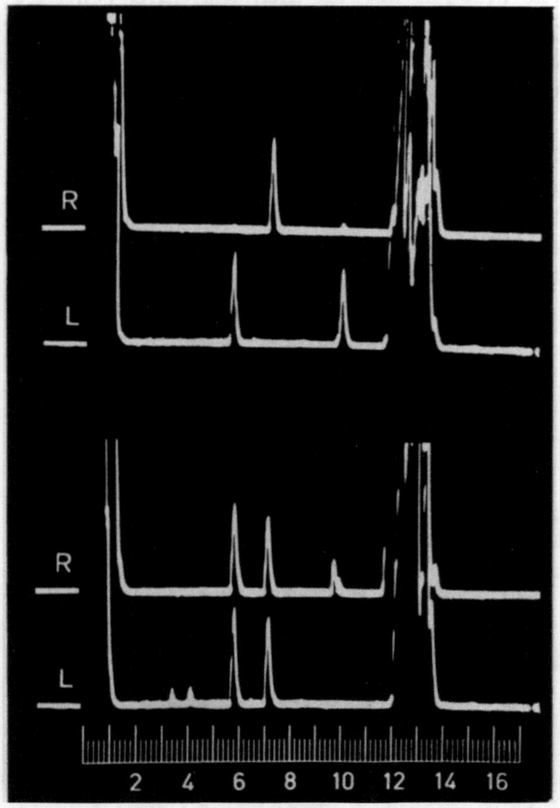

Fig. 81

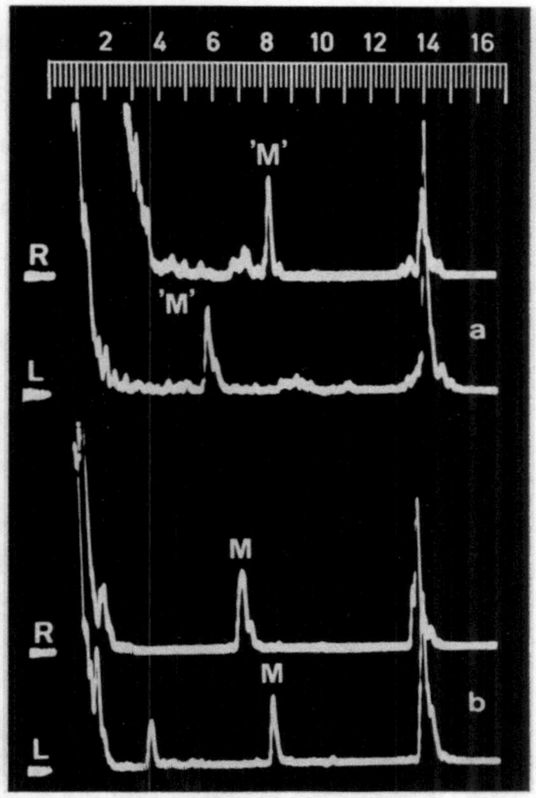

Fig. 82

Fig. 81. Factitious displacement of the midline echo in the presence of ventricular dilatation. a) The same wall of the dilated third ventricle was sounded from both sides and erroneously regarded as the midline echo thus simulating a midline echo shift. b) The correct echo-encephalogram shows the double echo of the third ventricle which is dilated to a diameter of 14.0 mm. Pat. Ch. G. 4 years of age. Echo-EG No. 1646/65.

Fig. 82. "Reversed shift" in the echo-encephalogram of a 44-year-old patient with a left frontal glioblastoma. a) During the first recording, a shift of the midline echo 'M', amounting to 10 mm from right to left was diagnosed. Since the sum of the two midline echoes was only 140 mm while the diameter of the head was 150 mm,there must be an error in measurement. b) A second echo-encephalogram which was recorded subsequently, showed that there was actually a 6.0 mm shift from left to right. The "midline echoes" which were erroneously regarded as such in the first record, probably originated from the lateral ventricles. Pat. E. F., Echo-EG No. 3688/68

which could be the midline echo. Here one must be satisfied with recognizing that the ultrasonic trace is pathological, but that it is impossible to make an accurate evaluation (Figs. 102 and 110).

Recently the width of the third ventricle is being measured by echo-encephalography more and more. One should beware of believing that this measurement is possible in all patients though. Excessive thickness of the bony skull in old age in some individuals and displacements of the midline structures are the most common causes for the inability to obtain echoes from the walls of the third ventricle (also p. 87 f. and 97 f.). If the third ventricle is midline and enlarged, however, the investigation is very simple. Nevertheless misinterpretations occasionally occur here too. For example it is entirely possible to regard the midline echo and one echo from the wall of the third ventricle as the double echo from the third ventricle. Only half the diameter of the third ventricle is measured in this way. This error can be avoided by considering the theoretical midline echo. *The reflections from the walls of the third ventricle must lie equally far from the midline echo,* otherwise there is an error in measurement (unless there is a displacement of the midline structures). The accuracy of determining the width of the third ventricle is greater if the temporal horn echo is also found and the echo-encephalographic brain mantle index is calculated, because the two values are clearly correlated (Fig. 116). If the brain mantle index is normal then the third ventricle cannot be greatly dilated.

In patients with ventricles of normal width, one sometimes observes numerous reflections from the midline region whose origin often cannot be accurately determined. In these cases it is often difficult to record the exact width of the third ventricle. Consideration must be given to the point that the width of the third ventricle can vary depending on the application point of the probe; thus the double echo from the third ventricle is not necessarily always equally wide (compare also GROSSMAN, 1966). Examinations from the right and from the left may produce a difference of up to 1.0—1.5 mm. Rarely an apparent dilatation of the third ventricle is observed. In such a case the reflections from the walls of the lateral ventricles at the level of the thalamus were regarded as the echoes from the lateral walls of the third ventricle (see also GARG and TAYLOR, 1968). Checking the position of the probe and calculating the BMI will help to avoid such misinterpretations.

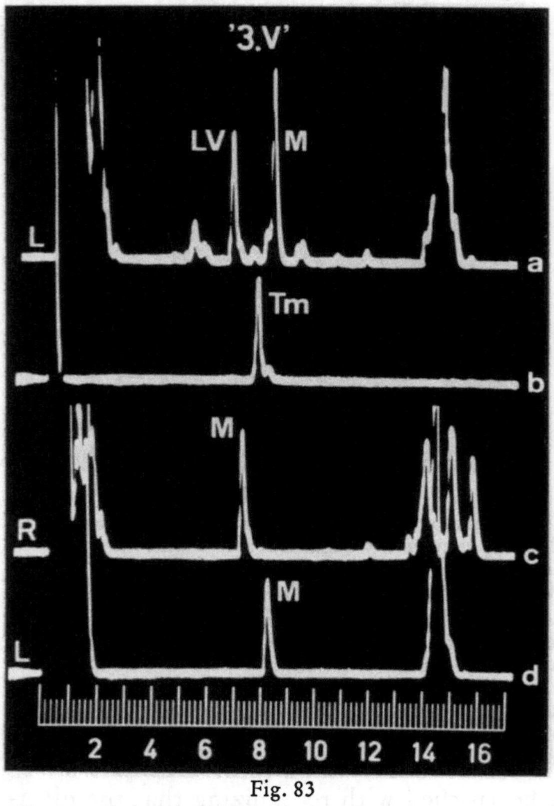

Fig. 83

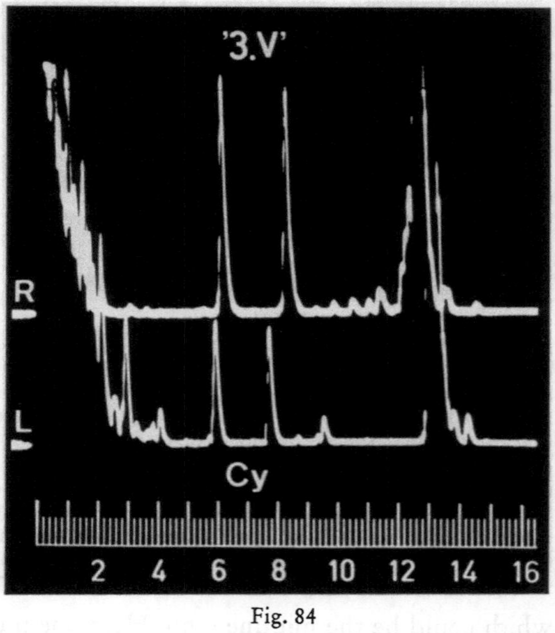

Fig. 84

Fig. 83. Apparent dilatation of the third ventricle in a 33-year-old man with a left-sided thalamic tumor. a) double echo in the midline which was regarded initially as a dilated third ventricle ('3. V'). b) The determination of the theoretical midline echo (Tm) shows that the double echo lies somewhat to the left of the midline. c) and d) In the course of further examinations, another echogram is performed which reveals a displacement of the midline structures 5.0 mm to the right which corresponds to the actual relations. The comparison with record a) indicates that one of the two echoes corresponds to the M-echo. The second echo probably originated from a wall of the lateral ventricles. Pat. G. B. Echo-EG No. 2017/65

Fig. 84. Double echo in the median region in a 27-year-old man which was erroneously regarded as the echogram of a markedly dilated third ventricle. Later investigations revealed a cystic spongioblastoma in the hypothalamus which protruded into the third ventricle from the front. Since the dilatation of the temporal horn in the echogram was only slight, doubts arose regarding the accuracy of the ultrasonic diagnosis of a dilated third ventricle. Pat. H. W. Echo-EG No. 1785/6/68

Basically every echo in the median region of the brain can naturally be regarded as the midline echo and every double echo as the echoes from the two lateral walls of the third ventricle. Thus it may happen that a ventricular dilatation is erroneously diagnosed in cases of midline displacements if a reflection from the lateral ventricles or from a tumor interface accidentally appears beside the displaced midline echo (Fig. 83). Even double echoes of various intervals can be recorded from the same patient in this way. Special mention must also be made here of *double echoes in the median area* which are *due to cystic tumors* e. g. craniopharyngiomas which cannot be distinguished from a dilated third ventricle. Fig. 84 shows the echogram of a 28-year-old man with a double echo in the median area whose peaks are 18 to 22 mm apart. The expected corresponding dilatation of the lateral ventricles was absent. Further investigations showed that a cystic spongioblastoma originating from the hypothalamus was the cause of this abnormal double echo.

Absolutely certain criteria for differentiating these reflections from other echoes are impossible to set up. *Great care must be taken not to interpret the echo-encephalographic findings carelessly or superficially.* The investigator will achieve correct and reliable results only if he continues to compare these with the neuroradiological findings and if he stays in constant practice. Above all, this concerns the determination of the width of the ventricles. The beginner cannot expect that he will be as successful in determining the width of the third ventricle as his colleague who has had several years' more experience. Recording and interpreting echo-encephalograms and avoiding misinterpretations is an art in some respects, but in our opinion it is an art that is not excessively difficult to acquire.

D. Accuracy of Echo-Encephalographic Findings in Comparison with the Results of Contrast Medium Studies

In examining the accuracy of echo-encephalographic findings the first question is, *with what degree of certainty can the position of the midline structures be determined.* From the literature at our disposal, we selected several papers where the results were grouped together and where neuroradiological control investigations were performed in a large proportion of the patients examined by echo-encephalography. Table 6 contains the results of 20 publications by 15 different authors. The total case material without our own patients consists of 4610 patients. If one author had several publications, the older results were not considered in calculating the error quota.

Table 6. *Reliability of midline determination using A-scan echo-encephalography* (literature)

Author	Year	No. of Cases	Midline Echo Shifted		Midline Echo Normal		Ecno-EG not Evaluable	Total Errors (Per Cent)
			Correct	Wrong	Correct	Wrong		
De Vlieger and Ridder	1959	47	21	—	24	2	—	4
Jefferson	1959	50	14	3	28	5	—	16
Jeppsson	1960	262	91	1	168	2	—	1
Lithander	1960	150	57	—	93	—	—	—
Jeppsson	1961	432	189	2	231	10	—	2.8
Lithander	1961	373	121	7	227	9	9	6.7
Taylor, Newell and Karvounis	1961	278	86	1	162	17	12	11
Jefferson	1962	229	60	7	146	13	3	10
Ford and Ambrose	1963	867	297	24	517	29	—	6.2
Schiefer, Kazner, Brückner	1963	520	126	1	391	—	2	1
Friedmann and Thun	1964	67	62	—	—	5	—	7.5
Barrows, Dyck, Kurze	1965	159	42	1	108	8	—	5.0
Brinker, King, Taveras	1965	287	51	7	228	1	—	2.8
Kessler	1965	103	23	3	75	2	—	4.8
Lapayowker and Christen	1965	358	44	7	302	5	—	3.4
Jefferson and Hill	1966	100	92	—	—	5	3	8
Schiefer and Kazner	1966	1603	382	6	1199	1	15	1.4
White	1966	310	11	4	268	—	27	10
Fischer, Revol, Munier et al.	1967	1000	244	4	741	8	3	1.5
Schiefer and Kazner	1967	2747	669	15	2044	3	18	1.25

Altogether a shift of the midline echo was registered 1410 times and in 1343 cases this was later confirmed by neuroradiology, operation or autopsy. *Thus the demonstration of an M-echo shift coincided with the actual relations in 95.2% of the cases.* In 67 cases, i. e. in 4,8% the displacement of the midline structures, suspected on echo-encephalographic examination could not be confirmed by later investigations. These are *false positives* which usually result in unnecessary contrast medium studies for the patient.

In 3143 patients, the *midline echo was at the normal site. In 3029 cases i. e. 96.4% this was correct.* However, in 114 cases an actual shift of the midline structures was present but failed to be recognized by echo-encephalography. The result of these *false negatives* would have been considerably more dangerous, had the clinical course not dictated further investigations at any rate.

If one observes all cases which actually showed a supratentorial mass shift (1457 patients) and if one calculates the quota of correct cases, *92.2% of all patients with a displacement of the midline structures were correctly diagnosed by echo-encephalography, while no less than 7.8% remained undetected. This last value is the percentage of errors.* However in the present summary we must consider that all the initial failures of all authors are included and that many series consist of only a few hundred cases. Today we know that at least 200 to 300 examinations are necessary for the physician to be truly in command of the procedure.

Besides the false positive and false negative findings, only a small percentage of echo-encephalograms are reported which are either not intelligible or in which the midline echo could not be accurately recorded. In the 4610 cases, such a non-interpretable echogram was recorded 57 times, approximately 1.2%.

Summarizing all erroneous measurements and all unsatisfactory results, we find a total of 238 such failures among the 4610 echo-encephalographic investigations. This is a *total error quota of 5.2%.* Larger series (with the exception of FORD and AMBROSE, 1963) generally have a much lower error quota (between 1.5 and 2.8%). Aside from the investigator's experience, the type of ultrasonic equipment also seems to influence the frequency of erroneous measurements.

The same procedure was followed for analyzing the accuracy of the midline echo measurement in 3140 patients in a series of our own (see Table 7). 844 times a midline echo shift was recorded which was wrong 15 times (1.8%). A normally situated midline echo was found 2278 times. However, in three of these patients (0.14%), the midline structures of the brain were shifted according to the dye studies which were performed later. *An error which could have endangered the patient thus occured in only 0.4% of the cases. In 829 of 832 cases, a displacement which actually existed was correctly recognized (99.6%).* 18 times no midline echo whatsoever could be picked up (bony layer in older patients too thick or pneumatized, tumors in the third ventricle or quadrigeminal region, serious head injuries, infantile hydrocephalus). The total error quota in our own series of 3140 patients was therefore 1.12%. Naturally contrast medium studies were not performed in all patients, especially if the neurological deficits were only transient. A control of the echo-encephalographic results by angiography, pneumoencephalography, ventriculography, operation or autopsy was carried out in 1870 cases. In reference to this number, *the total error percentage was 1.9.* A more detailed survey of our own series of patients is given in Table 7. Most erroneous measurements in the ultrasonic determination of the cerebral midline occured with tumors of the oral brain stem (17%) and with infantile hydrocephalus (3%).

Since the accuracy of the midline determination by means of ultrasound is 92 to 98%, echo-encephalography represents an absolutely useful diagnostic procedure of inestimable value.

Investigations on the *accuracy of the echo-encephalographic determination of third ventricular width* are reported only rarely. FORD and MCRAE (1966) for example, reported on 8 pneumoencephalographically controlled cases, all of whom showed good correlation. In the year 1966, we compared the echo-encephalographic and pneumoencephalographic or ventriculographic width of the third ventricle and found complete coincidence of the results when the enlargement due to exposure technique was adequately considered (Fig. 113). JACOBI and SCHUCH (1966) compared the results of ultrasonic measurements of the third ventricle with the results of pneumoencephalography in 145 children. Assuming a maximal deviation of ±1 mm, the findings were in agreement in 68% of cases. Using a correlation of ±2 mm, BETZ and HUBER (1967) found complete agreement in 82.4% of the cases.

Table 7.
Results of echo-encephalography in 3140 patients (Neurosurgical University Clinic, Erlangen 1962—1968)

Type of Lesion	No. of Cases	Midline Echo Shifted		Midline Echo Normal		Echo-EG not Evaluable	Errors (Per Cent)
		Correct	Wrong	Correct	Wrong		
Tumors of the cerebral hemispheres	443	417	1	22	—	3	0.9
Tumors of the oral brain stem and corpus callosum	58	12	4	36	—	6	17
Tumors at the base of the skull	71	13	—	57	—	1	1.4
Tumors of the posterior fossa	235	5	—	228	—	2	0.9
Suspected brain tumor (tumor excluded)	375	—	9	366	—	—	2.5
Head injury	1034	321	1	706	3	3	0.7
Stroke	154	52	—	102	—	—	—
Atrophic lesions	63	—	—	63	—	—	—
Infantile hydrocephalus	103	2	—	98	—	3	3
Epilepsy	183	—	—	183	—	—	—
Miscellaneous	421	7	—	414	—	—	—
Total	3140	829	15	2275	3	18	1.12

Our own series of investigations included 187 subjects who underwent both echo-encephalographic and pneumoencephalographic study. Assuming a deviation of ±1 mm, an agreement was noted in 183 cases = 97.8%. 4 times we suspected a ventricular dilatation from the echo-encephalogram which was not confirmed later. Obviously the reflections from the walls of the lateral ventricles were mistaken for echoes of the third ventricle. The opposite case, namely a normal third ventricle in the echogram and a dilatation on the pneumoencephalogram was not observed. *Determining the width of the third ventricle from the ultrasonic trace obviously requires more practice than the measurement of the midline echo.* Furthermore we must add that in contrast to the midline echo which was found in about 99% of the subjects, the reflections from the walls of the third ventricle can be recorded much more rarely. In a midline ventricular system of normal width, the demonstration of the echoes from the walls of the third ventricle is successful in about 90% of the examined cases, whereas if the ventricular system is enlarged, it is successful in 98 to 99% according to our own experience. However, if the midline structures are displaced, this percentage sinks considerably. To illustrate, we obtained a double echo from the walls of the third ventricle in only 24.7% of 352 tumors of the cerebral hemispheres. The reason for this poor result is due to angulation of the axis of the third ventricle in many cases of supratentorial space-occupying lesion. If the shift is more than 5 mm, the width of the third ventricle can be determined by echo-encephalography only in exceptional cases (compare also FORD and MCRAE, 1966).

According to FORD and MCRAE (1966) as well as UEMATSU (1966) the correlation between the echogram and pneumoencephalogram is good when *the width of the lateral ventricles is measured by echo-encephalography.* SJÖGREN (1967) also found good correlation between the two methods of measurement in 70 children. WEST (1967) was able to confirm the value of echo-encephalography in 30 hydrocephalic children as well. We have undertaken such comparative investigations only in several children with hydrocephalus where the lateral ventricles were measured correctly in every case (see also Fig. 63).

The determination of the position of the lateral wall of the inferior horn by echo-encephalography is most successful if the ventricles are dilated. In our own series of 114 patients, comparative investigations were carried out; assuming a deviation of ±2 mm, good agreement was found between the echogram and the pneumoencephalogram in 111 cases (97.3%). These were chiefly patients with marked hydrocephalus. In 3 patients another reflection was erroneously regarded as the temporal horn echo and on the basis of this echo a hydrocephalus was diagnosed while the width of the ventricles was actually normal. *If the ventricles are not dilated, the recording and accurate identification of the temporal horn is difficult at least in adults.* We did not study the reliability of the echo-encephalographic measurement of the temporal horn systematically in these cases. JACOBI and SCHUCH (1966) undertook such comparative examinations in children and were able to confirm the echo-encephalographic values in 119 of 145 such cases (=82%).

No literature is available on the *reliability of recording tumor echoes.* In our own case material, abnormal echo complexes or individual pathological echoes on the echo-encephalogram were found 116 times. In 113 cases, tumors, abscesses or cysts were found to be responsible for these reflections. However in three other cases, further investigations failed to confirm the presence of a tumor at the site suspected on echo-encephalography.

All in all tumor echoes are a relatively rare echo-encephalographic finding in our own series of patients (24.6% in 459 cases of supratentorial tumor). Special mention must be made of the fact that tumor echoes in children are recorded much more frequently (53% in our own series of 49 tumors). This is probably due to the lower absorption of ultrasound by the thinner infantile skull bone. PIA (1967) also found that the incidence of tumor echoes with supratentorial space-occupying lesions was only 20%. TANAKA et al., on the other hand, reported finding tumor reflections in 89% of such tumors (compare Table 3). However, the last named authors perform the ultrasonic investigation of the brain from application points other than the temporal area as well.

Turning finally to the *reliability of hematoma echoes,* we must also draw on our own case material, since exact figures have not been reported in the literature as yet. If we summarize all space-occupying intracranial hemorrhages (330 patients), we come to the following conclusions: altogether single pathological reflections or pathological echo complexes were found in 191 of 330 patients with hematomas (57.9%); further investigations or immediate operation (on the basis of the echo-encephalographic finding) showed that these were caused by hematomas. Beyond this however, reflections which were regarded as hematoma echoes turned out to have another cause in 9 patients (false positive results). Three of these were older patients with dilated subarachnoid spaces in whom a bilateral chronic subdural hematoma was suspected because of small echo peaks ahead of the end echo on both sides. Twice we regarded an echo from the lateral ventricle region as a hematoma echo. Both cases were acute subdural hematomas only 1.5 cm in diameter so that the reflection registered 3.5 and 4 cm ahead of the end echo respectively, could not have developed at the hematoma interface. In 4 other cases, high echo peaks were observed ahead of the end echo although there was no hematoma present (see also Fig. 135). In all 4 cases the patients had a contusion of the temporal lobe. Twice a midline echo shift was absent and the origin of the additional echo is unclear. The other two patients revealed marked shifts of the midline echo indicating massive cerebral edema. According to the operative and autopsy findings, the questionable hematoma echo developed at the site of an extensive crush injury (compare also BARROWS, DYCK and KURZE, 1968).

The echo-encephalographic diagnosis of hematoma is very accurate with unilateral chronic subdural hematomas and epidural hematomas in our experience (over 90% correct results) *while errors occur most frequently in cases of bilateral chronic subdural hematomas.*

The frequency of hematoma echoes with the different types of hematoma ranges from 28% with intracerebral and 89% with unilateral chronic subdural hematomas (see also Table 15).

This survey shows that *echo-encephalographic diagnosis can be very accurate in all areas in the hand of the experienced physician. However, the more differentiated the evaluation of the echo-encephalogram, the greater is the risk of misinterpretation.*

Part III. Special Echo-Encephalography

A. The Echo-Encephalogram in Healthy Subjects

1. Infants and Children

In order to evaluate an echo-encephalogram it is necessary to know the normal trace in all age groups. The interpretation of the cathode ray picture of an infant or a young child is often more difficult than that of an adult. The thickness of the head coverings in the temporal area (by this we mean scalp, temporal muscle and bone; FEUERLEIN speaks of the head wall thickness) according to our own measurements in new-borns is about 4 mm. In the course of the first months and years of life it rises steadily and reaches practically the same values (an average of 10 mm) as in adults at age 14 (Fig. 34). Thus the thickness of the scalp and skull in infancy and early childhood is much less. From the standpoint of ultrasound physics, however, the most decisive role is played by the *thickness of the skull bone,* as discussed extensively elsewhere. In infants and small children this thickness is rarely greater than 1.0 to 1.5 mm in the temporal area and in adults we found a thickness of 2.5 to 5 mm, corresponding to the reports of WHITE et al. (1965). The measurements obtained in the temporo-parietal area on cadaver skulls by DILLING and FEUERLEIN (1967) are even higher. Thus considerably less energy is absorbed by the twofold passage of ultrasound through the infantile skull and therefore considerably more reflections appear on the screen. However, many of these cathode ray deflections are so variable that they disappear and are replaced by new reflections whenever the position of the probe or the direction of the beam is changed minimally. Often a very confusing, at first seemingly uninterpretable picture develops this way. In children however, the same constant echo peaks are present as in adults and these can be used successfully for the diagnosis of intracranial lesions.

In order to study the reflections normally observed in the echogram and to determine the boundaries of normalcy and pathology, 189 children aged 3 days to 14 years were thoroughly investigated by means of echo-encephalography (see also M. JACOBI, 1966). It is important to point out that *sounding in children should always be carried out with as low a pulse power as possible since this makes the interpretation far easier.*

In agreement with the observations of DILLING and FEUERLEIN (1967) we found that it was not always possible to obtain an unequivocal midline echo from the usual application point above the ear. However, there was usually a double reflection from both walls of the third ventricle. The median plane of the brain in these cases was determined by taking the midpoint between the two echoes from the third ventricle. Nevertheless, it was possible to obtain an unequivocal midline echo in all 189 children when the application point was chosen more anteriorly or more posteriorly. In order to determine whether an echo-encephalogram contains signs of midline structure displacement, we compared the distances midline echo-end echo or median plane-end echo, sounding from both the left and the right.

The diameters of the two cerebral hemispheres are thus measured by echo-encephalography and related to each other. This procedure proves to be as simple as it is reliable because it eliminates factitious displacements due to variations in the thickness of the scalp and skull immediately. The evaluation of the echo-encephalograms of the 189 children examined in this manner reveals a surprisingly low difference between the results of sounding from the right and from the left. In our experience, this difference amounted to an average of 0.08 mm. Only 24 children (12.8%) showed a deviation of the midline structures by 0.5 mm and 2 (1.1%) a deviation of approximately 1.0 mm. Not a single healthy child revealed a shift greater than this (see Table 8). Therefore children with a midline echo displacement of 1.5 mm or more should undergo more precise neuroradiological investigation in our opinion.

Table 8. *Position of the midline echo (ME) in 189 children and 149 adults without neurological deficits and 283 patients with concussion*

Group	No. of Cases	ME Absolutely Midline	ME 0.5 mm Shifted	ME 1.0 mm Shifted	ME 1.5 mm Shifted	ME 2.0 mm Shifted	Mean Shift of the Midline Echo (mm)
Children 0 to 14 years	189 (100%)	163 (86.1%)	24 (12.8%)	2 (1.1%)	—	—	0.08
Adults	149 (100%)	88 (59.1%)	42 (28.2%)	15 (10.0%)	4 (2.7%)	—	0.28
Patients with concussion	283 (100%)	173 (61.1%)	75 (26.5%)	26 (9.2%)	8 (2.8%)	1 (0.4%)	0.27

Our measurements contradict those of Jeppsson (1961), who reports a greater range in normal midline echo variation in children than in adults. According to his calculations the probability of finding a space-occupying intracranial lesion in a child is 99% if the midline echo is displaced by 3.2 mm, while in adults this is thought to be true at 2.1 mm already. Jeppsson's assumption that a distinctly recognizable midline echo is often obtained only with difficulty in older children between the ages of 8 and 15 years, could not be verified by our investigations either.

Besides the position of the median plane of the brain, the normal width of the ventricle in the first years of life is of special interest. In 30 infants up to the age of 1/2 year, we found values between 3.0 and 4.5 mm where the 4.5 mm value was seen only 4 times. In 40 children between the ages of 6 months and one year, widths of 3.0 to 5.0 mm were obtained. Here 6 children revealed a width of 5.0 mm. From this we may draw the conclusion that a third ventricle width of 5.0 mm indicates slight ventricular dilatation in infants under the age of 6 months. Up to the age of 3 years, we normally do not find diameters in excess of 5.0 to 5.5 mm, while in older children a third ventricle width of 6.0 mm is regarded as just admissible (Fig. 39). Similar results were obtained by G. Jacobi and Schuch (1966), as well as by Feuerlein and Dilling (1967).

2. Adults

The same investigations as in infants and children were undertaken in 149 healthy adults. The age range extended from 18 to 90 years. The average deviation of the midline echo, 0.28 mm, was distinctly greater than in children. Only 88 persons (59.1%) had an exactly midline M-echo. A displacement of 0.5 mm was found in 42 (28.2%) subjects of approximately 1.0 mm in 15 (10.0%) and of 1.5 mm in 4 (2.7%) healthy adults (see Table 8). The echoes recorded in cases of simple cerebral concussion in 283 patients will serve as a comparison here. The two groups show remarkably good agreement. According to these results, a midline echo displacement of 2.0 mm must be regarded as a pathological finding which calls for further investigations.

In 136 of 149 persons (91.3%), the width of the third ventricle could be determined by echo-encephalography at all ages by the characteristic double echo from the walls of the third ventricle. The lowest values in 20- to 30-year-old women of slender build lay between 3.5 and 4.0 mm, the upper limit in this age group being 6.0 to 6.5 mm (see also Huber and Patiri, 1967). From the fourth to the sixth decade, values between 6.5 and 7.0 mm may be regarded a just admissible yet. In subjects over the age of 60, values between 5.5 and 7.0 mm were found. A third ventricle of 7.5 mm was noted in subjects over 70 years of age only two times. In these two cases an atrophic lesion could not be eliminated with certainty (see also Feuerlein and Dilling, 1967).

B. The Echo-Encephalogram in Brain Tumors

From 1962 until 1968, a total of 1182 patients suspected of having a brain tumor was examined by echo-encephalography at the Neurosurgical Clinic of the University of Erlangen. In 807 of these patients, there was actually a space-occupying lesion present, in the remaining 375 cases further investigations failed to confirm the suspected diagnosis. In the following sections the echo-encephalographic findings with the different types of tumors and in patients suspected of having an intracranial expansivity will be discussed separately. All patients who were admitted to the clinic up to February 1st, 1967 are included in this analysis (290 cases with suspected brain tumors, 630 brain tumors and abscesses). The 262 patients who have been investigated since then (85 with suspected brain tumors, 177 tumors) did not alter any aspect of this series with the exception of thalamic tumors which will be discussed in more detail here.

1. Suspected Brain Tumors

In 290 patients referred with the diagnosis of a suspected brain tumor, the presence of a tumor was not confirmed by further investigations or follow-up studies. Table 9 presents the echo-encephalographic findings of this group. Regarding the position of the midline

Table 9. *Echo-encephalographic findings in 290 patients with suspected brain tumor (still under observation or tumor excluded)*

Diagnosis	No. of Cases	Midline Echo Normal	Midline Echo Shifted	Third Ventricle in the Echogram		
				Normal Width	Dilated (> 7 mm)	Not Measurable
Suspect brain tumor (neuroradiologically investigated)	100	92	8	78	13	9
Suspected brain tumor (only follow-up study)	190	190	—	144	19	27
Total	290	282	8	222	32	36
Percentage	100	97.2	2.8	76.6	11.0	12.4

echo, the echo-encephalogram was unremarkable 282 times (97.2%). In 8 patients (2.8%), the midline echo deviated from the norm, 6 times by 2.0 mm and once each by 2.5 and 3.0 mm. Three times the explanation lay in an asymmetry of the skull, twice the pneumoencephalogram showed slight lateral displacement of the third ventricle and three patients with a 2 mm M-echo shift had entirely normal contrast medium studies. Whether these last 3 cases were echo-encephalographic misinterpretations must remain open to discussion, since even on control examination there was always a slight displacement of the midline echo. Possibly a midline echo displacement of 2 mm is normal in a very low percentage of the population. Such a finding is so rare however, that its occurrence is always an indication for contrast medium studies. As previously mentioned, the probability of finding a space-occupying lesion in the presence of such a midline echo shift is 99% according to JEPPSSON (1961).

In 254 of 290 patients, the width of the third ventricle could be determined by echo-encephalography. It was within the normal range in 222 patients; in 32 cases the third ventricle exceeded 7 mm in width. Most of the patients with a dilated third ventricle had cerebrovascular disease.

In 100 of the 290 patients, dye studies were undertaken. In 40 cases we carried out pneumoencephalography, in 23 cases carotid angiography and in 37 patients both investigations were performed. The results confirmed the echo-encephalographic findings regarding the position of the midline structures in 97 cases, in 3 patients the pneumoencephalogram showed a ventricular system of normal width where the echogram had raised the question of ventricular dilatation. Thus the reliability of echo-encephalography in determining the midline echo position in this group was 97% and the width of the third ventricle, 96%.

2. Tumors of the Cerebral Hemispheres

The value of echo-encephalography for the early recognition of mass shifts caused by tumors of the cerebral hemispheres is emphasized by many authors (DE VLIEGER and RIDDER, 1959; JEFFERSON, 1959, 1962; LITHANDER, 1960, 1961; JEPPSSON, 1960/1961; FORD and AMBROSE, 1963; SCHIEFER, KAZNER and Brückner, 1963; etc.). These tumors are one of the most important indications for echo-encephalography.

In supratentorial space-occupying lesions leading to a displacement of the midline structures of the brain, a shift of the midline echo towards the opposite side can be seen in the echo-encephalogram. This is especially important if homolateral neurologic signs are present.

Table 10. *Echo-encephalographic findings with 352 tumors and abscesses of the cerebral hemispheres*

Type of Lesion	No. of Cases	Midline Echo Normal	Midline Echo Shifted	Midline Echo Not Detectable	Error	Width of the 3rd Ventricle Normal	Width of the 3rd Ventricle Exceeding 7 mm	Width of the 3rd Ventricle Not Measurable	Tumor Echo
Meningioma	43	2	40	—	1	16	—	27	5
Glioblastoma	125	3	121	1	—	11	—	114	40
Astrocytoma	47	5	42	—	—	14	2	31	11
Oligodendroglioma	29	4	25	—	—	11	—	18	9
Other gliomas	21	—	20	1	—	7	2	12	7
Sarcoma	19	1	18	—	—	4	1	14	5
Metastases	34	3	30	1	—	8	2	24	5
Other tumors	9	2	7	—	—	3	2	4	3
Abscess, empyema	25	—	25	—	—	4	—	21	5
Total	352	20	328	3	1	78	9	265	90
Percentage	100	5.7	93.2	0.8	0.3	22.1	2.6	75.3	25.6

Case 3: Maria P., 39 years of age, Echo-EG No. 549/63. The patient, a medical secretary, had noticed a decrease in general performance and memory for some time already. 8 months prior to admission, she had a first generalized seizure, which later recurred. In the last months dizziness and repeated vomiting especially after a change in position. Clinical findings: obvious lack of motivation, cranial nerves including fundus unremarkable, slightly hyperreflexive on the right, walk uncertain, Romberg positive. Pneumoencephalography, performed at another hospital, showed incomplete filling of the relatively narrow ventricular system. The EEG pointed to a left-sided (!) frontal lesion.

The increased reflexes on the right also seemed to speak for o left-sided lesion. However, the echo-encephalogram showed a displacement of the midline echo 5 mm towards the left (Fig. 85 a). Right-sided carotid angiography which was subsequently performed, revealed a nearly fist-sized, homogeneously dye-tinged tumor in the anterior third of the sinus (Fig. 85 b). The large meningioma was completely removed on July 3rd, 1963, and the post-operative course was uneventful.

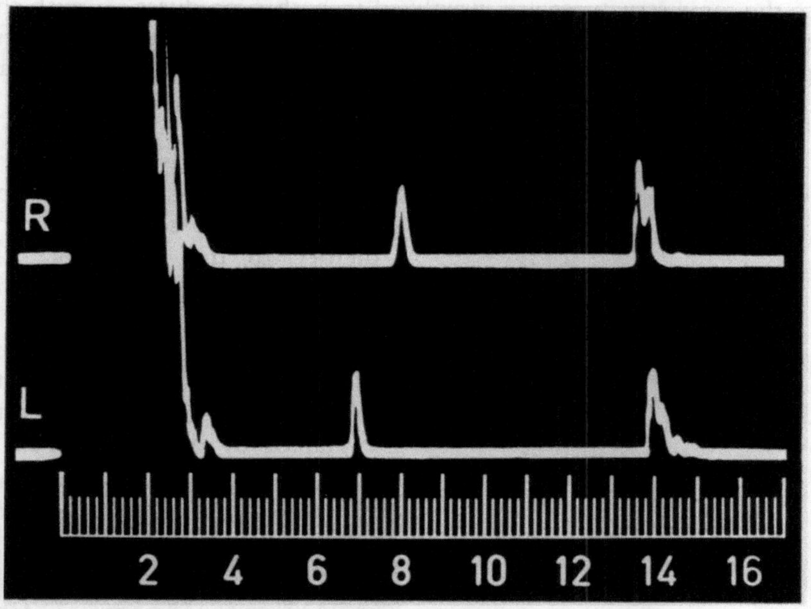

a

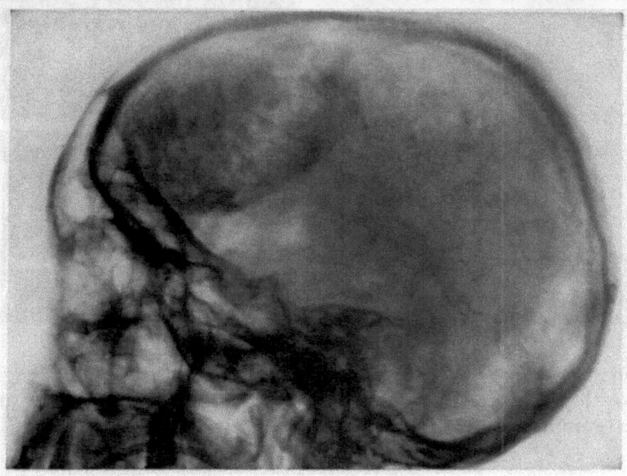

b

Fig. 85. a) Echo-encephalogram of a 39-year-old woman who was admitted with the tentative diagnosis of a left-sided frontal tumor. However, the ultrasonic trace revealed a 5 mm shift from right to left, so that a right-sided lesion must be assumed. b) The right-sided carotid angiogram indicated a large frontal meningioma. Pat. M. P., Echo-EG No. 549/63

Our own investigation series includes 352 tumors and abscesses of the cerebral hemispheres. In 351 cases, we possess neuroradiologic data, in one patient the operation was carried out on the basis of the echo-encephalographic finding alone without an x-ray. In 339 cases, an operation was performed which provided the histological diagnosis. In the remaining 13 patients, the clinical diagnosis was confirmed by autopsy 4 times, 9 times the tumors were inoperable and of unknown histology; x-ray therapy was given in these cases.

The echo-encephalographic findings of these 352 tumors and brain abscesses are grouped according to the histological diagnosis in Table 10. A displacement of the midline echo by 2.0 mm or more was found 328 times, in 20 cases a midline echo shift could not be demonstrated and once the echo-encephalogram was interpreted wrongly. The unsatisfactory results make up 6.8% of the total. These cases will be discussed in detail later.

Relationship between Midline Echo Shift and Tumor Localization

The degree of midline echo shift is strongly dependent on the localization of the tumor. The echo-encephalographic results can be compared best with the pneumoencephalographic results, because the same structures are visualized in both procedures — with the exception of the calcified pineal body. As expected, the least amount of lateral displacement in the region of the posterior part of the third ventricle is produced by the frontal, parasagittal and mediobasal tumors. The most marked shifts are caused by temporal tumors because the pressure of the tumor affects the midline structures directly here (Fig. 86). In our series, the occipital space-occupying lesions are in second place. This finding is contradictory to the results of other authors (PLANIOL et al., 1964, 1967; AKERMAN and GUIOT, 1965, BOSTEM et al., 1968) who often failed to find any midline echo displacement at all in tumors of the occipital lobe. Perhaps such an observation may be explained by the fact that an unequivocal

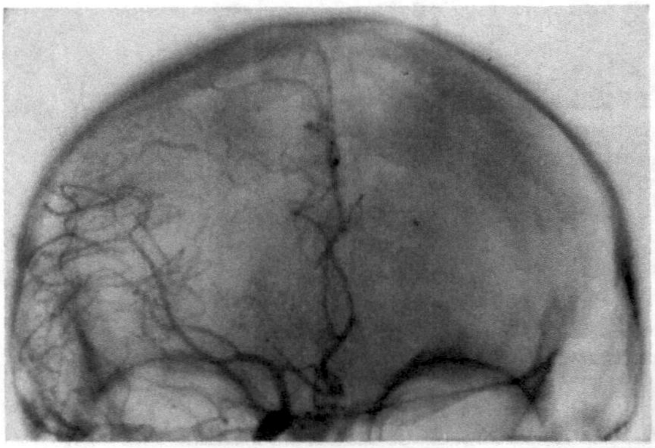

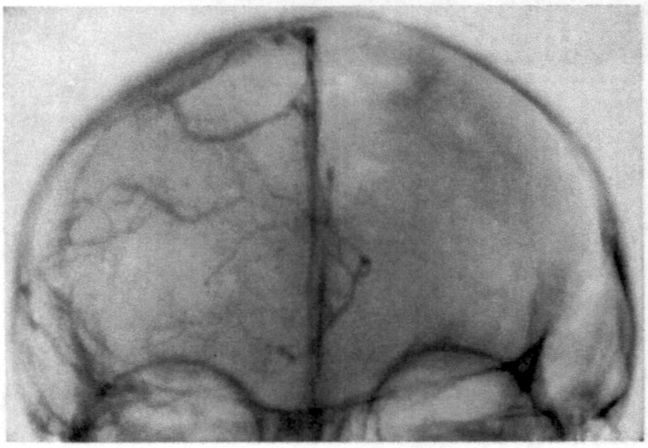

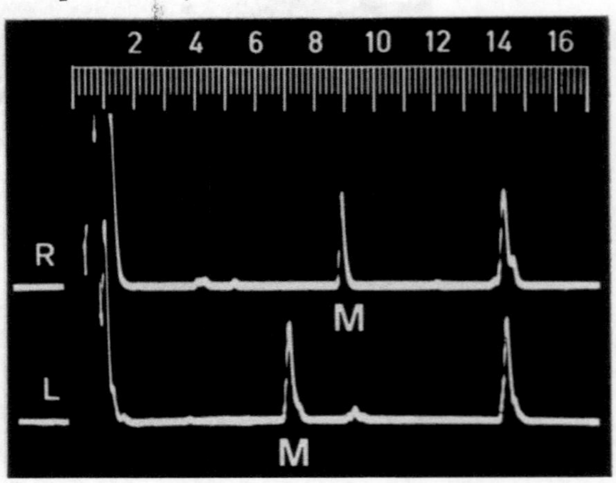

Fig. 86. Temporal tumors lead to the most pronounced average midline shifts. Right: shift of the midline echo by 9.0 mm due to a temporal glioblastoma. Left top: only slight shift of the anterior cerebral artery. Left bottom: considerable displacement of the internal cerebral vein towards the left. Pat. A. K., 47 years of age. Echo-EG No. 1739/65

midline echo is often very difficult to obtain with occipital tumors in our experience. The high degree of mass displacement with these tumors is due to the fact that an expanding lesion can escape only anteriorly because the occipital lobe is sqeezed between the falx and the tentorium which provide a firm barrier (TÖNNIS and SCHIEFER, 1953). On reviewing a large number of occipital tumors, the two authors also noted a lateral shift of the calcified pineal body, averaging 8 to 12 mm on AP plain films evaluated according to the method described by VASTINE and KINNEY (1927).

To bring about a *correlation between the angiogram and the ultrasonic measurement*, we must use the internal cerebral veins as the point of reference (JEPPSSON, 1960, 1961; LITHANDER, 1961). This applies especially to frontal and occipital tumors. The anterior cerebral artery and its branches are not suitable for a comparison because their course does not lie in the scanned area (Figs. 52 and 87).

The average displacement of the midline echo and its dependence on the site of the tumor is shown graphically in Fig. 88. The results are based on the ultrasonic findings of 328 tumors of the cerebral hemispheres, all of which are confirmed by neuroradiological methods.

Compared to our own earlier publications, an increase is noted in the midline echo displacement produced by frontal tumors from 3.1 to 4.1 mm. The reason for this is that we have observed several unusually large meningiomas recently which caused displacements of 7.5 to 8.0 mm and a glioblastoma which filled the entire fronto-precentral area and produced a displacement of 11.0 mm. In general frontal gliomas cause a midline echo displacement of only about 3 mm. The second deviation is seen with parietal tumors. The smaller average midline echo shift recorded here is due to the fact that we have grouped together the parietal and parasagittal tumors which had been separated in earlier publications. These space-occupying lesions showed an increase in the average midline echo shift of 0 mm with smaller tumors in the parasagittal area to a shift of 7.0 mm in the middle and low parietal area. In precentral, temporal and occipital neoplasms, we recorded the largest midline displacements namely 6.9, 8.2 and 7.9 mm respectively.

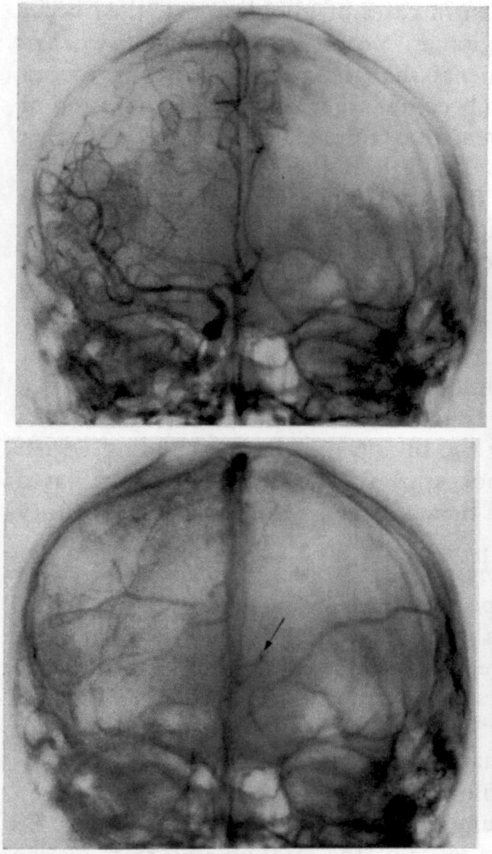

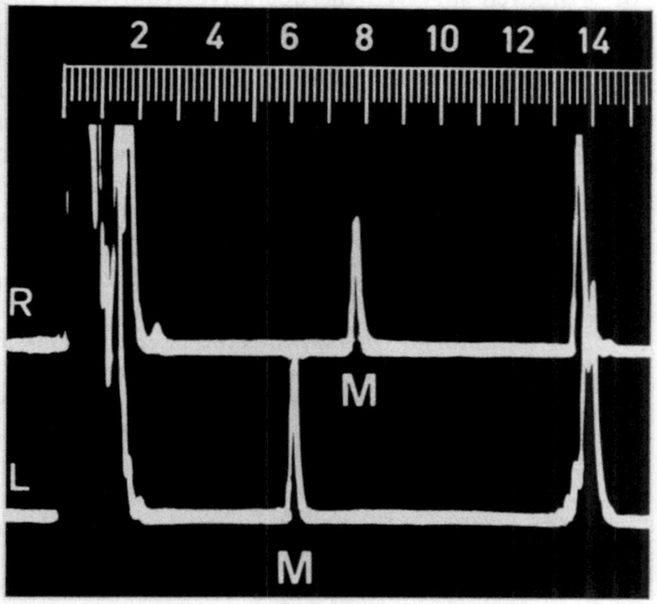

Fig. 87. Occipital tumors usually result in marked shifts of the midline structures. Right: displacement of the M-Echo by 8.0 mm due to an occipital glioblastoma. Left: carotid angiography confirms the extent of the displacement on the phlebogram (arrow), while the arterial phase fails to reveal any displacement. Pat. M. K., 64 years of age. Echo-EG No. 2852/67

The question now arises whether the midline echo shift can be used for more exact localization of the lesion in addition to determining the side on which the tumor is located. In his monograph, published in 1961, Jeppsson has already reported nearly the same mean deviation of the M-echo and noted a significant difference between frontal and temporal or parietal tumors. Nevertheless, we do not believe that the degree of midline echo shift can aid in the localization of a space-occupying lesion in the particular case. The only exception is probably the case of a tumor which causes an M-echo shift in excess of 10.0 mm. With very few exceptions, such a great mass displacement is found only with temporal tumors. This statement is not true for tumors of the cerebral hemispheres in infants and children however.

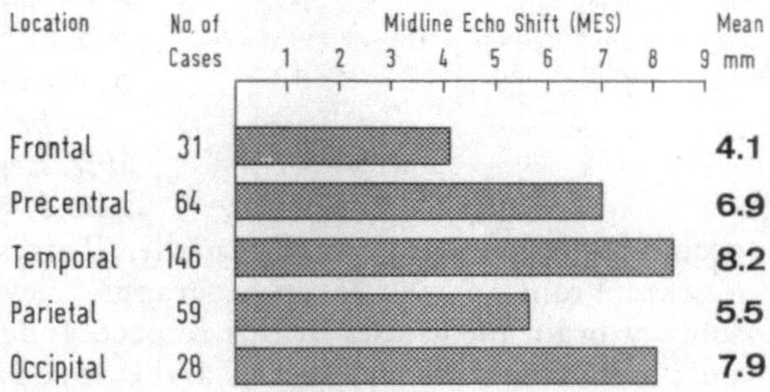

Fig. 88. Degree of midline echo shift in cerebral hemisphere tumors depending on the localization (328 cases)

In our experience the degree of midline echo shift also does not permit definite deductions about the *histological structure* of the tumor in question. The average M-echo shift produced by the various types of tumors ranges from 6 and to 8 mm. Although very marked mass displacements (up to 15.0 mm) are observed most often in cases of glioblastoma, they occur in other types of tumors as well, especially in meningiomas. *Supratentorial brain tumors in children* cause much more marked M-echo displacements than in adults according to our series. This is surprising because the infantile skull can become larger through a separation of the cranial sutures. However, the tumors of the children we examined, were nearly all larger than tumors in the corresponding location in adults. While half the adult patients with a midline echo shift exceeding 10.0 mm did not survive (mostly temporal glioblastomas), 8 out of 9 children survived the operative removal of their tumors despite the massive displacement. *The majority of adult patients with tumors causing an M-echo shift of more than 10.0 mm had disturbances in consciousness.* Such great lateral displacements of the midline structures of the brain will not affect the state of consciousness only if the tumor is a very slow-growing one e. g. a meningioma. The more acutely the tissue displacement occurs, the sooner we may expect a state of coma. However, in children with brain tumors, we observed M-echo shifts up to 15 mm without noting even somnolence. The infantile brain can obviously cope with mass displacements produced by the tumor and the surrounding edema much better so that gross neurological deficits do not appear. The only exception was a child with a large temporal ependymoma into which a hemorrhage had occurred. The history of this case is reported in detail as case 5 on page 101.

The behavior of the midline echo in cases of meningioma of the cerebral hemispheres will be discussed in more detail here since it shows some peculiarities. In our investigation series, there are 43 meningiomas in this location. This series is made up of 7 frontal meningiomas, 6 tumors between the anterior and middle third of the sinus, 12 tumors in the middle third of the sinus and 3 meningiomas of the posterior third of the sinus. 5 tumors originated from the falx and developed bilaterally. 7 meningiomas were found near the Sylvian fissure and 3 tumors in the lateral ventricles. Fig. 89 shows the dependence of the midline echo shift on the location of the meningioma. Falx meningiomas generally escape echo-encephalographic detection (displacements of 0.5 to 2.5 mm) while the mean value of frontal meningiomas is surprisingly twice the mean value of gliomas. The reason for this probably lies in the fact that these tumors generally become quite large before they lead to marked neurological deficits. The same is true for meningiomas of the posterior third of the sinus. In contrast to gliomas in the same location, parasagittal meningiomas of the middle third of the sinus exhibited a midline echo displacement in all cases (3.5 to 5.5 mm) and could thereby be diagnosed accurately as space-occupying lesions.

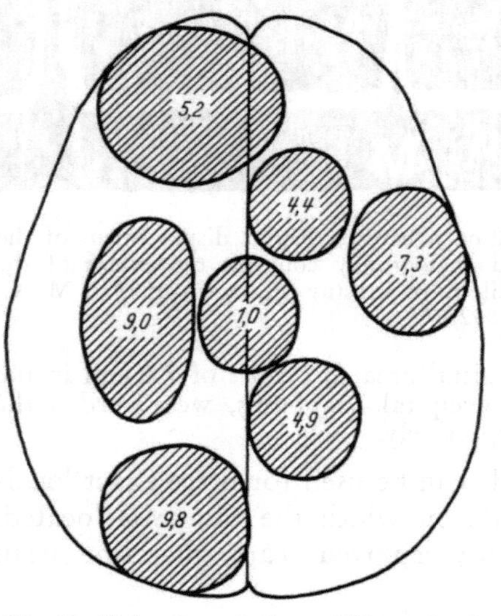

Fig. 89. Behavior of the midline echo in cases of meningioma of the cerebral hemispheres and of the lateral ventricles (43 cases). The figures indicate the average midline echo shift in mm. Only falx meningiomas do not produce a midline displacement

The degree of midline echo shift in cases of brain abscesses and empyemas almost reached the values obtained with glioblastomas although the actual space-occupying lesion was generally smaller. This is due to pronounced reactive edema around the abscess. From an echo-encephalographic viewpoint, the brain abscess can be compared to solitary brain metastases in this respect. Independent of location, the average midline echo shift in 25 cases of brain abscess and empyema was barely 8 mm.

From the amount of mass shift alone, however, there are no special diagnostic indications for a circumscribed inflammatory cerebral lesion. A chief characteristic, though, is the *rapid development of the midline echo displacement* which matches the often fulminating clinical course. A single echo-encephalographic investigation is not enough under any circumstances since the first echogram is frequently still normal when the first neurological deficits appear. We have observed a series of patients with brain abscesses, who either developed a mass displacement within a few days or even hours or in whom the shift increased rapidly. A similar observation was made by GROSSMAN (1964). Only in cases where a hemorrhage into a brain tumor occurs, can a similar rapid or even more rapid increase in the midline echo shift be recorded. This echo-encephalographic sign can aid in the differentiation between tumor and abscess.

The value of *echo-encephalographic follow-up studies* in cases of brain abscess is illustrated by the following history:

Case 4: Gabi I., 4 years of age, Echo-EG No. 2643/66. Child admitted to hospital on December 1st, 1966 because of a sudden left-sided hemiparesis following bacterial sinusitis. The first echo-encephalogram was unremarkable (Fig. 90 a). Right-sided carotid angiography revealed no signs of a space-occupying lesion. Since a 3 mm midline echo shift towards the left was observed in the echo-encephalogram several days later (Fig. 90 b), we undertook a pneumoencephalographic investigation which showed a slight displacement of the third ventricle towards the left and a downwards displacement of the body. Assuming the diagnosis of brain abscess in the upper parietal region, operation was performed on December 13th, 1966. Upon puncturing the abscess, 20 ccm pus were aspirated. Although the abscess cavity was flushed with Nebacetin, the echo-encephalogram revealed a further increase in the mass shift up to 10.5 mm (Fig. 90 c and d). From this finding we deduced that a further abscess must have developed; the assumption was confirmed first by angiography and then by operation.

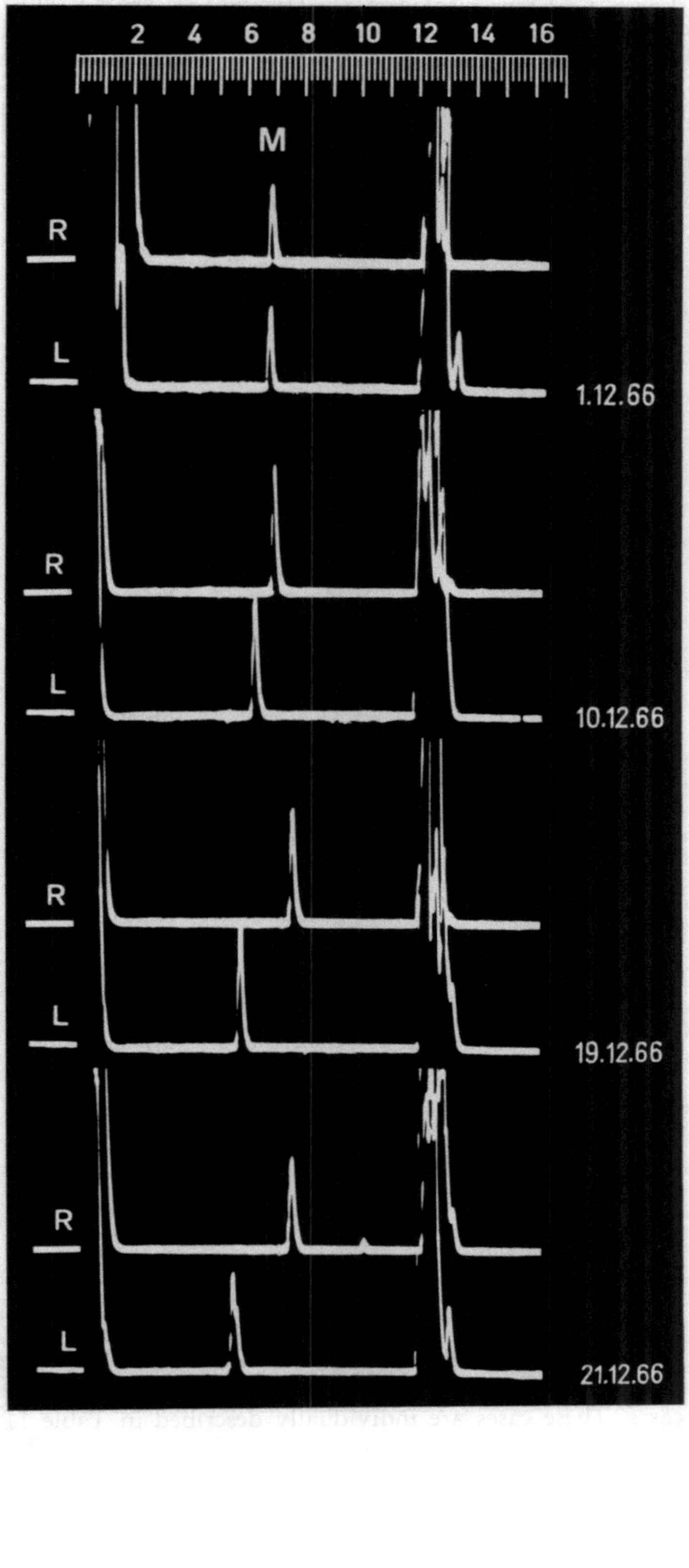

Fig. 90. Echo-encephalographic follow-up study of a 4-year-old child with a brain abscess. Pat. G. I., Echo-EG No. 2643/66

Tumors of the Cerebral Hemispheres without Midline Echo Displacement

Of the 352 tumors of the cerebral hemispheres we examined, 21 cases failed to produce a displacement of the midline structures on the echo-encephalogram. In one case, only the control investigation showed that the M-echo was midline while the initial examination had shown a 4.5 mm displacement. This was an error (Table 11). In all patients however, the

Table 11. *Tumors of the cerebral hemispheres without midline echo displacement (21 Patients)*

Echogram No.	Name	Age	Type of Tumor	Localization	Size of Tumor	Echo-Encephalogram
654/65	D. Sch.	58 y.	Meningioma (recurrence)	Falx, middle third of the sinus	Tangerine	Normal
670/66	G. M.	44 y.	Meningioma (recurrence)	Falx, middle third of the sinus	Tangerine	At first ME shift 4.5 mm, on control normal
855/63	G. S.	33 y.	Astrocytoma	Parasagittal parietal R	Chestnut	Midline, 3rd v. 8.5 mm
1004/64	H. Sch.	28 y.	Astrocytoma	Parasagittal parietal R	Chestnut	Midline, 3rd v. 7.5 mm
1430/64	R. R.	54 y.	Oligodendroglioma	Bilat. frontal	Pigeon egg	Normal
1463/64	F. L.	24 y.	Astrocytoma	Temporal L	Bean	Midline, abnormal echoes from tumor calcification
1557/65	D. D.	1 y.	Not classif.	Fronto-basal L	Walnut	Normal
1671/65	U. W.	11 y.	Cavern. hemangioma	Fronto-lateral R	Walnut	Normal
1699/65	M. T.	62 y.	Meningioma	Falx, middle third of the sinus	Tangerine	Normal
1753/65	W. Sch.	62 y.	Sarcoma of the dura mater	Temporal L	Flat, palm-sized	Midline, 3rd v. 9.0 mm
1772/65	Ch. H.	58 y.	Metastasis	Parasagittal parietal R	Chestnut	Normal
1930/65	E. C.	44 y.	Oligodendroglioma	Parietal L	Cherry	Midline, abnormal echoes from tumor calcification
1984/65	G. E.	45 y.	Oligodendroglioma	Frontal R	Chestnut	Normal
2143/66	H. O.	49 y.	Astrocytoma	Fronto-dorsal L	Cherry	Normal
2228/66	A. D.	25 y.	Oligodendroglioma	Parasagittal precentral L	Chestnut	Normal
2406/66	F. R.	57 y.	Multiple metastases		Up to chestnut size	Normal
2429/66	K. G.	63 y.	Glioblastoma	Parasagittal parietal R	Pigeon egg	Normal
2447/66	E. U.	55 y.	Glioblastoma	Sylvian fissure L	Cherry	Normal
2592/66	M. Sch.	55 y.	Metastasis	Parasagittal parietal R	Chestnut	ME shift 1.5 mm to the left
2598/66	G. P.	57 y.	Glioblastoma	Parasagittal parietal L	Walnut	Beside the ME on both sides high reflections from calcified choroid plexus
2694/67	E. F.	28 y.	Astrocytoma	Above the Sylvian fissure R	Not circumscribed	ME shift 1.5 mm to the left

neuroradiological investigations did not demonstrate a displacement of the midline structures or of the internal veins towards one side either. Particularly remarkable is the fact that the diagnosis of a space-occupying lesion could not be made even from the angiogram in 11 cases. These cases are individually described in Table 11; furthermore 15 of these tumors were drawn into a diagram according to scale (Fig. 91). This illustration shows us that *smaller tumors in the frontal and parasagittal area can escape echo-encephalographic detection.* Very small tumors in the temporal and parietal area sometimes do not produce a midline shift either. However, here it is simply a *question of the time* at which the patient goes to the doctor, who then orders further investigations. Therefore, although it is entirely possible that a broader use of echo-encephalography can lead to an earlier detection of many tumors, the number of tumors without midline echo shifts will also increase considerably. Thus PLANIOL (1967) found a displaced midline echo in only 69% of the cases of tumors of the cerebral hemispheres. Altogether, however, only 21 cases = 6.0% of our series of 352 tumors of the cerebral hemispheres failed to produce lateralization signs on the echo-encephalogram. But we must not forget that we are dealing with selected neurosurgical cases here.

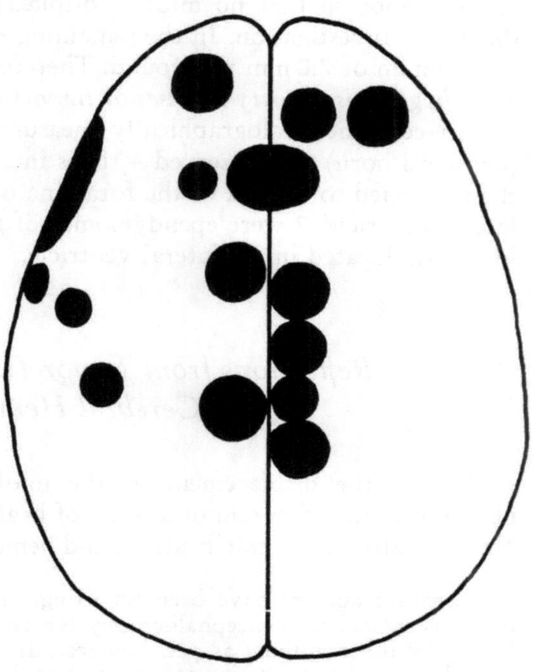

Fig. 91. Localization and size of 15 tumors of the cerebral hemispheres which did not lead to a midline echo shift. These were primarily smaller frontal and parasagittal gliomas.

Demonstration of the Third Ventricle in the Echo-Encephalogram with Tumors of the Cerebral Hemispheres

If the ventricular system is midline, the width of the third ventricle can be determined fairly accurately in 90% of the cases by means of a double echo. With tumors of the cerebral hemispheres on the other hand such a double reflection from the lateral walls of the third ventricle was recorded in only 24.7% of the cases. With few exceptions, these were patients with tumors which caused only slight displacements of the midline echo (up to 5.0 mm). If the M-echo shift was more marked, the width of the third ventricle could be determined only with difficulty or not at all. FORD and MCRAE (1966) arrived at the same results. The reason for this may be found in the fact that *echoes from the lateral walls of the third ventricle can be picked up only if there is a parallel shift of the walls of this ventricle.* Due to the pressure effects of the tumor, however, the third ventricle is no longer perpendicular to the direction of sounding in the overwhelming majority of tumors. This results in unfavorable conditions of reflection and if the axis of the third ventricle deviates as little as 10 degrees from the norm, the amount of ultrasonic energy reflected back to the probe is so little that a definite differentiation of the individual reflections is not possible any more because of the low amplitudes. While the width of the

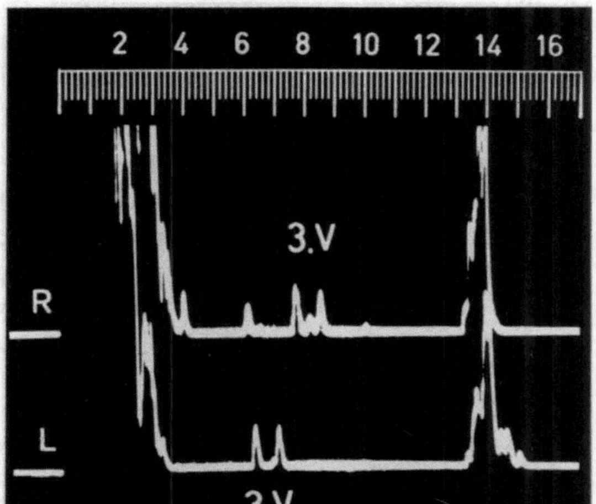

Fig. 92. Echogram of a 51-year-old man with a right temporal glioblastoma extending to the basal ganglia. The trace shows an 8 mm wide double echo from the walls of the slightly dilated third ventricle, shifted 7.0 mm to the left. Pat. H. F., Echo-EG No. 519/63

third ventricle lay within the range of normal in 78 patients (22.1%), a slight dilatation of this ventricle was seen only 9 times (2.6%). In 3 of the cases, the midline structures were not displaced (compare Tables 10 and 11). Twice there was an atrophic brain lesion in addition to the tumor so that no midline displacement could be demonstrated at the time of the diagnostic investigation. In the remaining 6 patients a midline echo shift of about 2.0 mm to a maximum of 7.0 mm was found. Therefore *a laterally displaced, dilated third ventricle in the echogram is a rarity in cases of tumors of the cerebral hemispheres* (Fig. 92).

An echo-encephalographically measurable dilatation of parts of the lateral ventricles (temporal horn) was observed 4 times in cases of tumors of the cerebral hemispheres where the tumor led to a block of the foramina of Monroe. One of these was a meningioma of the lateral ventricle, 2 were ependymomas of the lateral ventricles and one was a plexus papilloma also located in the lateral ventricle.

Reflections from Tumor Tissue and Cyst Walls with Tumors of the Cerebral Hemispheres ("Tumor Echoes")

Besides the displacement of the midline echo, a number of additional echoes were observed in the echogram of a series of brain tumors. These originated from tumors, cyst and abscess walls, tumor calcifications and hemorrhages into a tumor.

Japanese authors have been reporting on the possibilities of localizing brain tumors through one-dimensional echo-encephalography for years (see p. 65). Their good results have not been confirmed by other authors as yet, however. In the present investigation series, special emphasis was placed on the demonstration of tumor echoes because a positive result would be a valuable enrichment of echo-encephalographic diagnosis.

Such reflections were picked up in 90 of the 352 tumors of the cerebral hemispheres, a percentage of 25.6. Despite careful investigation technique, the proportion of tumor echoes has not risen significantly in the last 2 years when compared to the findings obtained till the end of 1964 (Schiefer, Kazner, Kunze, 1965) (25.6% now as contrasted to 22.4% in the earlier study). The frequency of tumor reflections with the different types of tumors is presented in Table 10.

Tumor echoes are found most frequently in temporal and temporo-occipital *glioblastomas*. 38 of the 83 glioblastomas observed since 1964 showed tumor reflections (= 45.8%). If this frequency is applied only to temporal, temporo-occipital and temporo-parietal glioblastomas (a total of 54 tumors) the surprisingly high proportion of 70.4% is obtained. The tumor echoes in cases of glioblastoma are seen primarily in the near field of the echo-encephalogram (between initial complex and midline echo) when the probe is applied on the side of the tumor. Sounding from the opposite side only rarely reveals tumor echo complexes because ultrasound which is reflected from the tumor must travel through the normal cerebral hemisphere as well; as a result much more ultrasound is absorbed. Fig. 93 shows a series of echograms with tumor reflections in cases of glioblastoma. These echo groups are in no way specific for glioblastomas, however. In cases of *ependymoma* and *oligodendroglioma* tumor echoes were recorded relatively frequently (50, i. e. 31%). Tumor echoes in cases of ependymoma were seen exclusively in children. In the echogram of these cases, the pathological reflections were usually recognized in both traces (Fig. 95). In 3 cases out of 5, the plain films of the skull already showed distinct calcifications. The tumor reflections in cases of oligodendroglioma were also caused primarily by the calcifications. 7 out of 9 patients with oligodendrogliomas and tumor echoes had calcified deposits in the area of the tumor on radiological or histological examination. Here we often saw complexes of unusually high pathological reflections in the echogram which could be recorded from both sides if the tumor was located in the temporal area (Fig. 70).

Besides these tumors, distinct tumor echo complexes were occasionally seen in cases of *astrocytoma* (Fig. 64) but much more rarely. However, the echogram of this type of tumor repeatedly revealed a single high reflection in the tumor area; this was later found to originate from a cyst wall (Fig. 66). In several cases a single reflection was picked up from the wall of a brain abscess; however, this did not enable us to make a specific diagnosis (see also Fig. 192 g).

Meningiomas caused two different types of tumor echoes. Firstly these echo complexes resembled those found with gliomas. These were probably echoes caused by calcium deposits not visible on x-ray. Secondly a single high echo was sometimes recorded which had obviously developed at the interface between tumor and brain tissue, when the ultrasonic beam hit the well-delineated tumor

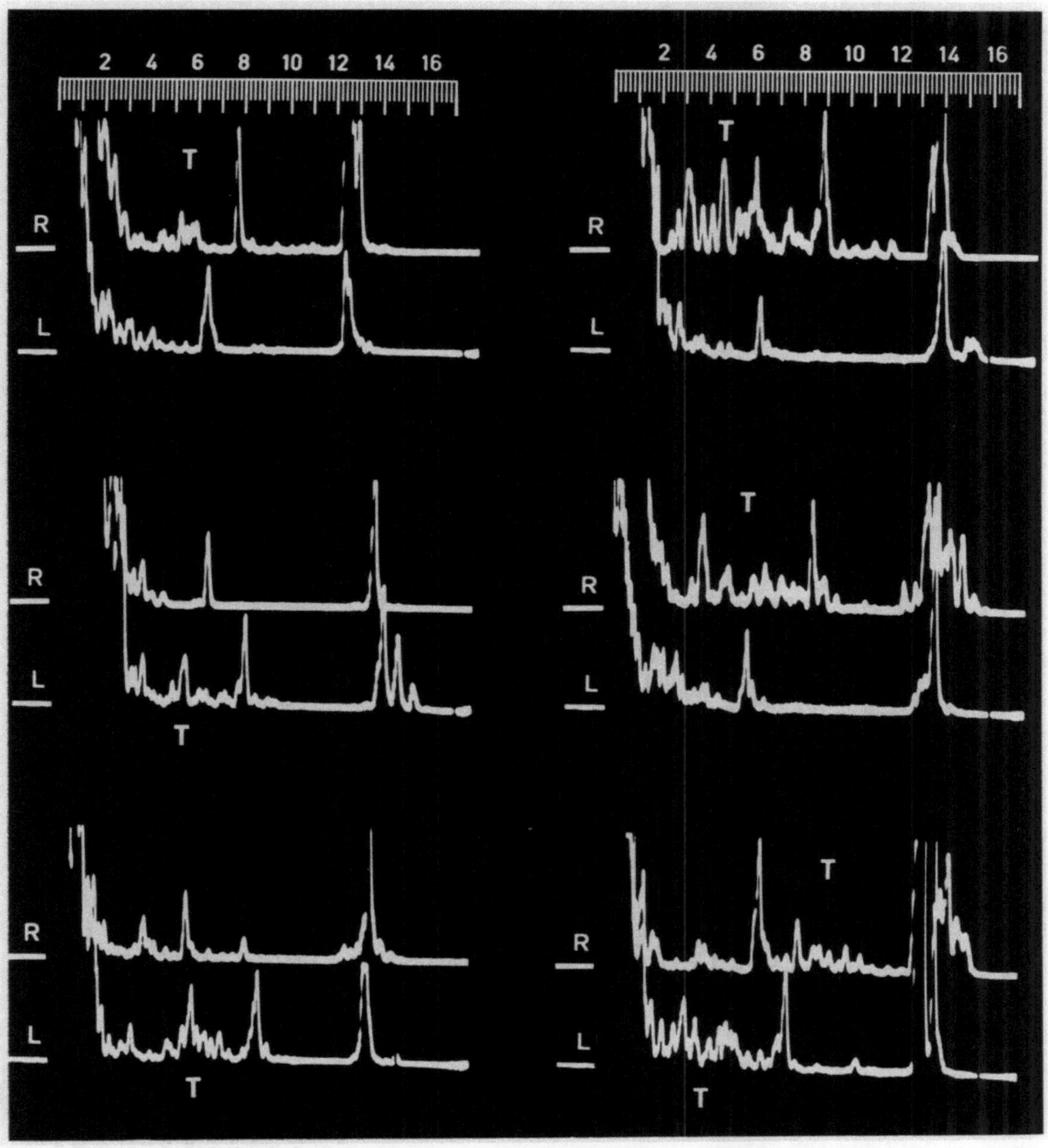

Fig. 93. Echogram series of temporal glioblastomas with tumor reflections. When sounding from the side of the tumor, the tumor echoes usually appear only as prolonged initial echoes. In all cases, there is a distinct midline echo shift. In order to make a correct interpretation, sounding must be carried out from both sides at the same adjustment level of the sensitivity time control. a) Pat. E. E., 59 years of age. Echo-EG No. 1779/65; b) Pat. K. H., 54 years of age. Echo-EG No. 1780/65; c) Pat. A. K., 47 years of age. Echo-EG No. 1739/65; d) Pat. M. U., 50 years of age. Echo-EG No. 1386/65; e) Pat. A. R., 52 years of age. Echo-EG No. 2360/66; f) Pat. J. R., 63 years of age. Echo-EG No. 1823/65

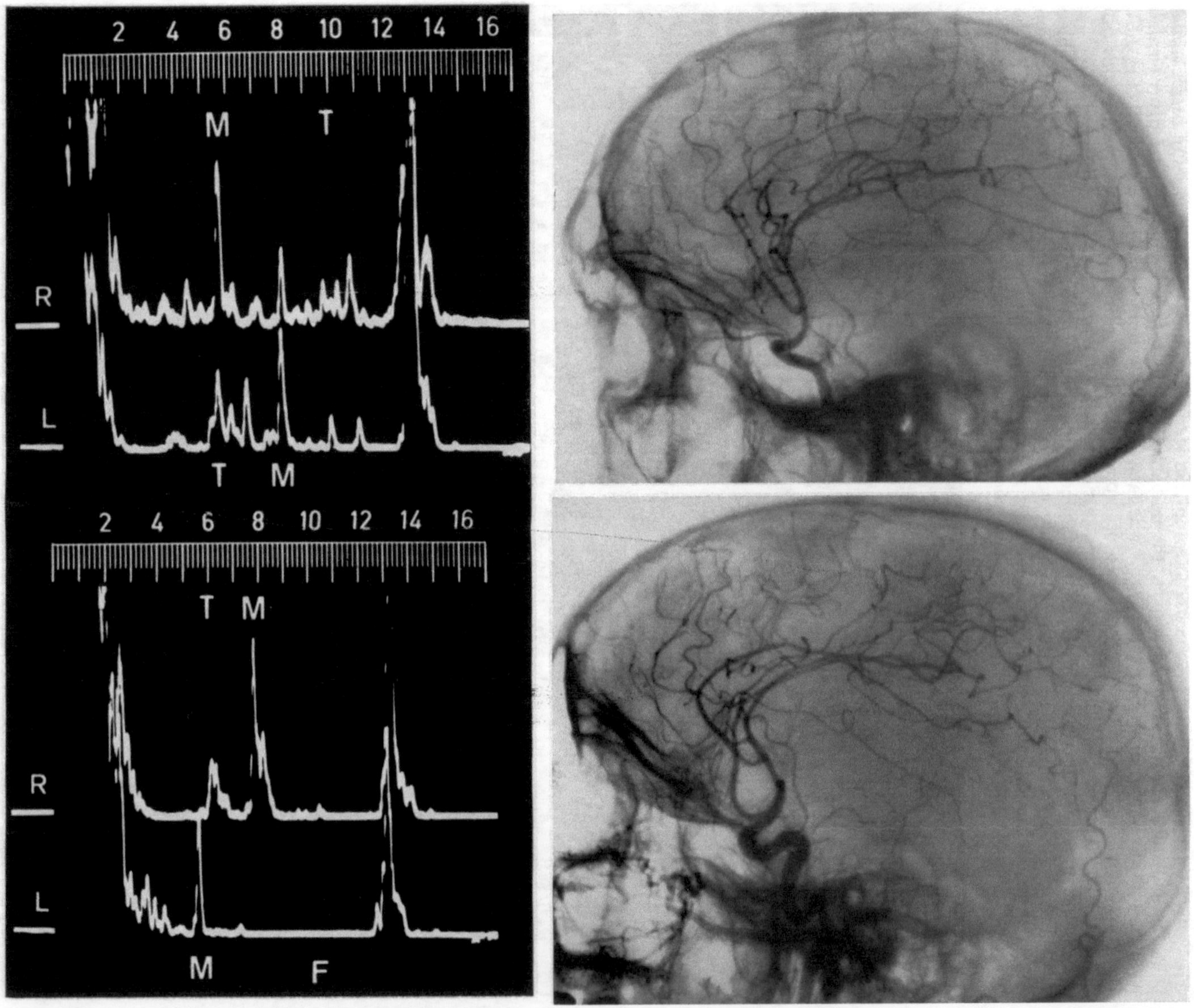

Fig. 94. Differentiation between cystic and solid tumors by echo-encephalography. In both angiograms, there are signs of a large avascular space-occupying lesion in the temporal area. The echograms on the other hand, are entirely different. Top: tumor reflections (T) in a glioblastoma. Pat. A. T. 39 years of age. Echo-EG No. 2756/67. Bottom: Echo-free zone (F) in a large cyst in the temporal lobe originating from a monstrocellular sarcoma. Pat. J. W., Echo-EG No. 2752/67

vertically. But the diagnosis of a meningioma could never be made from this reflection alone. With the exception of lateral sphenoid wing meningiomas, tumor reflections play a very secondary role because most of the tumors are not in the area scanned by the ultrasonic beam. All our investigations were made from a temporal application point, because all the other points did not result in satisfactory measurements. In an evaluation of the frequency of tumor echoes, this fact must be taken into consideration.

In 53 cases of sarcoma and metastases, tumor echo complexes were recorded 10 times (19%) but these did not differ remarkably from the echoes obtained from glioblastomas or other irregularly structured tumors either.

In recording pathological echo complexes, the type of ultrasonic apparatus used, plays a decisive role. Echo-encephalographs which automatically suppress smaller reflections (e. g. by raising the sound level) are unsuitable.

Even if one-dimensional echo-encephalography in its present form possesses only limited value for the exact localization of tumors by means of pathological echo complexes, still it provides valuable information in selected cases. Not infrequently the growth of a malignant tumor towards the midline can be determined by means of ultrasound. Echo-encephalography can also contribute to clarifying the differential diagnosis between cystic and solid tumors (see Figs. 67 and 94 in this respect).

However, we do not think it correct to do without dye studies and to operate brain tumors on the basis of the ultrasonic finding alone, aside from very few exceptions. Thus we performed a trepanation on a child admitted in a moribund state in one case only. The history is briefly presented here.

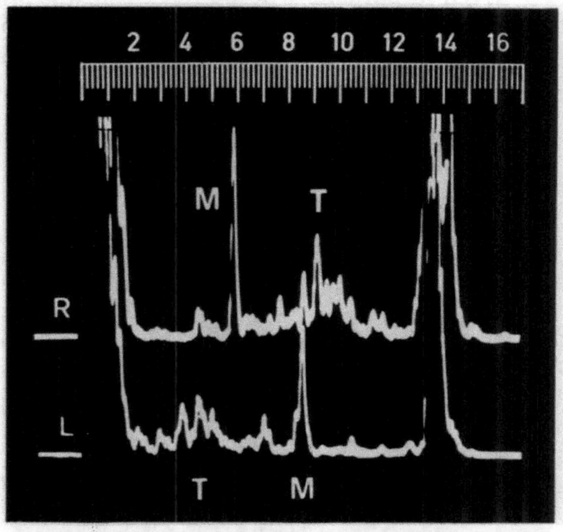

Fig. 95. Echo-encephalogram of an 8-year-old boy with an ependymoma in the left temporal lobe. Exact localization of a space-occupying lesion due to a tumor echo complex (T). Pat. R. B., Echo-EG No. 1707/65

Case 5: Rainer B., 8 years of age. Echo-EG No. 1707/65. Vomiting for four weeks. Unconscious since 6:30 p. m. on June 24th, 1965. Hospital admission on the same day at 11 p. m. Clinical finding: deep unconsciousness. Left pupil fully dilated and unresponsive to light. The right pupil is initially normal; however, within a few minutes it is also dilated and fails to react to light. Massive papilledema with fresh retinal hemorrhages bilaterally. Clonic leg reflexes, painful stimuli produce decerebrate rigidity. Echo-encephalogram: displacement of the midline structures 14 mm from left to right. In addition, a complex of pathological echoes 4 cm in width originating from a tumor or a hemorrhage in the left temporal area (Fig. 95). Immediate operation without further radiological investigation: ependymoma in the left lateral ventricle containing a fresh hematoma of about 100 cm³.

Postoperative Findings and Long-Term Observations

Echo-encephalography is an ideal method for the postoperative follow-up of brain tumors, because it is safe, rapid painless and repeatable at will. Jeppsson (1961) dealt with the use of echo-encephalography following neurosurgical procedures in great detail and emphasized a number of characteristic findings. Brückner (1965) also studied the postoperative complications using the echo-encephalogram. Further investigations on this topic have been published by Hotermans and Thiry (1967). In the last 5 years, 282 patients in the postoperative phase (following removal of tumors of the cerebral hemispheres) were investigated it the University of Erlangen. Here the results reported by Jeppsson were confirmed in every detail.

a) Acute Phase Following Operation

Localized cerebral edema and postoperative bleeding are the most important complications following surgery of cerebral hemisphere tumors. The clinical picture and the follow-up study (neurological findings, control of the state of consciousness, of blood pressure, pulse, temperature and respiration) do not permit us to make the decision necessary for the right course of treatment with the certainty that would be desirable.

In cases of tumors of the cerebral hemispheres where the midline echo was shifted before the operation, generally no decrease in the M-echo shift appeared within the first few days after surgery. A 1 to 2 mm increase in the M-echo shift is not an exceptional finding and is due to *reactive cerebral edema.* Normally the decrease in mass displacement and consequently in the midline echo shift begins 4 to 7 days after operation. Depending on the degree of the preoperative M-echo shift, complete normalization may take as long as 20 to 60 days. We have not been able to determine a difference in the speed of normalization between benign and malignant tumors. Following subtotal resection of a tumor, the midline echo shift is frequently even markedly increased at first (4 to 6 mm) and never normalizes completely.

Even in patients who do not show a primary displacement, a transient midline echo shift of 2 to 5 mm may appear. This is also true for unilateral chiasma exposure.

While the slight increase in the midline echo shift due to circumscribed cerebral edema may be registered usually only on the second and third postoperative day, a midline echo shift caused by postoperative bleeding can develop within the first 24 or 36 hours after surgery already. The extent of the M-echo shift is also much higher in the majority of *postoperative hemorrhages* than with cerebral edema (+4 to 10 mm). Special reflections, pathognomonic of a hematoma, can be obtained only in cases of postoperative hemorrhage into the epidural space.

However, a dura echo 8 to 10 mm ahead of the end echo is no rarity after temporal and parietal exposures and generally does not make reoperation necessary. An epidural accumulation of blood up to 1 cm in thickness is produced by the fact that the dura which is originally curved along the inner table of the skull now forms a nearly flat surface because it is tightly stretched by the sutures. The space created in this way slowly fills with blood after the operation if no drainage tube is inserted. The resorption of such a blood accumulation within a period of 3 to 4 weeks can be recognized in the echo-encephalogram by the progress of the dura echo towards the end echo and finally by its disappearance.

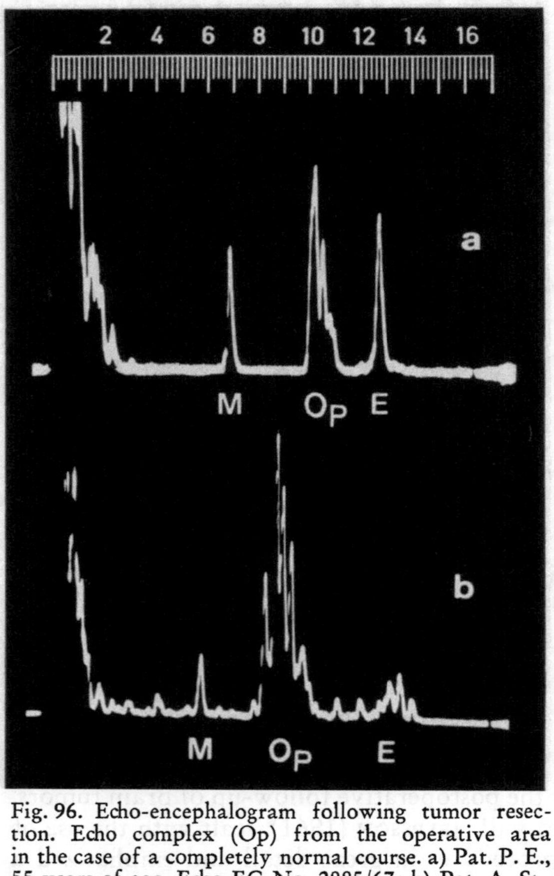

Fig. 96. Echo-encephalogram following tumor resection. Echo complex (Op) from the operative area in the case of a completely normal course. a) Pat. P. E., 55 years of age. Echo-EG No. 2885/67; b) Pat. A. St., 48 years of age. Echo-EG No. 1387/64

In 7 cases of echo-encephalographically diagnosed postoperative epidural hematomas, varying in thickness from 12 to 28 mm, the wound was reopened in order to evacuate the hemorrhage. Aside from the considerable increase in the midline echo shift we saw no specific hematoma echo complexes in the ultrasonic trace of postoperative intracerebral hemorrhages. Even if the postoperative course is completely normal, we often find series of pathological echoes originating from the walls of the tumor resection cavity, from metal clips and from small sponges inserted for coagulation purposes; these cannot be differentiated from hematoma echo complexes (Fig. 96). Thus the echo-encephalographic diagnosis of an intracerebral postoperative hemorrhage has to be made chiefly on the basis of an increase in the midline echo shift. According to JEPPSSON, the suspicion of a postoperative hemorrhage must also arise when the midline echo shift fails to return to normal in the postoperative phase.

Our own experience is limited to 4 cases of intracerebral postoperative hemorrhage. The great advantage of echo-encephalography here is that an increase in the midline echo shift was noted even before circulatory and respiratory disturbances could be detected. In cases where trepanation was performed in the temporal area, the investigation can generally be carried out only from the healthy side. Finally it must be pointed out that *postoperative cerebral edema* can be so massive on occasion that the increase in the midline echo shift makes it impossible to differentiate cerebral edema from a hemorrhage. We saw such a complication on the second postoperative day in 3 cases of basal meningioma; the complication necessitated the transient removal of the bony flap.

b) Long-Term Follow-up Studies, Recognition of Recurrences

In 257 patients echo-encephalographic control examinations were performed from 4 weeks to 10 years after the removal of a tumor of the cerebral hemispheres. Normally the midline echo is found at the normal site again, 1 to 2 months after the operative procedure. If the resection was very extensive, especially in children, a midline echo shift towards the side of the lesion ranging from 2 to 5 mm could be observed. The characteristic finding is a *slight dilatation of the third ventricle amounting to 7.5 to 9 mm.* This sign is seen very frequently after a glioblastoma operation with subsequent radiation therapy, but it is also common after the removal of large meningiomas of the convexity.

The first sign of a recurrence in the echo-encephalogram is the decrease in the width of the third ventricle. Only after a period of weeks or months do neurological deficits and a new midline echo shift appear. Altogether a total of 42 tumor recurrences were recognized in the echo-encephalogram. The only exception is that of a 40-year-old man with a frontal astrocytoma where the ultrasonic trace failed to show any signs of renewed tumor growth. Carotid angiography, carried out somewhat later, showed a distinct displacement of the anterior cerebral artery at that time, however. PLANIOL (1967) recorded less favorable results with tumor recurrences.

A distinct decrease in the M-echo shift was observed in 4 cases of inoperable glioma following radiation doses of 4000 to 5000 R; twice the midline echo even returned to normal.

3. Tumors of the Oral Brain Stem and Corpus Callosum

45 of the 630 patients with brain tumors examined by echo-encephalography revealed a tumor in the oral brain stem or corpus callosum. The ultrasonic examination of these tumors in our first series resulted in widely different and often unsatisfactory findings. The echo-encephalographic peculiarities in a series of these tumors made it necessary to group them separately, not together with lesions of the cerebral hemispheres.

At the beginning of our echo-encephalographic research, we always tried to find an unequivocal midline echo in all our patients. In those with tumors of the thalamus and of the third ventricle, this resulted in great difficulties. Not infrequently a whole series of reflections appeared in the median area so that a clear interpretation of the echogram was not possible or errors occurred. The reason for this may be that such tumors directly affect the structures responsible for the midline echo and that they produce their own reflections.

The histology of many of these tumors is unknown because surgical treatment of the tumor itself was rarely carried out, and the majority of the patients who had been treated with ventriculo-cisternostomy (Torkildsen) and radiotherapy is still alive.

The location of the tumors and/or their histology is shown in Table 12. In the majority of cases, 31 of 45 tumors of the oral brain stem and corpus callosum (69%), the midline echo was found at the normal site. In 9 patients, corresponding to a quota of 20%, the determination of the midline echo was erroneous or it could not be recorded accurately. Thus these tumors showed the highest rate of error in the entire field of echo-encephalographic diagnosis (see Table 7). Midline echo displacements are quite rare here (5 cases = 11%).

Table 12. *Echo-encephalographic findings with 45 tumors of the oral brain stem and the corpus callosum*

Type of Tumor / Localization	No. of Cases	Midline Echo Normal	Midline Echo Shifted	Midline Echo Not Detectable	Error	Width of the 3rd Ventricle normal	Width of the 3rd Ventricle Exceeding 7.0 mm	Width of the 3rd Ventricle Not Measurable	Temporal Horn Dilated, BMI Exceeding 2.3	Tumor Echo
Thalamus	7	—	2	2	3	—	2	5	1	2
Hypothalamus and anterior third ventricle	7	7	—	—	—	3	—	4	—	2
Posterior third ventricle, quadrigeminal region	12	9	1	2	—	1	5	6	9	6
Craniopharyngioma	11	9	1	1	—	3	4?	4	8	6
Corpus callosum	8	6	1	—	1	2	2	4	2	1
Total	45	31	5	5	4	9	13	23	20	17
Percentage	100	69	11	11	9	20	29	51	44	38

The most important echo-encephalographic finding in this group of tumors is the demonstration of a dilatation of the lateral ventricles and the registration of tumor reflections (44 and 38%).

In the last 3 years, decisive diagnostic pointers were obtained from the echo-encephalogram in these patients also. For it was noted that a tumor echo complex is often found near the midline in cases where the midline echo was missed, when a large area of the brain stem is scanned with slow to and fro movements of the probe. In tumors which cause a dilatation of the lateral ventricles by an occlusion of the foramina of Monroe or of the aqueduct, this enlargement can be deduced from the echo-encephalographic measurement of the position of the echo from the lateral wall of the opposite temporal horn in nearly all cases. This reflection which lies about half-way between the midline echo and end echo in the healthy subject and which normally possesses only a slight amplitude, approaches the end echo and attains a greater height if the ventricle is dilated because the reflecting surface is enlarged. By calculating the echo-encephalographic brain mantle index (BMI), the degree of ventricular dilatation can be determined even without measuring the width of the third ventricle. Fig. 97 shows the degree of ventricular dilatation to be expected with different BMI values in cases of midline tumors.

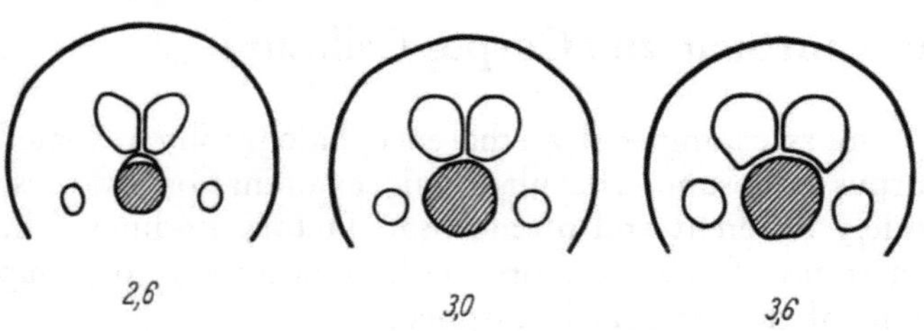

Fig. 97. The brain mantle index allows us to draw conclusions about the degree of ventricular dilatation, even if the width of the third ventricle cannot be measured because a tumor is growing there

Altogether enlargements of the lateral ventricles were detected and confirmed by neuroradiology in 20 patients (44%, see Table 12). These were chiefly tumors of the posterior part of the third ventricle and craniopharyngiomas.

While these echo-encephalographic findings are indirect signs of a tumor, the registration of tumor echo complexes makes it possible to pick up directly and localize exactly a considerable proportion of the oral brain stem tumors. In the 45 tumors of the oral brain stem and corpus callosum, such tumor echoes were detected in 17 patients, i. e. in 38%. In the following section, the echo-encephalographic findings with the different groups of tumors will be discussed in detail.

Thalamus and Basal Ganglia

Of the 45 tumors of the oral brain stem in our first series 7 were located chiefly in the thalamus and basal ganglia. The echo-encephalographic results were very unsatisfactory. A midline echo shift was seen 5 times, but the findings coincided with the results of contrast medium investigations only twice. Thus there was an error in measurement 3 times. Particularly remarkable is the fact that although echo-encephalography indicated the correct side in each case, the extent of the midline echo shift in the echogram was considerably higher than in the pneumoencephalogram. Till now we have no explanation for this. In the remaining two cases, it was impossible to obtain an unequivocal midline echo. The width of the third ventricle could also be determined only twice, while a dilatation of the lateral ventricles was found once. Tumor reflections were seen in two patients. Once a brain abscess was the cause and in the second patient as in the remaining 5 cases, we do not have a histological or pathological-anatomical diagnosis. The echogram of the last mentioned patient is shown in Fig. 72.

All in all the findings in cases of thalamic tumors of our first series were often misleading, probably because the structures responsible for the midline echo were directly affected by the tumor. After the first series of investigations, we were able to observe 9 more tumors in the area of the thalamus and basal ganglia. The M-echo was displaced 7 times, once the echoencephalogram did not show any deviation from the norm, in 7 cases an exact localization of the tumor was possible through tumor reflections (Fig. 98 and 99) and one echo-encephalogram could not be evaluated. In these patients, the evaluation of the ultrasonic trace was also sometimes very difficult due to numerous pathological echoes near the midline echo.

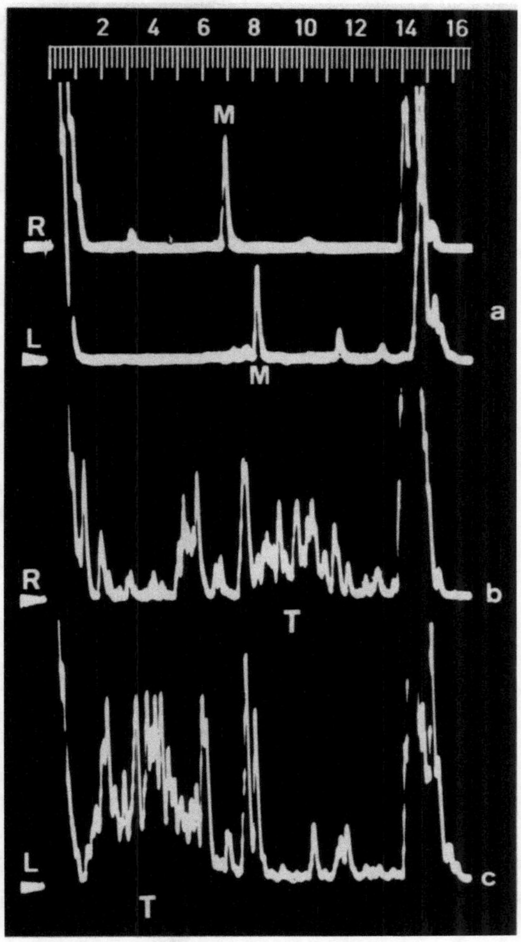

Fig. 98. a—c. Echo-encephalogram of a 7-year-old boy with temporal lobe epilepsy. a) 4.5 mm displacement of the M-echo towards the right. b)—c) broad complex of abnormal echoes in the left cerebral hemisphere extending to the midline. Echo-encephalographic diagnosis: tumor in the basal ganglia on the left. Pat. G. P., Echo-EG No. 3529/68.

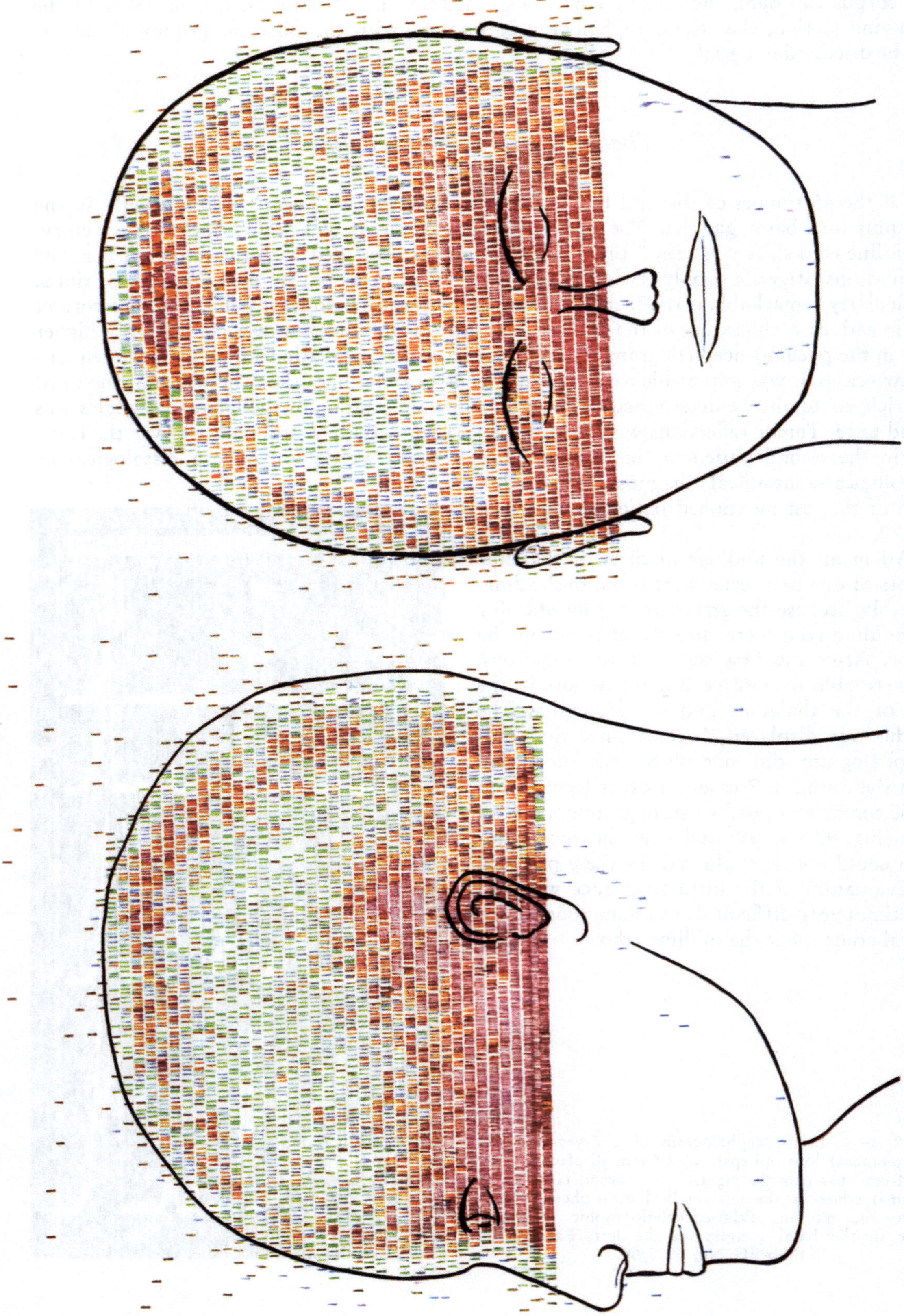

Fig. 98 d and e. Scintigraphy with technetium 99m confirmed the presence of a lesion in the basal ganglia left. (Picker colorscan, with the kind permission of Priv. Doz. Dr. Wolf, Dept. of Nuclear Medicine, Medical Clinic, University of Erlangen-Nuremberg). In this case contrast medium studies provided no further diagnostic information. Exploratory trepanation revealed an astrocytoma (same patient as in Fig. 98 a—c)

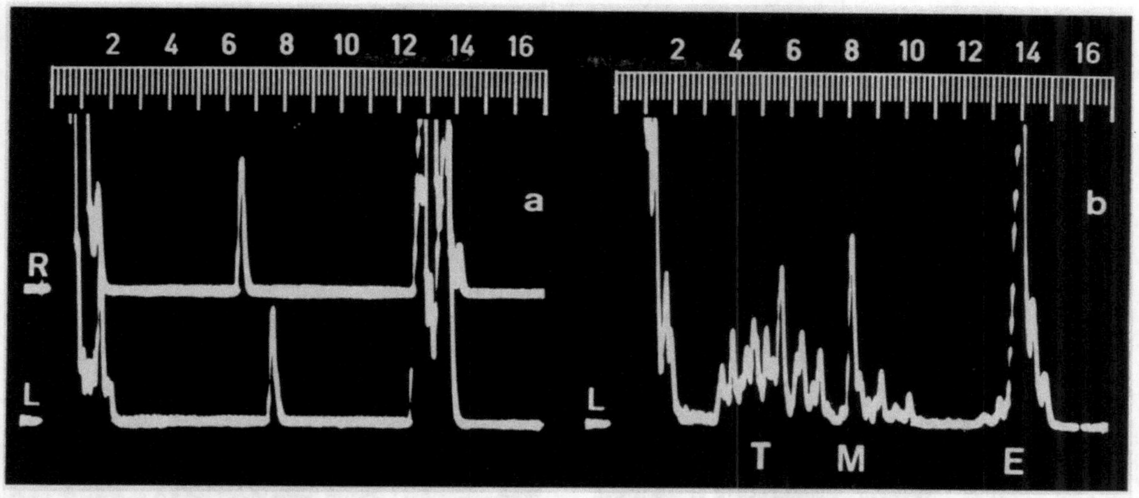

Fig. 99 a and b. Echo-encephalogram of a 35-year-old woman with a left-sided tumor of the basal ganglia. a) M-echo shift by 6.0 mm. b) Tumor echo complex on recording from the left near the midline. Pat. I. H., Echo-EG No. 3072/67

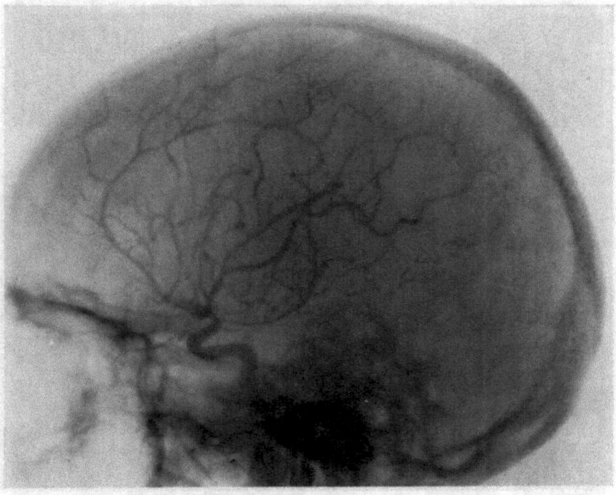

Fig. 99 c

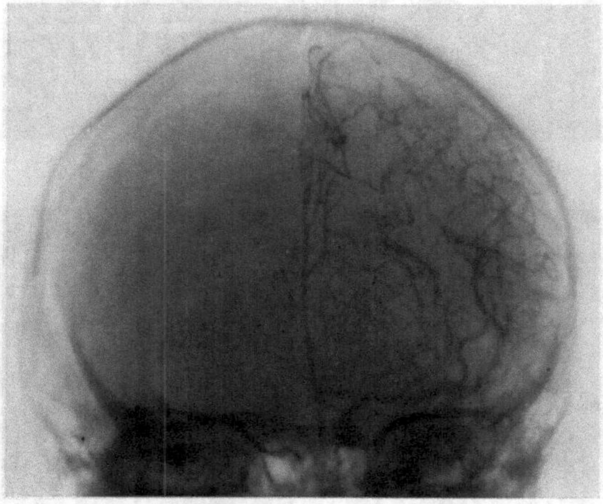

Fig. 99 d

Fig. 99 c and d. Angiograms of the same patient as in Fig. 99 a and b. The tumor is clearly delineated by pathological marginal vessels in the anterior basal ganglia

Hypothalamus and Anterior Part of the Third Ventricle

7 tumors in our series of investigations lay in the area of the hypothalamus and the anterior part of the third ventricle. 3 of these were spongioblastomas of the optic chiasm, one was a neurinoma of the optic chiasm and in 3 cases the histology of the tumor was unknown. The latter protruded into the third ventricle from below. All echo-encephalograms revealed a normal M-echo. The third ventricle was of normal width in 3 cases, while the other 4 cases failed to show unequivocal echoes (Fig. 100). Once tumor reflections of extraordinary height were registered in the median area on sounding from either side (Fig. 101). In a second patient we saw a pathological echo complex whose amplitude was less marked. Unfortunately we do not possess a histological report in either case.

Thus we can expect a pathological echo-encephalogram only in exceptional cases of tumors in the hypothalamic area and in the anterior part of the third ventricle.

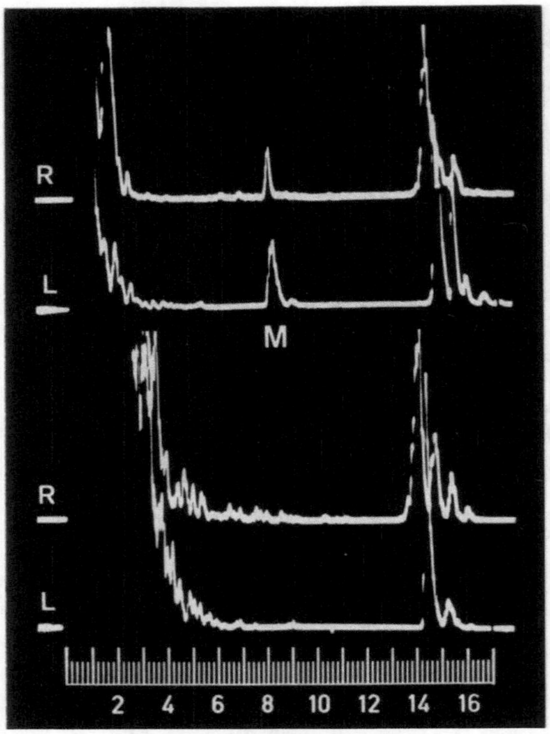

Fig. 100

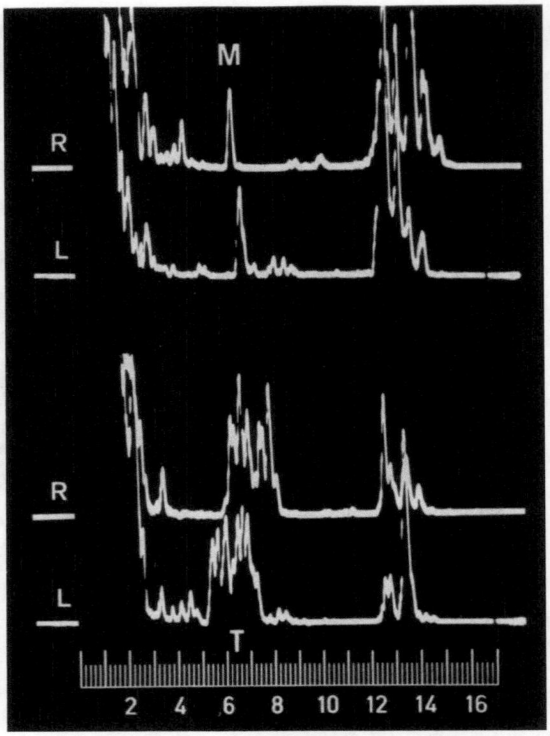

Fig. 101

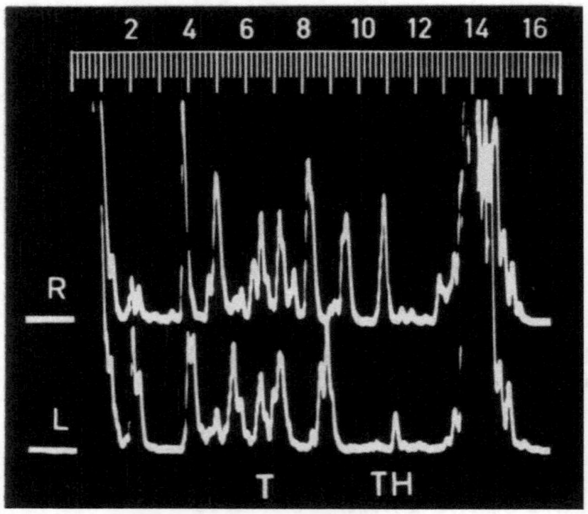

Fig. 102

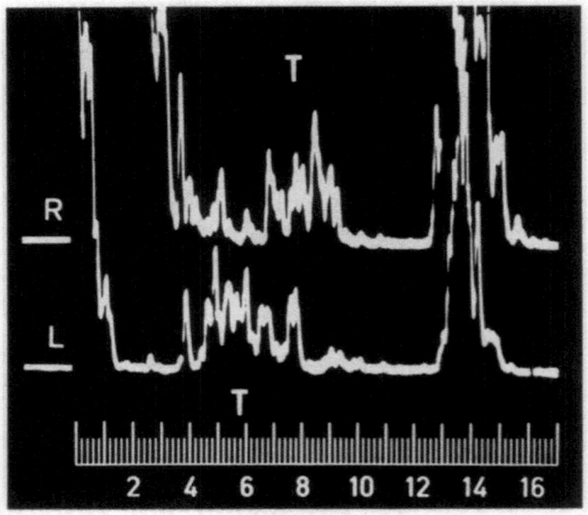

Fig. 103

Fig. 100. Echogram of a patient with a tumor in the anterior part of the third ventricle. Top: midline echo at the usual site. Below: on directing the ultrasonic beam at the third ventricle, we can find only a generalized undulation of the base line, and the echoes from the ventricular walls are absent. Pat. A. K., 37 years of age. Echo-EG No. 234/63

Fig. 101. Echogram of a patient with a tumor in the anterior part of the third ventricle. Top: midline echo shift of 1.5 mm still in the normal range. Bottom: when pointing the ultrasonic beam at the anterior part of the third ventricle, a tumor echo complex of high amplitude is found at the midline. Pat. A. K., 38 years of age. Echo-EG No. 1773/65

Fig. 102. Echo-encephalogram of a 4-year-old boy with a pineoblastoma the size of an apple. The ultrasonic trace is altered markedly by tumor reflections (T) and prevents a clear interpretation. Pat. N. M., Echo-EG No. 1625/65

Fig. 103. Echo-encephalogram of a 16-year-old girl with a tumor in the posterior part of the third ventricle more developed on the left side. Absence of the midline echo, broad tumor echo complex (T). Pat. E. E., Echo-EG No. 2050/66

Posterior Part of the Third Ventricle and Quadrigeminal Region

In this series of tumors, there are 12 tumors of the posterior part of the third ventricle and quadrigeminal region. The histology is known in 4 cases only: one pineoblastoma, one ependymoma, one teratoma and one meningioma which underwent malignant degeneration. In two other cases the diagnosis is probably pinealoma, while the other 6 tumors are primarily unilateral and of unknown histological structure. 9 times the midline echo was in a normal position, in one case it was displaced. In all 10 cases the neuroradiological results could be correlated with the results of echo-encephalography. Twice it was impossible to record an unequivocal midline echo. This was a pineoblastoma in one case (Fig. 102) and a

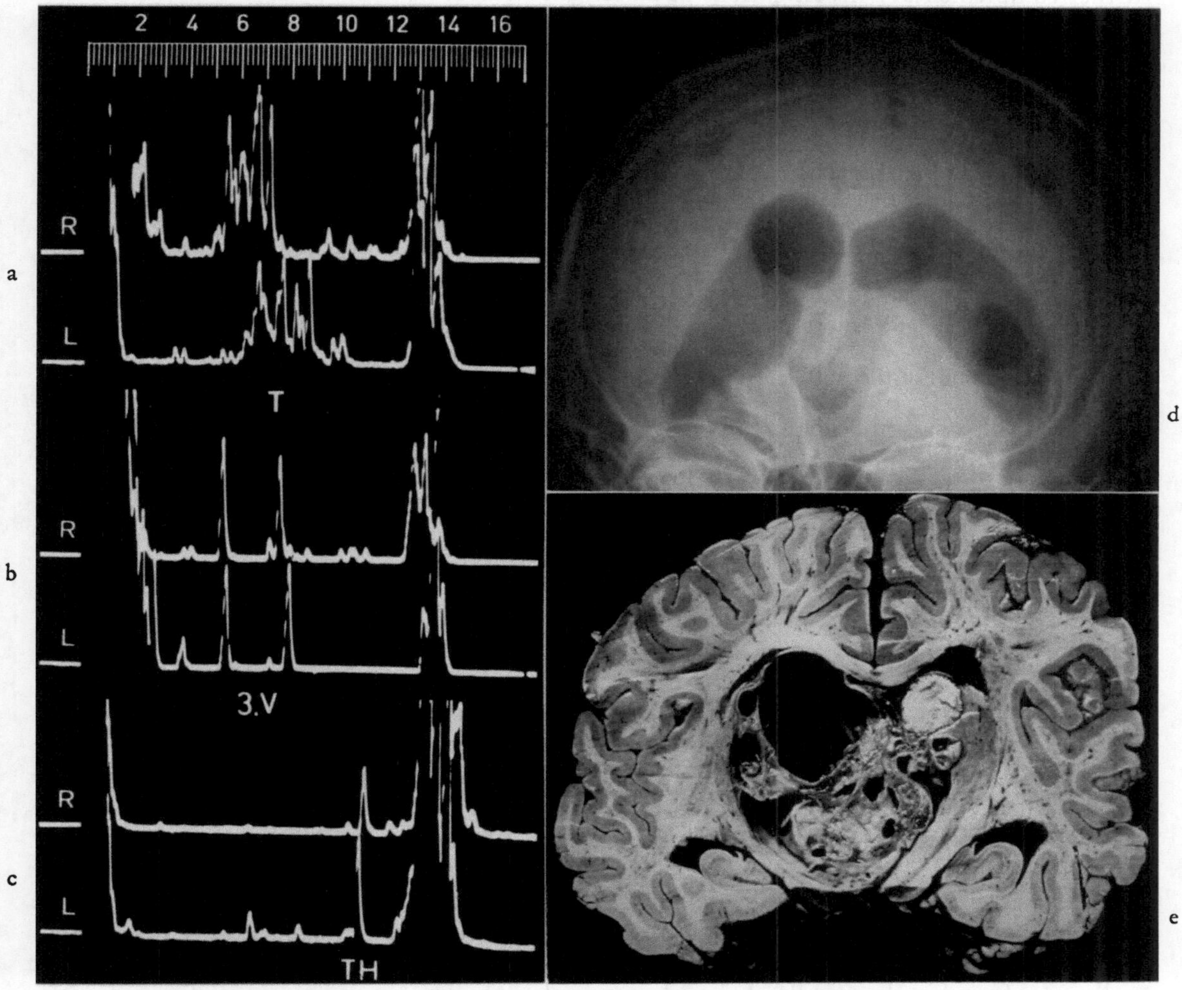

Fig. 104. Echo-encephalographic diagnosis of a tumor in the posterior part of the third ventricle. a) Tumor echo complex (T) in the midline region using normal temporal sounding. b) Abnormal double echo of the third ventricle 23 mm apart when sounding from a more anterior application point. c) The location of the temporal horn echo also indicates marked ventricular dilatation BMI 3.0. d) Confirmation of the echo-encephalographic findings by means of ventriculography. e) The autopsy specimen shows a large midline teratoma. Pat. H. E., 14 years of age. Echo-EG No. 2472/66

tumor of the posterior part of the third ventricle developed more on the left side in the other case (Fig. 103). In contrast to the often unsatisfactory results obtained with echo-encephalography of tumors of the thalamus, the hypothalamus and the anterior part of the third ventricle, the diagnosis was correctly set on the basis of abnormal ventricular echoes and tumor reflections with most of the tumors in the posterior part of the third ventricle.

5 times a double echo was picked up from the third ventricle, indicating a dilatation of this ventricle. However, the application point of the probe usually had to be displaced towards the front in order to record an unequivocal double reflection (see Fig. 104 b). In 6 cases, the width of the third ventricle could not be determined. 9 times however, there was a dilatation of the lateral ventricles whose extent was determined echo-encephalographically by measuring the position of the lateral wall of the temporal horn. The echo-encephalographic brain mantle index ranged from 2.5 to 4.5, the mean value being 3.2. In all cases the echo-encephalographic findings were confirmed by ventriculography. Furthermore tumor reflections were picked up 6 times. Thereby the exact localization of the tumor was possible even before further complementary investigations had been performed.

Case 6: Heinz E., 14 years of age. Echo-EG No. 2472/66. Sporadic seizures for the past 4½ years. At the beginning of August 1966 the boy attracted attention because his behavior was slowed and he was nauseated. Occasional headache. Referral to a neurological clinic. Here increasing somnolence and finally unconsciousness on August 21, 1966. On the basis of the clinical examination a brain tumor in the quadrigeminal area was suspected (bilateral papilledema, complete vertical ophthalmoplegia hypotonia, hyperreflexia). During a consultation on August 21, 1966, an echogram was performed which revealed marked dilatation of the inferior horns (BMI = 3.0), and the midline structures were not displaced. On directing the beam at the posterior part of the third ventricle, a pathological echo complex of unusual size was found (Fig. 104 a). Diagnosis: tumor of irregular structure in the posterior part of the third ventricle with marked dilatation of the anterior part of the third ventricle and the lateral ventricles. Following admission of the patient, the echo-encephalographic diagnosis was confirmed by immediate ventriculography. Despite Torkildsen drainage and radiotherapy of the tumor, the boy died on October 18, 1966. At autopsy an apple-sized teratoma was found in the posterior part of the third ventricle (Fig. 104 e).

Subsequent neuroradiological investigations practically never brought more information in such cases. The simplification of the diagnostic process by echo-encephalography is an important step forward in the recognition of these tumors. Further improvement in the results of ultrasonic investigation should be possible here, especially with two-dimensional echo-encephalography.

Craniopharyngiomas

What was said above is also true for the echo-encephalographic diagnosis of craniopharyngiomas which frequently fill out the third ventricle and lead to obstructions in the flow of CSF. The position of the midline echo can be determined in most cases, being midline in the majority (9 of 11 cases); only once did we find a 3 mm shift of the M-echo, and in one case the M-echo could not be definitely recorded. Of the 11 craniopharyngiomas in our series, 3 did not lead to a dilatation of the ventricles so that the echo-encephalograms of these patients were unremarkable. In the remaining 8 cases, the echo-encephalographic brain mantle index indicated considerable dilatation of the lateral ventricles. In these cases, it varied from 2.7 to 5.4. In addition, pathological echo complexes originating from tumor calcifications or cyst walls were found in the middle of the echogram when the ultrasonic beam was directed at the region of the third ventricle in 6 cases (Figs. 105 and 106). In Table 12 there are 4 cases of dilated third ventricles and the number 4 is marked with a question mark. In these cases, the echo-encephalogram showed a wide double reflection in the middle of the echogram, resembling a markedly dilated third ventricle. Later pneumoencephalographic investigations and/or operation suggested, however, that this double reflection originated from the walls of a cyst within the craniopharyngioma. Control examinations indicated that the distance between the two reflections varied, probably depending on the state of fulness of the tumor cyst. However, together with the history and plain films of the skull, the diagnosis was established in every case.

Corpus Callosum

8 tumors were localized in the corpus callosum or developed from this structure. 4 of these were glioblastomas, the histology of the other 4 tumors is unknown. In 6 cases the midline echo was at the normal site, twice it was displaced. In one of these, an error in

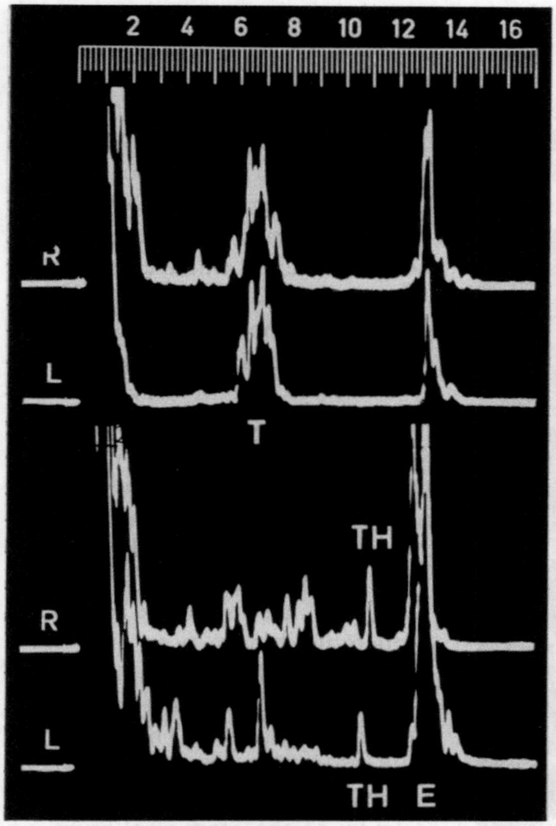

Fig. 105

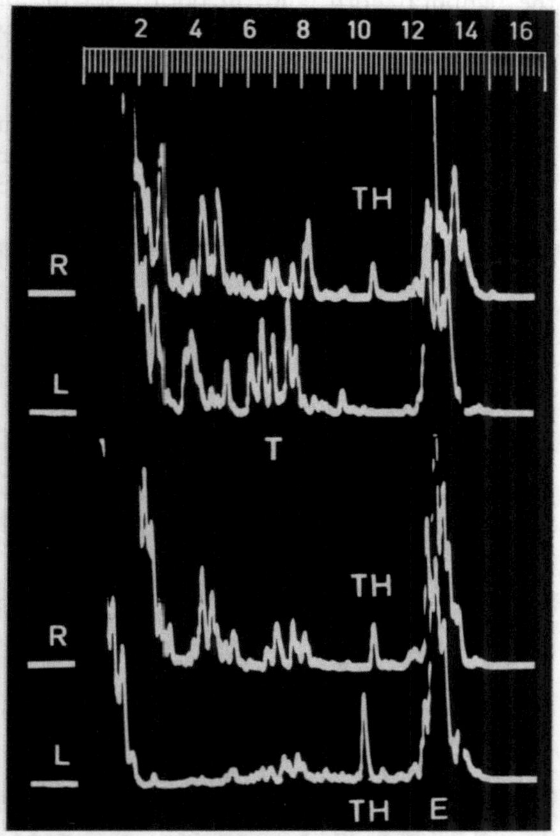

Fig. 106

Fig. 105. Echo-encephalogram of a 50-year-old patient with a calcified craniopharyngioma. a) Tumor echo at the centre (T). b) Echoes from the dilated temporal horn (TH). Brain mantle index 3.5. Pat. R. J., Echo-EG No. 2892/67

Fig. 106. Echo-encephalogram of a 12-year-old girl with a suprasellar craniopharyngioma which led to a block of the foramina of Monroe. a) Multiple reflections in the midline (T) originating from the tumor. b) Recording a high reflection from the lateral wall of the dilated temporal horn (TH). Brain mantle index 2.7. Pat. R. S., Echo-EG No. 1757/65

measurement was discovered at autopsy. This was a glioblastoma where the echo-encephalogram had indicated a 2.5 mm displacement of the midline structures. At autopsy however, there was no definite shift of the midline structures. The corpus callosum was infiltrated with necrotic masses of tumor and the posterior part of the third ventricle was also filled with tumor material.

The width of the third ventricle was determined 4 times, i. e. in 4 cases a double echo was recorded at the midline; we assume that these were reflections from the lateral walls of the third ventricle. In 2 cases of tumor, the width of the third ventricle was within the normal range, a result which was confirmed by pneumoencephalography. In another patient, we saw a double reflection 14 mm apart. The clinical picture was that of an acute stroke. The echo-encephalogram led us to suspect a ventricular hemorrhage and this was confirmed by autopsy. It came from an astrocytoma that had undergone malignant degeneration after growing out of the corpus callosum and into the roof of the lateral ventricle. In the last patient of this group, we obtained a double reflection 13 mm apart which we thought had originated from a dilated third ventricle. The pneumoencephalogram, however, showed a widening of the septum pellucidum while the third ventricle was not filled with air. Therefore we must assume that the two reflections originated from the walls of the dilated septum.

Twice a marked dilatation of the lateral ventricles was diagnosed from the position of the temporal horn echo; in both cases the brain mantle index was 3.0. Tumor reflections were seen only in the case of a gliablastoma which had invaded the corpus callosum as well as the entire oral brain stem.

A critical review of the results of echo-encephalography with tumors of the oral brain stem and corpus callosum reveals that a tumor-dependent CSF obstruction had been detected in all cases by recording the position of the echo from the lateral wall of the temporal horn. Craniopharyngiomas and tumors of the posterior part of the third ventricle were localized exactly by means of tumor echoes in half the cases. Echo-encephalography demonstrated its value for postoperative observation in these two tumor groups as well. Following ventriculo-cisternostomy with subsequent radiotheraphy, a decrease in the dilatation of the lateral ventricles could be observed by locating the temporal horn echo (Fig. 107). The increase in the size of an inoperable tumor near the third ventricle or its environment may be measured by echo-encephalography on occasion. The danger of an error in measurement is particularly great with thalamic tumors. Tumors of the corpus callosum, of the hypothalamus and in the anterior part of the third ventricle, generally do not produce significant changes of the echo-encephalogram.

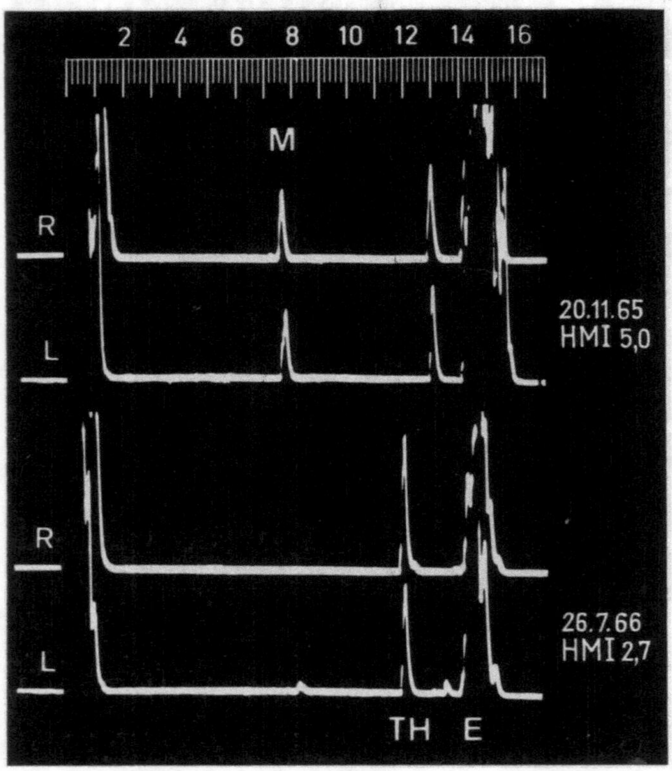

Fig. 107. Normalization of lateral ventricle dilatation following Torkildsen drainage in a craniopharyngioma. The increase in the thickness of the brain mantle can be determined from the position of the temporal horn echo (TH). Pat. G. M., 15 years of age. Echo-EG No. 1970/3/65 and 1970/6/66

The echo-encephalographic findings with tumors of the oral brain stem and corpus callosum show very clearly that *echo-encephalography alone in its present form is never capable of excluding a space-occupying intracranial lesion*, especially if only the midline echo is determined. Negative or erroneous M-echo findings occured in 89% of this group of tumors (Table 12). Here the limitations of this procedure become obvious.

The occasionally confusing ultrasound traces obtained with tumors of the third ventricle or its surroundings require special experience for correct interpretation. Still there will always be a few cases in which the echo-encephalogram cannot be interpreted correctly. However, in general the absence of an unequivocal midline echo and the appearance of multiple reflections mean that the echo-encephalogram is pathological even if there is no midline echo shift or ventricular dilatation. And a pathological echo-encephalogram is the indication for contrast medium studies.

4. Tumors at the Base of the Skull

From an echo-encephalographic point of view, tumors located at the base of the anterior and middle fossa assume a special position among the intracranial tumors because such neoplasms lead to a lateral displacement of the ventricular system only very rarely. Altogether a total of 62 tumors of our series was located in this area. 40 of these were pituitary adenomas, 19 were meningiomas, 2 were osteomas and one was a sarcoma (see Table 13).

Table 13. *Echo-encephalographic findings with 62 tumors at the base skull*

Type of Tumor / Localization	No. of Cases	Midline Echo Normal	Midline Echo Shifted	Midline Echo Not Detectable	Error	Width of the 3rd Ventricle Normal	Width of the 3rd Ventricle Exceeding 7.0 mm	Width of the 3rd Ventricle Not Measurable	Tumor Echo
Pituitary adenoma	40	38	1	1	—	30	6	4	1
Meningioma of the tuberculum sellae	3	3	—	—	—	3	—	—	—
Olfactory meningioma	3	3	—	—	—	2	—	1	—
Meningioma at the base of the middle fossa	4	2	2	—	—	—	1	3	1
Lateral sphenoid wing meningioma	6	—	6	—	—	1	—	5	4
Medial sphenoid wing meningioma	3	1	2	—	—	1	1	1	—
Other tumors at the base of the skull	3	1	2	—	—	2	—	1	—
Total	62	48	13	1	—	39	8	15	6
Percentage	100	77	21	2	—	63	13	24	10

The echo-encephalogram of pituitary adenomas is nearly always normal. Only once did we observe a 2.5 M-echo displacement. This was a recurrent adenoma which had developed towards the front on the left and become a goose-egg-sized tumor in the left frontal lobe. In one patient, it was impossible to record a midline echo. Here a recurrent adenoma had grown into the hypothalamus and completely deformed and elevated the midline structures.

The width of the third ventricle lay within the normal range 30 times, in 6 cases the third ventricle exceeded 7.0 mm in width (see Table 13). However, the width of the third ventricle exceeded this limit only by 1.0 to 1.5 mm. Perhaps this is due to a compression of the third ventricle by the tumor which projects into the hypothalamus. Certainly the age of the patients also plays a role here, since these were mainly older men and women in whom an additional atrophic lesion must be drawn into consideration. In 4 cases it was impossible to obtain an unequivocal double echo from the walls of the third ventricle.

Tumor reflections were found only in one case of a recurrent adenoma which had grown to the size of an hen's egg thereby projecting into the cone of the ultrasonic beam (Fig. 108).

A more detailed discussion of the 19 *meningiomas* originating from the anterior and middle fossa will follow. *Olfactory meningiomas, suprasellar meningiomas and meningiomas located near the midline at the base of the middle fossa* generally do not produce a shift of the midline structures and therefore *fail to be recognized in the echo-encephalogram.* The average displacement in cases of sphenoid wing meningiomas of the medial third was 2.3 mm (1.5 to 3.5 mm); therefore they are often missed in the echo-encephalogram. *Meningiomas of the middle and lateral portion of the sphenoid wings on the other hand, cause a clear displacement of the midline structures.* The average M-echo shift in the echo-encephalogram was 6.5 mm. Meningiomas situated laterally at the floor of the middle fossa lead to unequivocal midline echo shifts (mean shift = 6.7 mm).

The behavior of the midline echo in cases of meningioma at the base of the skull is summarized in the diagram in Fig. 109. Of the 19 meningiomas located at the base of the anterior and middle fossa 9 (= 47%) had a normal echo-encephalogram. Here the neuroradiological findings in regard to the midline echo shift agreed with the results obtained by ultrasound completely.

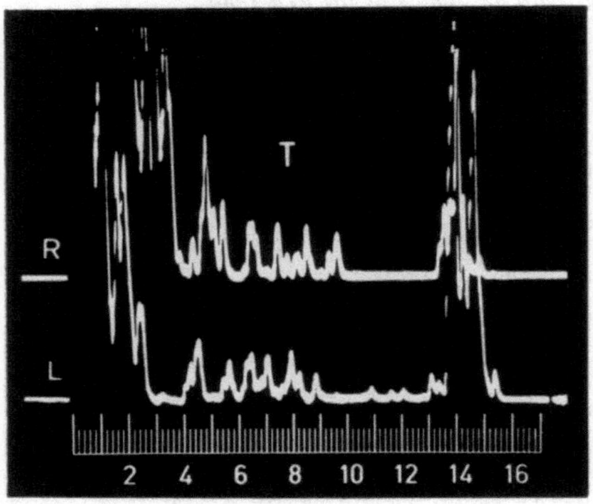

Fig. 108

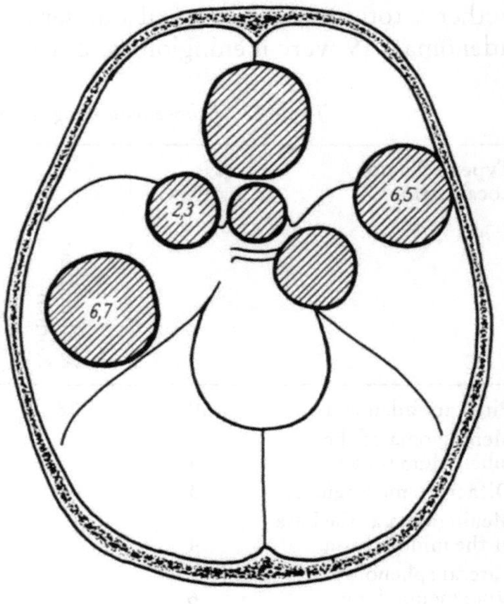

Fig. 109

Fig. 108. Tumor reflections (T) in a pituitary adenoma recurrence. Pat. J. B., 21 years of age. Echo-EG No. 1544/64

Fig. 109. Behavior of the midline echo in meningiomas of the base of the skull (19 cases)

Tumor reflections play a secondary role in meningiomas of the base of the skull (5 cases). Only in cases of large lateral sphenoidal wing meningiomas are tumor reflections in the near field of the echogram i. e. preceding the midline echo, a relatively frequent occurrence. However, this does not allow us to distinguish them from other tumors (Table 13 and Fig. 64).

This group also makes it clear that echo-encephalography is not a method for excluding tumors. It can only demonstrate a mass shift which is already present. However, the results obtained in the latter case are very reliable.

Two somewhat larger osteomas of the sphenoid wing caused a slight M-echo shift in the echo-encephalogram. A sarcoma at the base of the middle fossa did not lead to alterations in the echo-encephalogram.

Thus we see that echo-encephalography is unable to detect tumors located in the central area of the base of the skull. Only meningiomas of the lateral sphenoid wing and lateral part of the base of the middle fossa can cause a displacement of the cerebral midline structures to one side and thereby an M-echo shift. Here too, half the cases produce tumor reflections.

5. Tumors of the Posterior Fossa and Caudal Brain Stem

Since its inauguration by Leksell in 1955, echo-encephalography has found wide use for the localization of supratentorial space-occupying lesions. The possibility of diagnosing tumors of the posterior fossa and chronic occlusions of the aqueduct with this method was hardly noticed for years, however. As indicated in the part on general echo-encephalography, it is possible to determine the exact width of the individual parts of the ventricles with the help of the ultrasound pulse echo technique. In cases of space-occupying lesions of the posterior fossa, a method of investigation which can detect the presence of an internal hydrocephalus and its extent in a short time without stressing the patient, is especially important.

In his first publication (1955/56), LEKSELL already suggested that ultrasonic reflections could probably be picked up from the walls of the ventricles.

In 1961 LITHANDER reported on ventricular dilatations in children which were diagnosed by ultrasound; two of these had cerebellar tumors. We reported our own first echo-encephalographic findings on 24 tumors of the posterior fossa in 1963. In 20 of these cases, the echogram showed a dilated third ventricle (SCHIEFER, KAZNER, BRÜCKNER, 1963).

Also in 1963, DREESE and NETSKY showed an echo-encephalogram of a dilated third ventricle. AMBROSE published such an echogram in 1964. The results of investigations of space-occupying lesions in the posterior fossa have been confirmed by CALATAYUD-MALDONADO, GELETNEKY and LORENZ (1965) in the meantime. In 26 such neoplasms, the authors found 17 dilatations of the third ventricle in the echo-encephalogram.

While our method deals with the demonstration of occlusive hydrocephalus due to tumor, other authors have attempted to detect tumors of the posterior fossa directly by ultrasound. TANAKA et al. (1966) reported finding tumor reflections in 17 (=45%) of 38 such tumors. The examination was performed chiefly in a sagittal direction from an application point on the forehead. In several cases the tumor could be demonstrated only by means of a special probe applied to the posterior wall of the larynx. Results of this method by other authors have not been published as yet.

171 of the 630 tumors in our series were located in the posterior fossa and in the caudal brain stem. In every case, either dye studies, operation and/or autopsy revealed a space-occupying infratentorial lesion. The echo-encephalographic results obtained *before further investigations were undertaken* in these 171 tumors, are shown in Table 14. In 169 cases it

Table 14. *Echo-encephalographic findings with 171 space-occupying lesions of the posterior fossa*

Lesion	No. of Cases	Midline Echo Normal	Midline Echo Shifted	Midline Echo not Detectable	Error	Width of the 3rd Ventricle Normal	Width of the 3rd Ventricle Exceeding 7.0 mm	Width of the 3rd Ventricle not Measurable	Temporal Horn Dilated, BMI Exceeding 2.3	Average Width of the 3rd Ventricle (mm)
Cerebellar tumor (68)										(14.4)
Astrocytoma	16	16	—	—	—	—	16		14	15.3
Medulloblastoma	28	28	—	—	—	—	28	—	20	14.8
Ependymoma	15	15	—	—	—	—	15	—	12	14.0
Others	9	9	—	—	—	—	9	—	6	13.9
Posterior fossa meningioma	9	7	1	1	—	2	6	1	4	11.0
Cerebellar metastasis	9	7	2	—	—	—	9	—	6	11.6
Cerebellar abscess	4	4	—	—	—	1	3	—	2	12.5
Acoustic neurinoma	45	42	2	1	—	6	36	3	22	10.0
Caudal brain stem tumor	7	7	—	—	—	1	6	—	4	11.1
Aqueduct stenosis/occlusion	20	20	—	—	—	—	20	—	18	17.1
Arachnitis	9	9	—	—	—	2	7	—	7	9.7
Total	171	164	5	2	—	12	155	4	115	
Percentage	100	95.9	2.9	1.2	—	7.0	90.6	2.4	67.2	

was possible to determine the position of the midline structures from the *midline echo*. The oscilloscope showed a midline M-echo in 164 cases, and a displacement up to a maximum of 4.0 mm in 5 cases.

Of the cases with M-echo displacements, one had a cerebellar metastasis from a hypernephroma. A few months earlier, an occipital metastasis had been removed from the same patient. In another patient, a metastasis from a breast cancer was removed from the cerebellum, while another metastasis was located in the cerebrum. In 3 further cases, tumors were found which originated from the cerebello-pontine angle and expanded into the supratentorial region as well (2 acoustic neurinomas, 1 meningioma). Twice it was impossible to pick up a clear midline echo (1 meningioma of the clivus [Fig. 110], 1 acoustic neurinoma).

The width of the third ventricle was determined by echo-encephalography 167 times (=97.6%). In 12 cases it was within the range of normal, chiefly in cases of cerebellopontine angle tumors. 155 times (=90.6%) the third ventricle was dilated, i. e. the distance between the two reflections of the double echo as calculated for tissue was more than 7.0 mm.

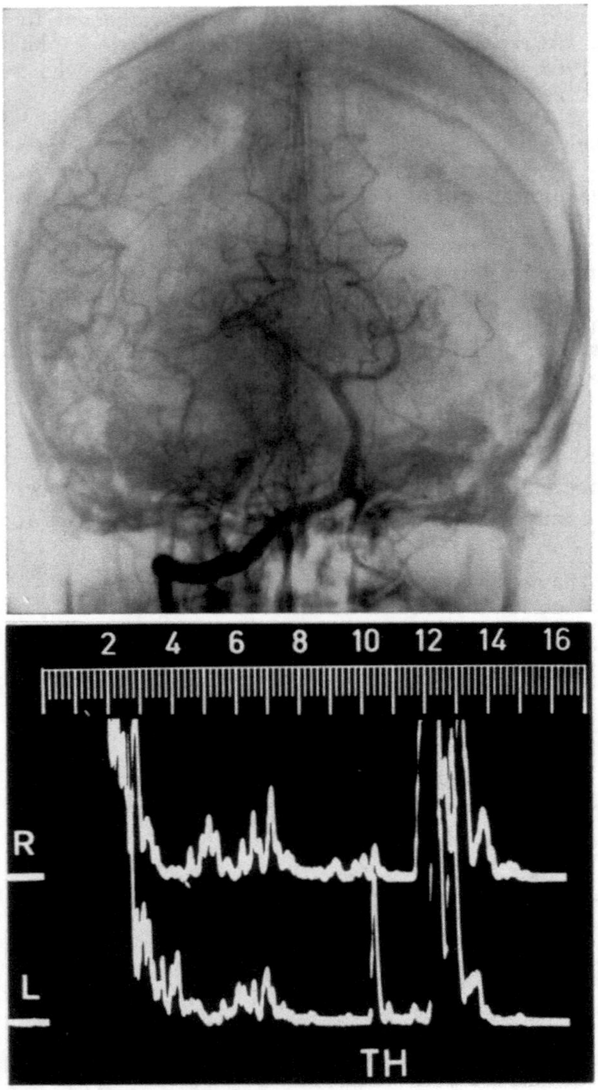

Fig. 110. Vertebral arteriogram and echogram of 59-year-old man with a clivus meningioma. Recording the midline echo was impossible. However, the echo of the external wall of the temporal horn pointed to a dilatation of the ventricles. Pat. L. B., Echo-EG No. 1846/65

The measurability of the third ventricle width by echo-encephalography was by no means limited to infantile or young patients with thin skull bones. The energy of the ultrasonic beam which is reflected at the interface between the ventricular wall and the CSF when the ultrasonic beam hits the interface perpendicularly, is quite sufficient to penetrate the skull bone of an adult especially if the ventricular system is dilated; an intelligible reflection then appears on the screen of the cathode ray tube. In many cases, the reflections from the walls of the third ventricle even have a greater amplitude than those reflected from the calcified pineal body. The age distribution curve in Fig. 111 shows that adults predominate in our investigation series.

The technique of sounding has been described in the section on general echo-encephalography already. Fig. 112 presents a graph of the measurements of the third ventricle in 169 patients grouped into cerebellar lesions and acoustic neurinomas. The width of the third ventricle attained a maximum of 25 mm in cases of cerebellar tumors and occlusions of the aqueduct, the average amount of enlargement being 14.5 mm. In acoustic neurinomas, the average diameter, 10.0 mm, was much lower.

A further subdivision of average ventricular enlargements is seen in Table 14. The differences between the various types of cerebellar tumors are not marked, although it is noteworthy that extreme enlargements of the third ventricle (18.0 mm and more) occur more frequently with cerebellar astrocytomas than with the other types of tumor.

By far the greatest dilatation, however, was regularly found in cases of occlusion of the aqueduct due to chronic inflammation. Cerebellar metastases and abscesses were accompanied by much smaller diameters of the third ventricle (11.6 and 12.5 mm) than the other lesions located in the cerebellum. This is probably the result of the shorter period of illness. In meningiomas of the posterior fossa and tumors of the caudal brain stem, we found average enlargements of the third ventricle of 11.0 or 11.1 mm respectively. In our experience a complete block of the CSF flow is a rarity in these tumors. Probably these partial CSF blocks also account for the fact that arachnitis about the posterior cistern and arachnoid sarcomas or sarcomatoses produce only slight dilatations of the third ventricle (9.7 mm).

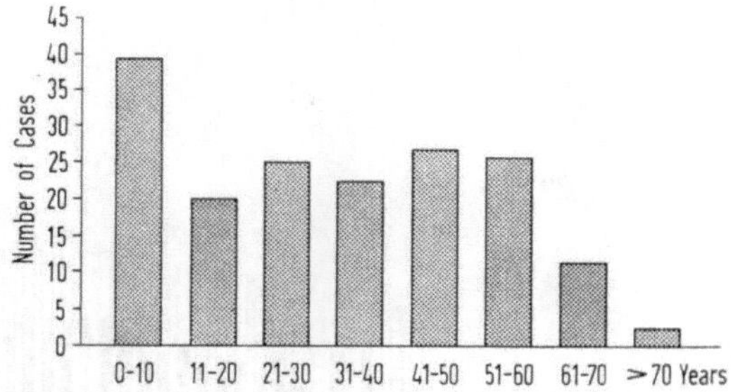

Fig. 111. Age distribution in 171 cases of space-occupying lesions of the posterior fossa

The ventricular widths measured by echo-encephalography, are obtained in living subjects without introducing contrast medium into the CSF spaces. They are not distorted or enlarged by the x-ray technique either. The great reliability of these findings was seen in a comparative study between ultrasonic and pneumoencephalographic measurements in 56 patients with cerebellar tumors, reported by us in 1966 (Kazner and Schiefer). In all these cases, the third ventricle seemed to be considerably larger in the air study than in the echo-encephalogram. This enlargement is due to the divergence of the x-rays and is dependent on the distance between the film and the focal point. The amount of enlargement can reach 40% when the common x-ray technique is employed (Fig. 113).

If this fact was drawn into consideration, there was complete agreement between the two techniques. Furthermore we want to point out that the echo-encephalographic measurement was generally carried out 1 to 2 days before ventriculography so that no great alterations in the width of the ventricle could have taken place in the meantime.

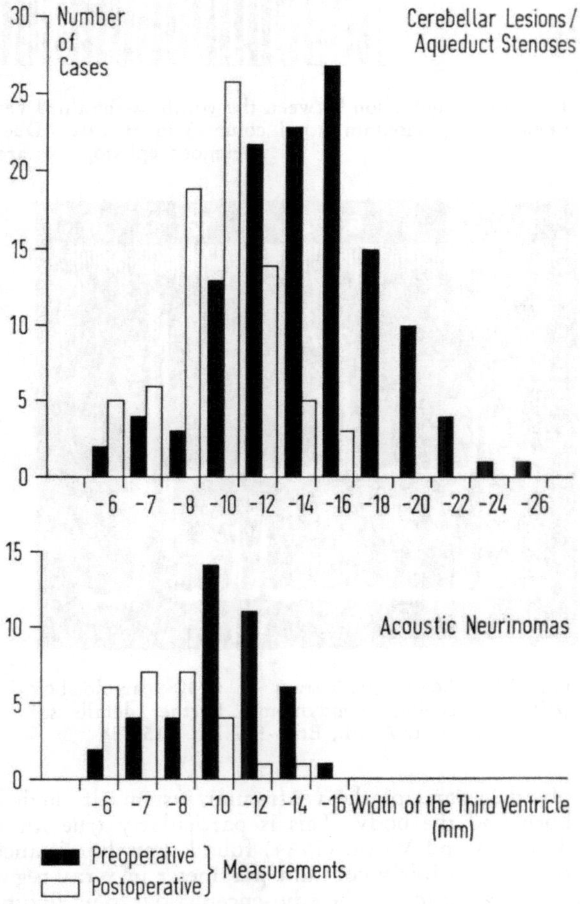

Fig. 112. Frequency and degree of third ventricle dilatation in space-occupying lesions of the posterior fossa (pre- and postoperative measurements).

In order to confirm the echo-encephalographic diagnosis of an internal hydrocephalus, the reflections from the temporal horn may also be used. The correct method of investigation has been discussed in detail previously. Fig. 114 shows two ultrasonic records of a cerebellar ependymoma in a 10-year-old boy. The upper trace represents the double echo of the third ventricle. In the far field of this echogram, there is a small deflection of the cathode ray preceding the end echo which originates from the lateral wall of the temporal horn. The lower trace was recorded in the same patient by using a probe which was also applied in the left temporal area and directed somewhat towards the base. High amplitude echoes are reflected from both walls of the dilated inferior horn. From the position of the *echo from the external wall of the temporal horn* alone (which usually has a great amplitude on the basis of the hollow mirror principle) certain deductions can be made on the degree of internal hydrocephalus even if the double echo from the walls of the third ventricle cannot be picked up for some reason (e. g. in tumors of the oral brain stem).

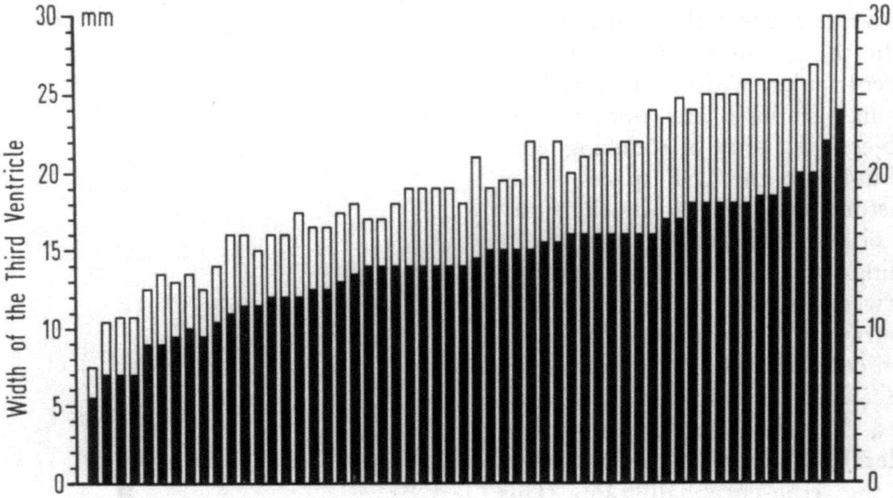

Fig. 113. Comparison between the width of the third ventricle in the echogram (solid black column) and in the pneumoencephalogram (total column) in 56 cases. Due to the divergence of x-rays, the values obtained in pneumoencephalograms are as much as 40% higher

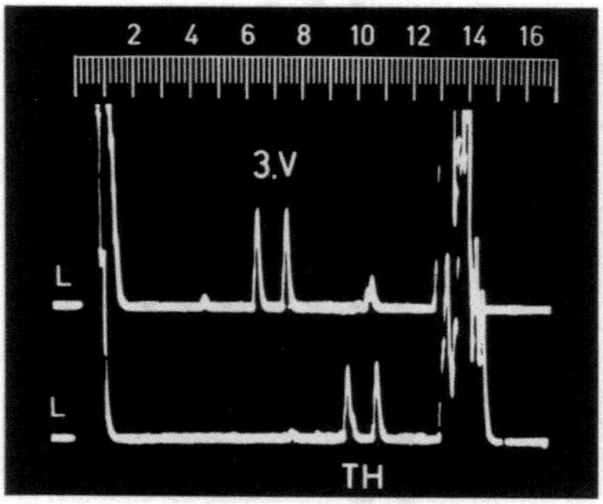

Fig. 114. Echo-encephalogram of a 10-year-old boy with a cerebellar ependymoma. Further details see text. Pat. A. H., Echo-EG No. 1153/64

As described in the section on general echo-encephalography, the echo from the lateral wall of the temporal horn has a very low amplitude in healthy subjects, so that it cannot be definitely identified in many cases. Normally it lies half-way between the midline echo and the end echo. When the ventricular diameter increases, however, the conditions of reflection improve because the reflecting surfaces are enlarged. For this reason, temporal horn echoes can be recorded without difficulty. Occasionally the amplitude of this echo can exceed that of the end echo. As the hydrocephalus increases, the position of the echo from the lateral wall of the temporal horn is displaced towards the end echo.

In order to use the position of the temporal horn echo as a standard for generalized ventricular dilatation, we must determine whether an internal hydrocephalus is always accompanied by a dilatation of the inferior horns. After studying over one hundred ventriculograms, we found that the enlargement of the temporal horns generally parallels a dilatation of other parts of the ventricular system although it rarely assumes the proportions of the anterior horn and the body. This is particularly true for the initial stage of the occlusive hydrocephalus. Taveras and Wood (1964) found that the distance between the skull and the external wall of the trigone is a fairly constant parameter in x-ray regardless of the level at which it is measured.

In a survey of the echo-encephalographic findings obtained in cases of space-occupying lesions in the posterior fossa, we noted that a measurement of the temporal horn was possible only 115 times in 171 patients (=67.2%). This relatively low quota is due to the fact that we were interested chiefly in a measurement of the midline echo and the third ventricle at first. The constancy of this additional reflection in cases of hydrocephalus occured to us only later; we then recorded it routinely. Thus the position of the lateral wall of the temporal horn was recorded in all of the last 96 cases of tumor-dependent ventricular dilatation.

In order to simplify the interpretation of echo-encephalograms, performed in hydrocephalic patients, an echo-encephalographic brain mantle index (BMI) was described by us

(SCHIEFER, KAZNER, KUNZE, 1965). Its calculation has been described above already and is shown diagrammatically in Fig. 60. In healthy persons, this index ranges from 2.0 to 2.2. In cases of ventricular dilatation due to cerebellar tumors, the brain mantle index ranges from 2.6 to 3.2 (Fig. 115). Values above this level are rare, being found only in small children and in patients with occlusion of the aqueduct.

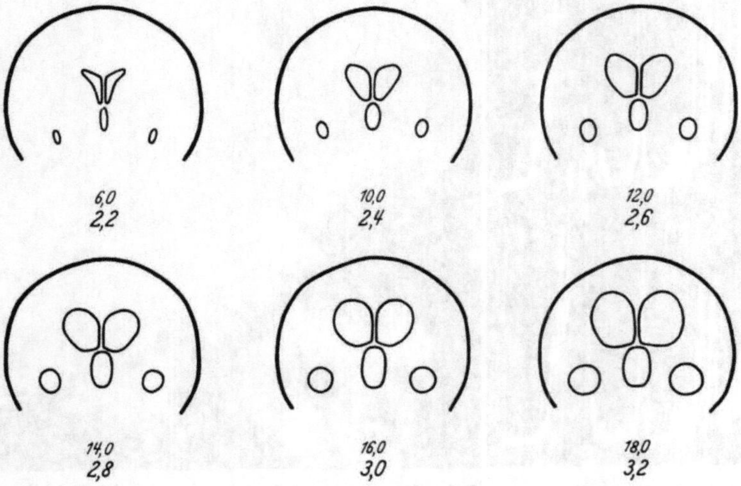

Fig. 115. Diagrammatic illustration of ventricular width at different brain mantle index values (large numbers). The small figures indicate the width of the third ventricle in mm

The brain mantle index is also a much more accurate measurement of the degree of ventricular dilatation than the width of the third ventricle. This is particularly true for smaller children. As Fig. 116 demonstrates, there is a clear parallel between the width of the third ventricle and the brain mantle index between 2.0 and 3.4. However, beyond this boundary, there is considerable scattering. This is due to the fact that children with a skull diameter of 13.0 cm or less have a more marked degree of hydrocephalus at a certain width of the enlarged third ventricle than adults do. Generally we do not set up a relation between the width of the third ventricle and the diameter of the head. This error is compensated by the echo-encephalographic brain mantle index. Fig. 116 is based on results obtained in 163 pre- and postoperative echoencephalograms in 96 patients.

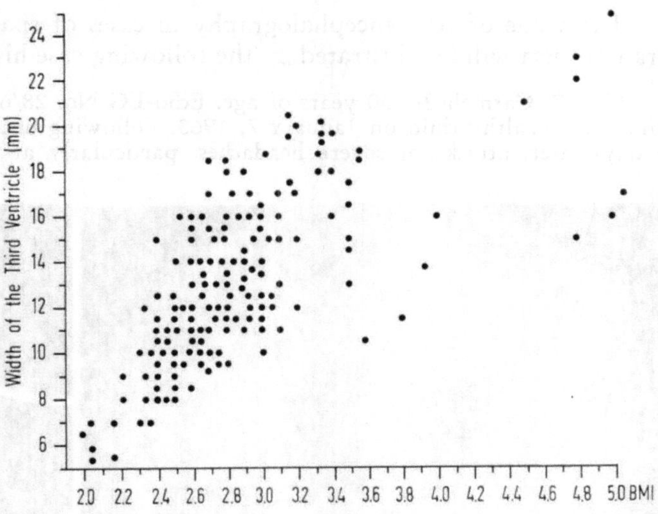

Fig. 116. Relationship between third ventricle width in the echogram and brain mantle index (163 measurements)

Tumor echoes were recorded only in exceptional cases by the sounding technique we used (temporal sounding). Altogether we have been able to observe an echo complex which could be said to originate from an infratentorial tumor only 4 times. These were one clivus meningioma (Fig. 96), one glioma of the pons, one ependymoma of the fourth ventricle which extended very high up and one exceptionally large cerebellar astrocytoma. In the last two cases, the actual midline echo was absent. The tumor echo complex appeared on the screen whenever the ultrasonic beam was directed posteriorly and downwards (Fig. 117).

a 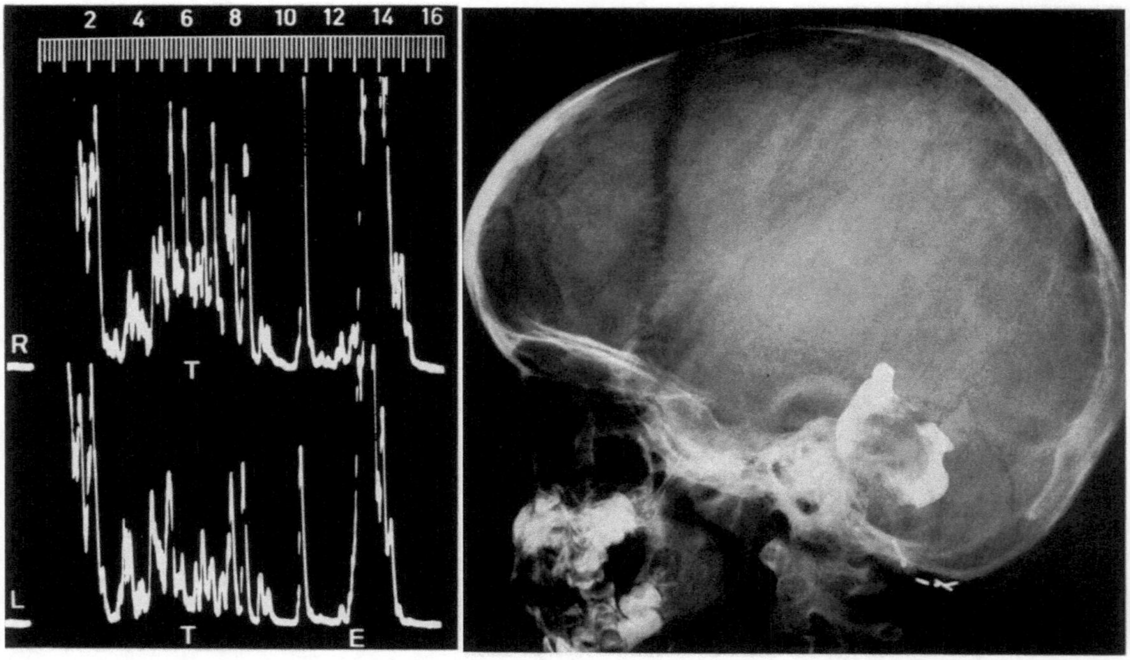 b

Fig. 117. a) Recurrence of an ependymoma of the fourth ventricle in a 4-year-old girl diagnosed by a tumor echo complex (T). The probe was applied behind the ear. Pat. J. L., Echo-EG No. 3208/3/68. b) Pantopaque ventriculogram of the same patient as in Fig. 117 a. Confirmation of the echo-encephalographic diagnosis. The tumor appears as a recess in the contrast medium which fills the markedly dilated fourth ventricle

Case Histories

The value of echo-encephalography in cases of space-occupying lesions in the posterior cranial fossa will be illustrated by the following case histories.

Case 7: Carmela B., 30 years of age. Echo-EG No. 28/63. No previous serious illnesses. Normal birth of a healthy child on January 7, 1963. Following uncomplicated delivery she was discharged. 3 days later, attacks of severe headaches, particularly at the back of the head. Examination on

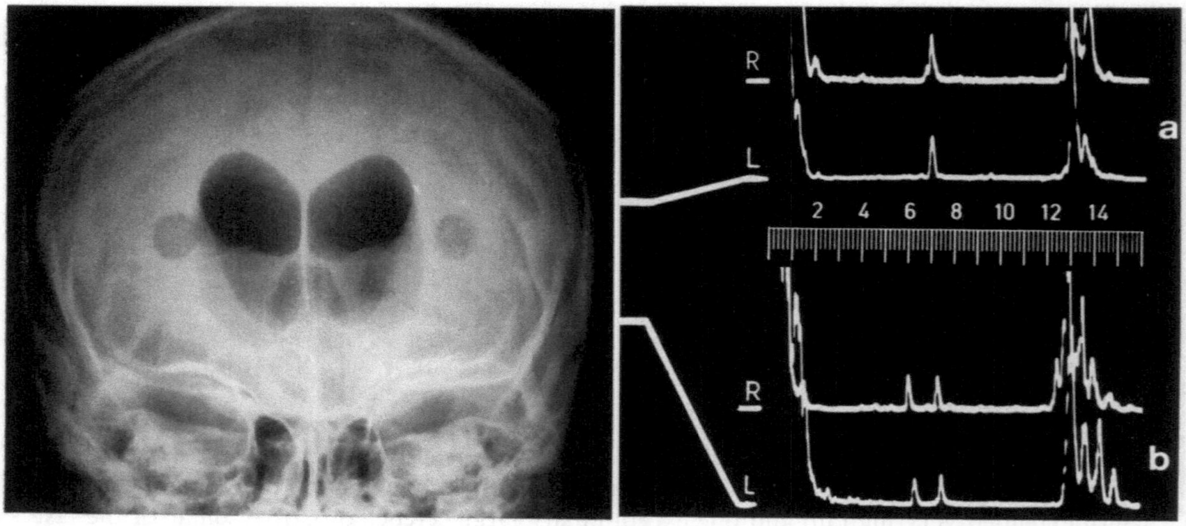

Fig. 118. Echo-encephalogram and ventriculogram of a 30-year-old woman with a cerebellar angioblastoma. a) normal midline echo using a high temporal probe application point; b) double echo from the walls of the dilated third ventricle. Pat. C. B., Echo-EG No. 28/63

hospital admission on January 25, 1963: beginning neck stiffness, left-sided abducens nerve paralysis, no obvious disturbances in co-ordination, bilateral papilledema of 2 to 3 diopters with fresh hemorrhages. CSF 39/3 cells, protein was 26 mg⁰/₀, slightly xanthochromic. On the basis of the history and CSF findings, we suspected a postpartum cerebral vein thrombosis. Echo-encephalography, however, revealed a double midline echo 14.0 mm apart (Fig. 118). We ourselves had neither seen such a finding until then, nor was it reported in the echo-encephalographic literature. We suspected that the two reflections originated from the walls of a dilated third ventricle and assumed that this was an occlusive hydrocephalus. Later operative ventriculography was carried out and it confirmed this suspicion (Fig. 118). The displacement of the aqueduct anteriorly suggested a space-occupying lesion in the posterior fossa. At the operation on February 2, 1963, a cystic angioblastoma was removed from the vermis of the cerebellum. No complications in the post-operative course.

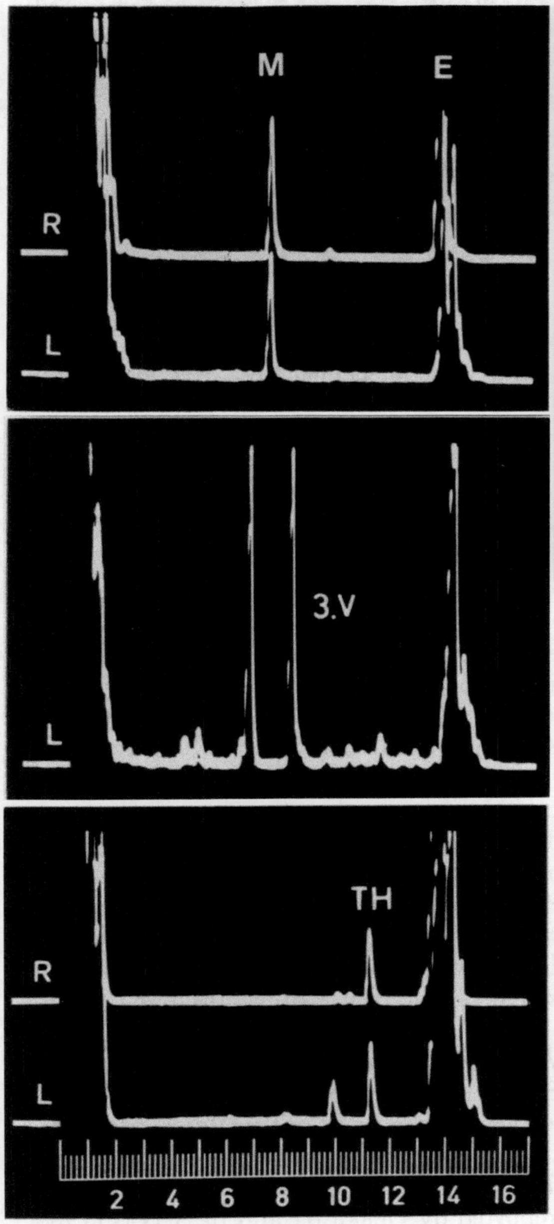

a
b
c

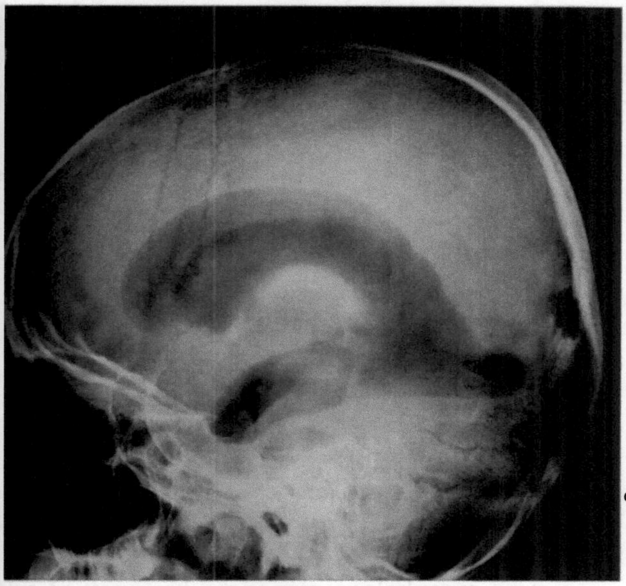

d

Fig. 119. Echogram and ventriculogram of a 9-year-old boy with a left cerebellar medulloblastoma. a) Normal midline echo, b) abnormal double echo from the walls of the third ventricle which is dilated up to 15 mm (3. V), c) using the position of the temporal horn echo, a BMI of 2.72 is calculated, d) ventriculogram of the same patient typical angulation and forward displacement of the aqueduct. Pat. K. L., Echo-EG No. 1977/65

Case 8: Klaus L., 9 years of age. Echo-EG No. 1977/65. Vomiting for 4 weeks. Several days before admission to hospital, occipital headaches and peculiar tilted head position. Neurological findings on day of admission (24. 11. 1965) within normal limits, fundi normal. On the echo-encephalogram (Fig. 119a to c) the midline structures at the normal site. Double echo from the walls of the dilated third ventricle, 15.0 mm apart. On sounding from both the left and the right, high reflection from the lateral wall of the temporal horn which is displaced towards the end echo. When calculating the brain mantle index from the position of the temporal horn echo, we find a value of 2.72, indicating the presence of an internal hydrocephalus. Confirmation of the echo-encephalographic findings using operative ventriculography which indicates a cerebellar tumor (Fig. 119). At the subsequent operation (26. 11. 1966), a medulloblastoma is removed from the left cerebellar hemisphere.

Case 9: Ida Sch., 59 years of age. Echo-EG No. 2046/66. The patient was admitted to the clinic on January 1, 1966 because of an unsteady gait and occasional vomiting. The neurological investigation revealed occasional spontaneous nystagmus towards the right, poor hearing on the left ear, slight ataxia and beginning dysdiadochokinesia bilaterally. Papilledema was not present. However, there was considerable dilatation of the ventricular system in the echogram (Fig. 120 right) which left no

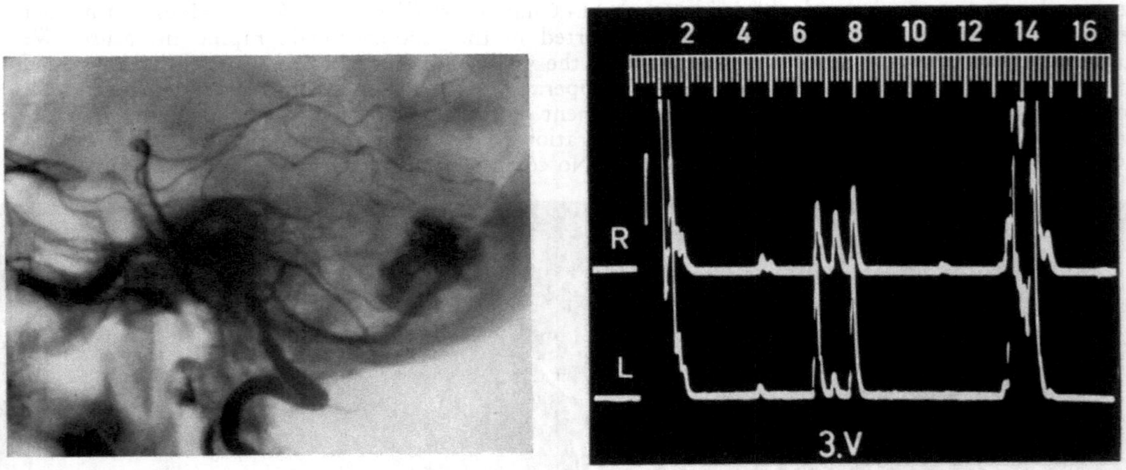

Fig. 120. Echogram of the dilated third ventricle (diameter of 15 mm) and vertebral angiogram of a 59-year-old woman with an angioblastoma. Pat. I. Sch., Echo-EG No. 2046/66

doubt about the presence of space-occupying lesion in the posterior cranial fossa. Vertebral angiography which was carried out subsequently, showed an angioblastoma in the left cerebellar hemisphere (Fig. 120 left).

Postoperative Examinations

Echo-encephalography is the *first method which gives us the opportunity of checking the width of the ventricular system following surgery in the posterior fossa at any point in the postoperative phase.* Postoperative ultrasonic examinations were performed in 104 such patients. A number of patients had echograms during the first few days. The other patients were examined between 4 weeks and 6 years after surgery. In 91 of these patients, we have a preoperative ultrasonic finding so that changes in the ventricular width following tumor exstirpation could be picked up and compared.

The results obtained in this way were surprising. Until now only a few scattered cases on the normalization of a dilated ventricular system have been reported in the literature, and these were usually check-ups after a Torkildsen procedure (Torkildsen, 1941, 1960; Tönnis, 1948; Fincher, Strewler and Swanson, 1948; Pia, 1953; Lang and Pia, 1961; Vogelsang, 1964). However, the earliest follow-up studies were all performed 1 to 3 months after the CSF flow had been re-established. In most cases, the interval between operation and follow-up study was even longer. No exact details have existed on the course of ventricular decrease over a given period of time until now.

Determining the width of the third ventricle has proved to be a good standard for checking the normalization of ventricular dilatation. Fig. 121 presents two characteristic progress charts of the first two weeks after removal of a cerebellar tumor or a Torkildsen procedure in a case of occlusion of the aqueduct. It is remarkable how much the hydrocephalus decreases, i. e. how narrow the third ventricle becomes within the first 72 hours after operation already. In the meantime, similar observations have been reported by Ulbricht and de Seixas (1966, 1967), Geletneky (1967) and Pia and Geletneky (1968). The dependence of third ventricle width on CSF pressure can be demonstrated by means of lumbar punctures as well. Following the removal of 50 to 70 cm^3 CSF, we were able to observe a narrowing of the third ventricle from 1 to 2 and in one case even 3.0 mm.

In 23 patients, comparative examinations were carried out over an extended period of time following the removal of a cerebellar tumor. According to Fig. 122 there was a distinct, although frequently variable, decrease in the width of the third ventricle of all these patients in the echogram. The average decrease was nearly 6.0 mm. The oberservation period for the group varied from 2 to 20 months. In the diagram, the initial values were connected to the control results by a straight line; naturally this does not mean that the decrease in the width of the ventricle was linear.

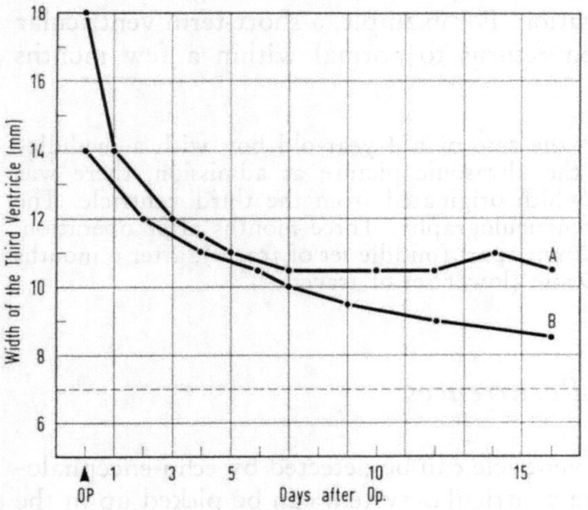

Fig. 121

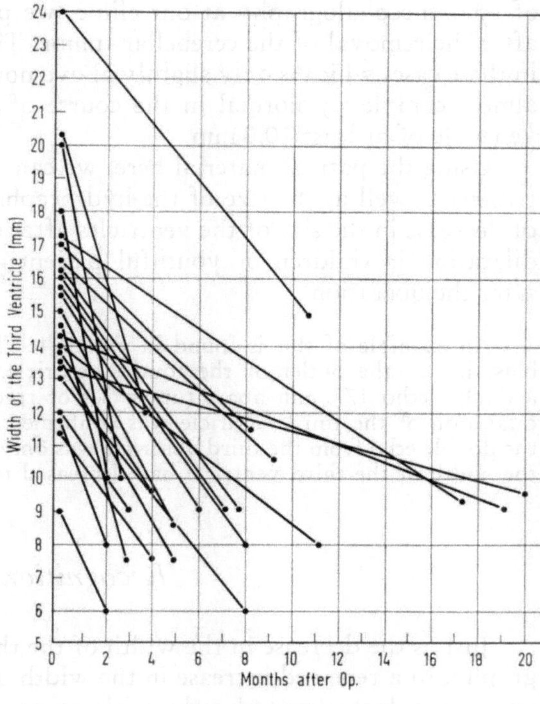

Fig. 122

Fig. 121. Echo-encephalographic follow-up of the width of the third ventricle which decreases in the first two weeks after operation; A. occlusion of the aqueduct due to chronic inflammation (Pat. D. K., 36 years of age. Echo-EG No. 1747/1—10/65); B. cerebellar tumor (Pat. K. G., 37 years of age. Echo-EG No. 1721/1—8/65)

Fig. 122. Decrease in the dilatation of the third ventricle in the echo-encephalogram over a period of two to 20 months following resection of a cerebellar tumor. 23 cases

In the meantime, the number of patients in whom follow-up studies were carried out, was so large that control charts could be drawn of the width of the third ventricle following the exstirpation of cerebellar tumors or of acoustic neurinomas over a period of 5 years. The two curves in Fig. 123 were drawn on the basis of 322 individual records. They show *that the decrease in ventricular dilatation comes to an almost total standstill after a period of half a year as a rule.* Up to this point, the width of the third ventricle has decreased by an average of 5.5 mm in cases of cerebellar tumors and by 2.7 mm in acoustic neurinomas. In subsequent years, a further decrease of 0.8 and 0.3 mm respectively, was observed. Using the methods of investigation we know, such insight into the behavior of compressed brain following decompression was not possible.

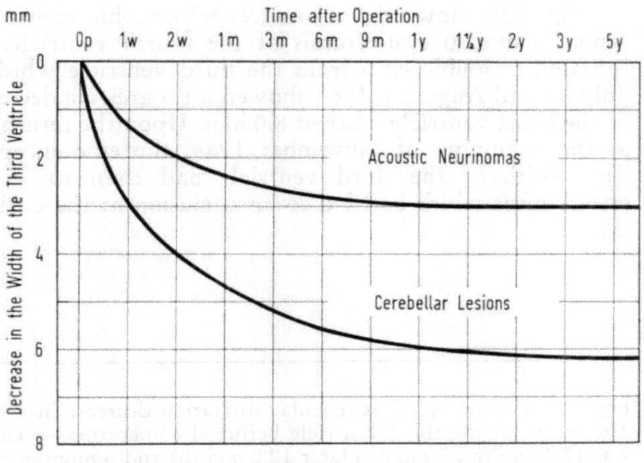

Fig. 123. Decrease of ventricular dilatation following tumor resection. Curve of mean values of 322 echo-encephalographic follow-ups

Fig. 112 shows post-operative measurements of the third ventricle in 104 patients. On comparing these with the preoperative echo-encephalographic results, a distinct shift in the frequency peak towards the left is noted. Altogether 24 patients were now found to have a third ventricle of normal width. In 13 patients, who were operated before the introduction of echo-encephalography at our clinic, we performed ultrasonic measurements 3 to 6 years after the removal of the cerebellar tumor. The width of the third ventricle averaged 0.8 mm in these cases which is only slightly above normal. The infantile patients in this group became almost completely normal in the course of time, while the adults usually retained a third ventricle of at least 10.0 mm.

Using the patient material here, we can say that the length of the history, the age of the patient as well as the size of the hydrocephalus are of importance for the degree and speed of decrease in the size of the ventricles after operation. For example, a short-term ventricular dilatation in children or youthful patients often returns to normal within a few months after the operation.

An example of this is found in Fig. 124. This is the case of a 4-year-old boy with a medulloblastoma at the outlet of the fourth ventricle. In the ultrasonic picture at admission, there was a double echo 17.0 mm apart (upper set of traces) which originated from the third ventricle. The dilatation of the third ventricle was confirmed at ventriculography. Three months after operation, the double echo from the third ventricle was only 12.0 mm apart (middle set of traces); after 6 months the width of the third ventricle had decreased to 7.0 mm (lower set of traces).

Recognition of Recurrences

Just as the decrease in the width of the third ventricle can be detected by echo-encephalography, so a renewed increase in the width of the ventricular system can be picked up in the outpatient department already, without stressing the patient. If the distance between the two peaks of the double echo increases after a transient period of decrease, then there must be a new obstruction of the path of the CSF-flow. Tumor recurrences or other impediments to the flow of CSF which may be due to chronic inflammatory processes can be recognized rapidly in this way. Till now we have observed a renewed dilatation of the ventricular system in the echogram seven times following an operative procedure in the posterior fossa. In 3 cases the echogram was checked and confirmed by ventriculography before further surgery was carried out; in the remaining 4 cases the second operation was undertaken without further diagnostic investigation.

Fig. 125 shows the echo-encephalographic course of a 44-year-old-man with occlusive hydrocephalus due to cysticercosis of the fourth ventricle. Preoperative measurement on Mai 11, 1964 revealed a double echo from the third ventricle which was 15.0 mm wide. Postoperative checks on July 10 and August 4, 1964 showed a progressive decrease in ventricular dilatation until the diameter of the third ventricle reached 8.0 mm. Upon the return of new signs of increased intracranial pressure at the beginning of November 1964, the echo-encephalogram showed the corresponding finding. The width of the third ventricle had risen to 15.0 mm again. Reoperation revealed granular ependymitis which had led to an occlusion at the outlet of the fourth ventricle.

Fig. 124. Follow-up of ventricular dilatation decrease in a child following the resection of a medulloblastoma. The width of the third ventricle before the operation — corresponding to the pneumoencephalogram (top) — was 17.0 mm (a), 3 months later 12.0 mm (b) and 6 months later 7.0 mm (c). Pat. G. A., 4 years of age, Echo-EG No. 1385/1/4/64

Fig. 125. Detection of a renewed occlusion of the CSF flow by using echo-encephalography. Pat. A. L., 44 years of age, Echo-EG No. 1143/64 (for details see text)

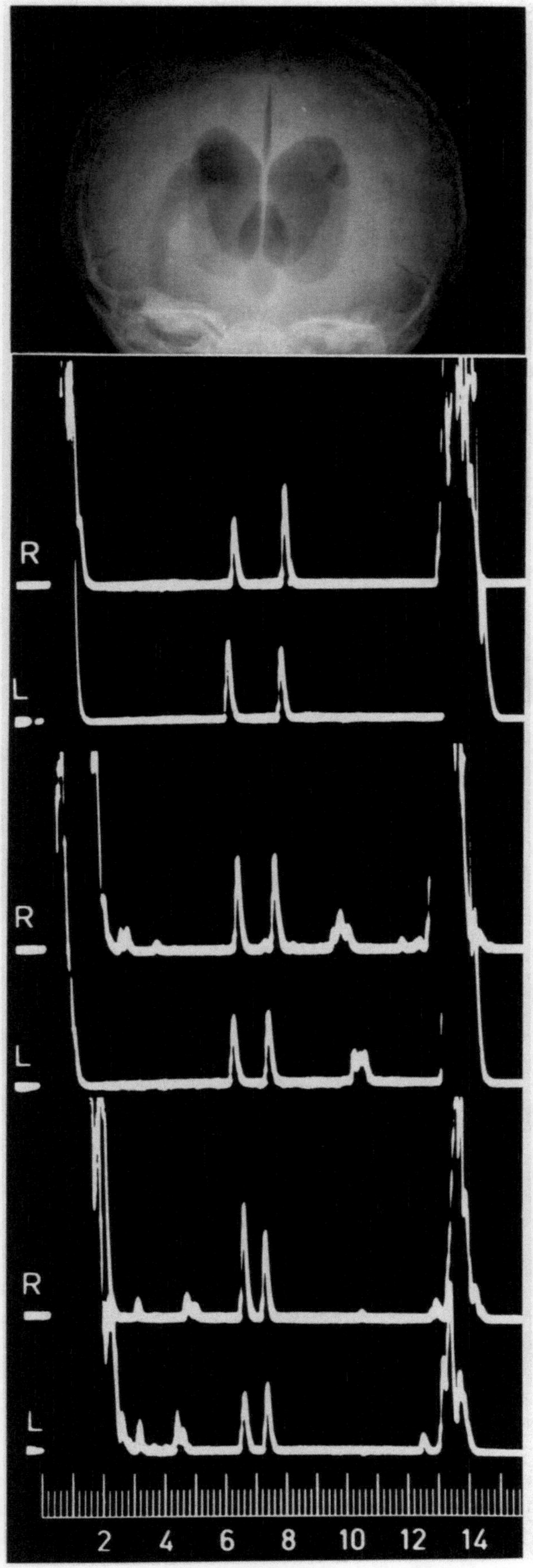

a

b

c

Fig. 124

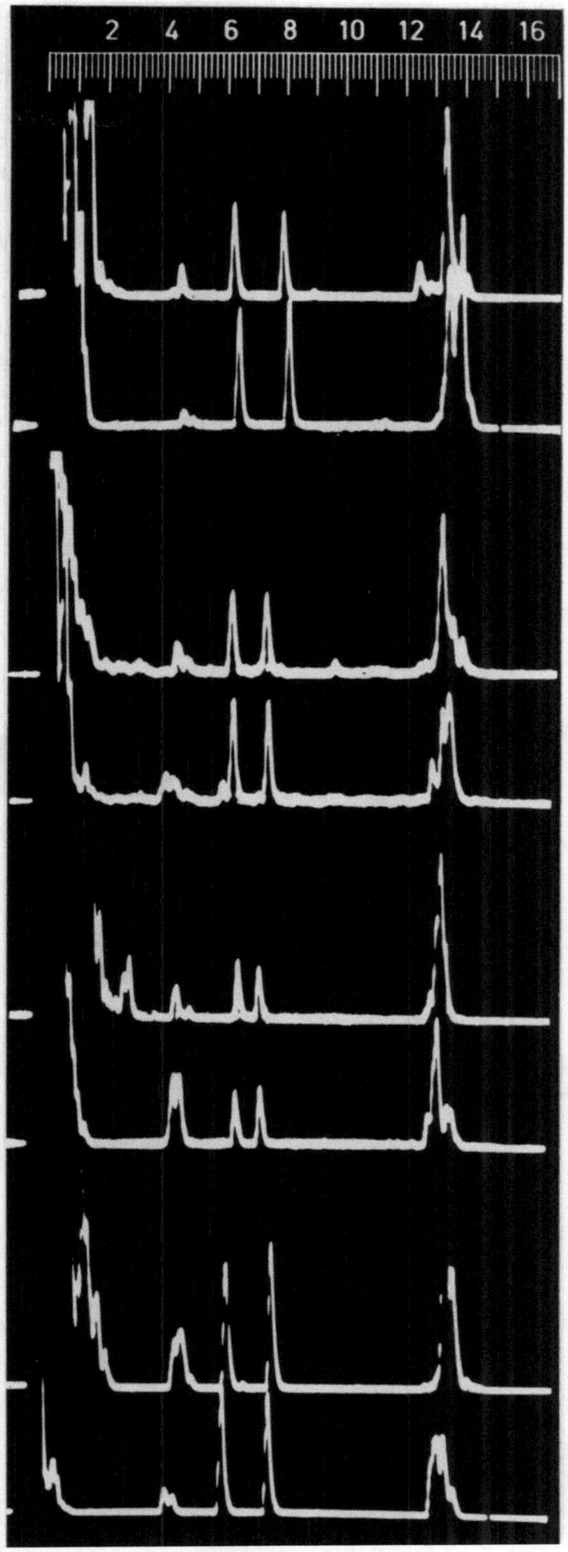

11. 5. 64

10. 7. 64

4. 8. 64

5. 11. 64

Fig. 125

6. Summary of the Echo-Encephalographic Results in Brain Tumors

A pathological echo-encephalogram can be expected in cases of brain tumor only if this lesion produces an alteration in the area scanned by the ultrasonic beam and this alteration is demonstrable. If the information provided by the echo-encephalographic examination were limited to a determination of the *midline shift* alone, a promising application of this method would be practically limited to lesions of the cerebral hemispheres (Fig. 126). Unfortunately many investigators are statisfied with the determination of the midline echo alone and thus arrive at understandably disappointing results.

Our own series of about 630 brain tumors and abscesses shows that only 55.7% (or 351 cases) produced such a midline echo shift. However, according to the pathologic-anatomic findings such a displacement of the midline echo could hardly be expected in the remaining cases because these tumors did not cause a lateral displacement of the midline structures.

A decisive improvement in the application field of echo-encephalography occurred only after ventricular dilatation was detectable. This possibility enables us to gain information on space-occupying lesions in the posterior fossa and on some of the tumors of the oral brain stem (Fig. 126). Thereby the proportion of pathological echo-encephalograms in

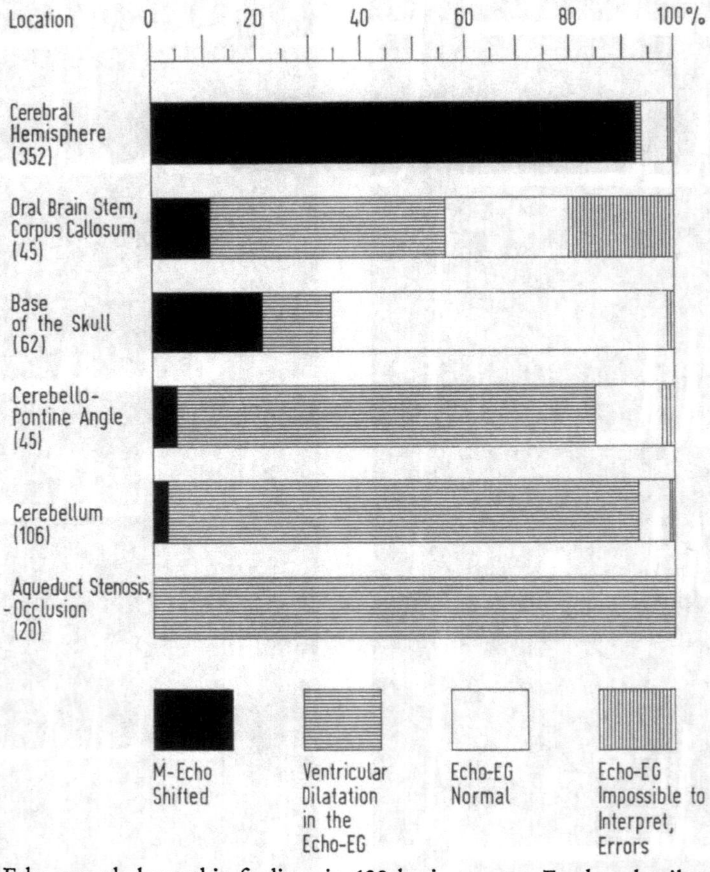

Fig. 126. Echo-encephalographic findings in 630 brain tumors. Further details see text

our own case material rose to 84.6%. The demonstration of *tumor reflections* does not lead to a decisive improvement of this quota because there is usually a midline shift or a ventricular dilatation present as well.

It is interesting to make a comparison of the results of echo-encephalographic investigations of brain tumors in children and adults. While only 323 of the 410 cases (77.8%) of the supratentorial tumors in adults had a pathological ultrasonic trace, no less than 45 of the 49 children (92%) with tumors and abscesses in this localization, revealed an abnormal echogram. Still more striking is the difference in the frequency of tumor echoes. Such reflections were observed in only 22.2% of adults with supratentorial tumors, children on the other hand, evidenced 53%. The same tendency is apparent in cases of space-occupying lesions of the posterior fossa. Of 115 adults with such a tumor, only 103 had a pathological echo-encephalogram (89.6%). However, all 56 children with a cerebellar tumor had an abnormal echogram.

In summary, we may say that *echo-encephalography in children with brain tumors is one of the most successful screening methods at present*, the frequency of detecting intracranial expansivities being 94%; furthermore it can usually be established whether the tumor is in the cerebral hemispheres, in the oral brain stem or in the posterior fossa. In addition, more than half of the supratentorial tumors can be localized exactly by means of tumor echo complexes.

Drawing 16 erroneous interpretations of the ultrasonic trace (2.5%) into consideration, there remains a total of 81 cases (12.9%) out of 630 brain tumors in our own series where the echo-encephalogram was completely normal. These were chiefly pituitary adenomas, acoustic neurinomas and smaller gliomas in the frontal and parasagittal region. While the two first-named groups can generally be excluded in the evaluation of echo-encephalography as a screening method, because they can be diagnosed on the basis of neurological findings alone, a number of frontal and parasagittal tumors pose a real diagnostic problem. The angiogram was frequently negative here too and only pneumoencephalography made the correct diagnosis possible. To regard these normal echo-encephalograms which accurately represent the relationships within the area examined, as failures of the method, would be a *gross misjudgement of the possibilities and limitations of the procedure.*

The fact that echo-encephalography is not an investigation for the definitive exclusion of a tumor must be strongly emphasized here. Only the positive demonstration of pathological changes is an important contribution to the diagnosis. But this is just as true for other technical diagnostic aids, which often enable us to make the final diagnosis only if they are all taken together.

C. The Echo-Encephalogram in Head Injuries

The marked increase in the number of traffic accidents over the last few years has resulted in a corresponding rise in the proportion of head injuries admitted to surgical and neurosurgical clinics in relation to the total number of patients. Besides simple skull fractures and concussions requiring little diagnostic endeavor, there are posttraumatic intracranial hemorrhages and cerebral edemas due to contusions, whose timely recognition often poses great problems even for the experienced surgeon. In a large series of patients, Tönnis, Frowein and Euler (1963) demonstrated that it is impossible to exclude or confirm a hematoma in two-thirds of all suspicious cases using the common methods, i. e. without the use of dye studies, unless one decides to perform exploratory trepanations or to make burr-holes. Numerous hospitals do not have the facilities for contrast medium investigations however, and exploratory trepanations do not have satisfactory long-term results because many hematomas are atypically located and contusions can often simulate the clinical picture of a space-occupying hemorrhage. Therefore it is frequently necessary to transport patients with head injuries over great distances to special hospitals. This loss of time greatly worsens the prognosis in many cases. In the search for a simpler and better method of diagnosing space-occupying hemorrhages following head injuries, Leksell developed "echo-encephalography". His first publication on this new method in the year 1955/56 deals chiefly with the recognition of posttraumatic hematomas by means of ultrasound.

Today *head injuries are the chief indication for echo-encephalographic investigation* since the position of the cerebral midline structures can be determined within a few minutes by this method so that space-occupying hematomas which produce a mass shift are immediately recognized. Beyond this, frequently recordable special reflections (hematoma echoes) enable us to determine the type, localization and dimensions of the intracranial hematoma without further diagnostic effort (KAZNER, KUNZE, SCHIEFER, 1965). In the following section, the diagnostic possibilities on the one hand and the limitations of echo-encephalography with head injuries on the other hand, will be discussed on the basis of our own experience with 1034 patients.

1. Brain Concussion — Mild Closed Head Injuries

In patients with a simple concussion syndrome, we do not expect a displacement of the midline structures of the brain. Only one patient in our series of 283 cases of cerebral concussion reached the borderline value of 2.0 mm; this was a 41-year-old woman (Echo-EG No. 2408/66). Repeated checks continued to show this slight M-echo shift. A 1.5 mm displacement of the midline echo was found 8 times (2.8%). The average displacement of the midline echo from the ideal midline was only 0.27 mm which is almost the same value as was obtained in a control group of 249 healthy subjects (see also Table 8). Occasionally *factitious displacements of the midline structures were observed in cases of subgaleal hematoma in the temporal area.* However, in these cases a parallel displacement of the end echo appears on the echo-encephalogram (see also p. 78 and Fig. 77 b). The distance midline echo—end echo is the same, regardless of whether one is sounding from the left or from the right; here a clear differentiation from a genuine shift can easily be made. Fig. 77 a shows an example of such an echo-encephalogram. On sounding from the healthy side, a distinct reflection may be recognized behind the end echo; this is the "egress" echo, which develops at the skin-air interface. In healthy subjects on the other hand, i. e. in patients without an injury to the scalp, this "egress" echo usually cannot be separated from the end echo complex. These peculiarities which can easily lead to echo-encephalographic misinterpretations, were first pointed out by JEPPSSON (1961). CHADDUCK and CRUTCHFIELD (1964) have devoted a detailed discussion to this problem (see also p. 78 f.).

In 240 patients (84.8%) with concussion, the width of the third ventricle could be determined by echo-encephalography. This was within the normal range in 224 cases, i. e. the diameter of the third ventricle was less than 7.5 mm here. The 16 patients, who revealed a dilatation of the third ventricle in the echo-encephalogram, had cerebrovascular disease or chronic epilepsy which explains the dilated third ventricles. Here the width of the third ventricle ranged from 7.5 to 12.0 mm.

2. Brain Contusion — Moderate and Severe Open and Closed Head Injuries

Echo-encephalographic investigation was performed in 302 patients with cerebral contusion. In most cases we carried out repeated check-ups during the first days after the trauma in order to detect complications at an early stage. In 207 injured patients, the midline echo remained within the normal range (68.5%). In 3 cases, no clear midline echo could be obtained. This was either due to extensive subgaleal hematomas so that increased amounts of ultrasound were absorbed which resulted in inadequate energy returning to the probe or the skull was so severely battered that it was impossible to apply a probe. In one instance, an error in measurement occurred. The echogram showed the midline structures in the normal place, but the internal cerebral veins were displaced by 3 mm on the angiogram. The echogram continued to remain unremarkable even at control investigation. Altogether the sum of unsatisfactory results was only 4 out of 302 investigations which is 1.3% (see also Table 15).

91 patients with severe closed or open head injuries exhibited a midline echo displacement of 2.0 mm or more in the echo-encephalogram during the posttraumatic phase (30.1%). On the average, the midline echo shift was only 3.4 mm, thereby being considerably below the mean value for space-occupying hemorrhages (PIA and GELETNEKY, 3.5 mm, 1968).

Table 15. *Results of echo-encephalography in 1034 patients with head injury*

Type of Lesion	No. of Cases	Midline Echo Normal	Midline Echo Shifted (2.0 mm and more)	Hematoma Echo	Width of the Third Ventricle Exceeding 7.0 mm	Error or Echo-EG not Interpretable
Concussion	283	282	—	—	16	1
Contusion	302	207	91	— (4?)	22	4
Sequelae of head injury	204	199	5	—	84	—
Posttraumatic intracranial hematomas (245)						
Extradural hematoma	76	4	72	59	—	—
Acute subdural hematoma	82	1	80	38	—	1
Intracerebral hematoma	32	3	29	9	—	—
Unilateral chronic subdural hematoma	39	1	37	33	—	1
Bilateral chronic subdural hematoma	16	9	7	12 [a]	1	—
Total	1034	706	321	155	123	7

[a] In 2 cases a hematoma echo could be detected on one side only.

Immediately after the trauma (i. e. during the first admission to the clinic), a midline echo shift was found in only 26 cases out of 41 within the first 24 hours. Patients, who showed distinct midline echo displacements shortly after the accident, nearly all had extensive unilateral brain contusions without an additional hemorrhage. Particularly remarkable is the fact that early midline echo displacements occurred chiefly in children up to the age of 10 years (see Table 16).

Table 16. *Behaviour of the midline echo in severe open and closed brain injuries (contusion). Results in 298 patients*

Midline Echo		No. of Patients in the Age Groups from 0—10	11—20	21—30	31—40	41—50	51—60	61—70	over 70 years
Not displaced at any time after the trauma	(207)	47	55	38	22	14	16	11	4
Displaced on day of trauma already	(41)	18	6	10	3	1	2	1	—
Displaced later	(50)	5	10	14	12	2	2	5	—
Total	(298)	70	71	62	37	17	20	17	4

In the majority of cases (50 patients), the displacement of the midline echo was found only during control examinations 24 to 48 hours after the trauma, sometimes not before one week. The greatest time intervals between trauma and the initial appearance of a midline echo shift was 10 days (3 times), 14 days (twice) and 19 days (once). In every case the increase or decrease in the M-echo shift was parallel to the worsening of the clinical picture: deeper unconsciousness, more marked or first appearance of neurological deficits, seizures.

The *point of maximal displacement occurs between the third and fifth day after the injury* according to our investigations so far. A midline echo shift due to a contusion generally recedes within 2 to 14 days and is easily kept under surveillance by repeated echo-encephalograms (Fig. 127). Only in 3 patients did the mass shift caused by a contusion take 18 to 22 days to normalize completely.

While hematomas usually produce midline echo shifts of 7 to 8 mm and more, contusions and posttraumatic vascular occlusions only rarely cause such pronounced mass shifts (see Table 18). Therefore we will briefly describe the injuries of the 6 patients, who failed to show a hematoma in spite of an 8 to 10 mm M-echo shift.

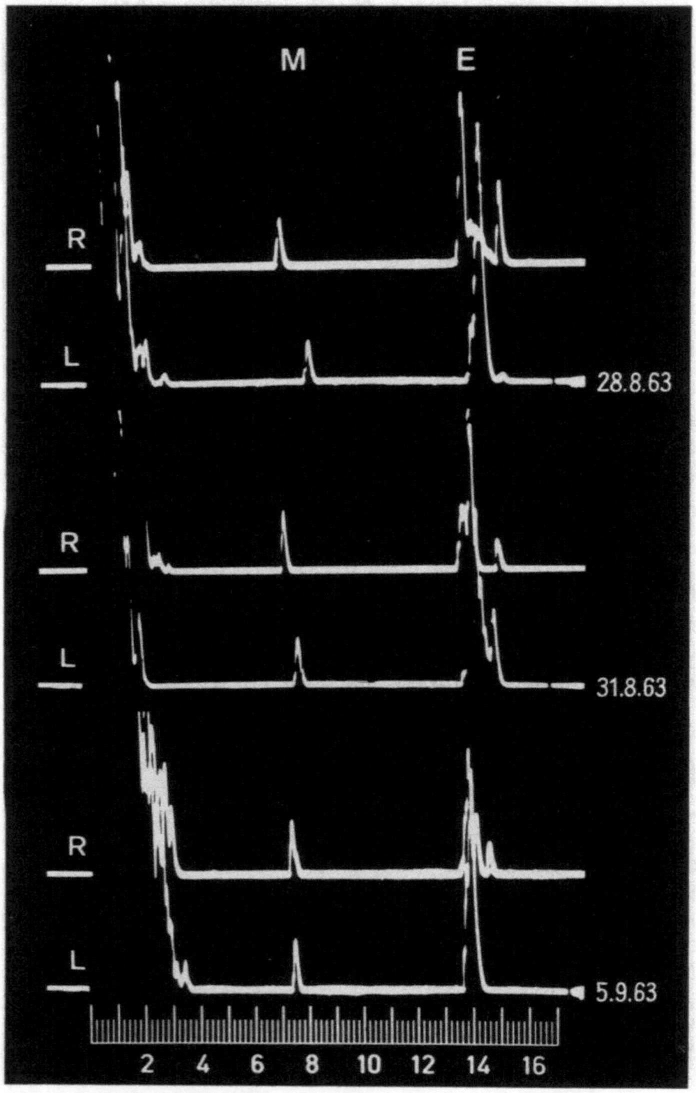

Fig. 127. Echogram of a patient with a contusion in the left temporal lobe; a few days after the trauma a midline echo shift of 4.0 mm towards the right appeared (August 28, 1963). Within a week complete normalization of the echo-encephalogram (September 5, 1963). Pat. M. J., 34 years of age, Echo-EG No. 672/1—3/63

1. A. D., 26 years of age. Echo-EG No. 1177/64. M-echo displacement of 10.0 mm 2 hours after the trauma already. Death occurred within 5 hours. The entire left cerebral hemisphere was crushed on account of a palm-sized depressed fracture of the temporal and sagittal bone.

2. H.-E. H., 13 years of age. Echo-EG No. 1403/64. 8.0 mm displacement of the midline structures within a few hours after the trauma. Confirmation of the ultrasonic findings through left-sided angiography. Death occurred one hour later. Autopsy: massive postcontusional cerebral edema with a marked increase in the volume of the left hemisphere. Thin subdural film of blood on the left without other space-occupying hemorrhage.

3. A. Sch., 39 years of age. Echo-EG No. 1870/65. 8.0 mm displacement of the midline structures 8 days after injury. Exploratory trepanation in the left temporal area, no hematoma, massive cerebral edema. Death set in 3 days after the operation. Autopsy: left frontotemporal contusion extending far into the white matter accompanied by softening of the entire left hemisphere.

4. J. Sch., 18 years of age. Echo-EG No. 2335/66. 8.0 mm displacement 5 days after trauma. Left-sided angiography showed partial occlusion of the carotid artery and thrombosis of the middle cerebral artery resulting in a considerable amount of displacement. Decompression trepanation and removal of the bony flap. Marked cerebral edema so that closure of the dura mater becomes impossible. Decrease of the midline echo shift within 2 weeks after surgery and subsequent development of ventricular dilatation.

5. G. R., 18 years of age. Echo-EG No. 3036/67. On admission to hospital shortly after a traffic accident, the M-echo was displaced by 6.0 mm. Left-sided angiography: diffuse increase in volume of the left cerebral hemisphere without circumscribed expansivity. Increase in the midline echo shift up to 9.5 mm on the 4th day after the accident. Subsequently normalization of the midline echo within 3 weeks. Development of distinct ventricular dilatation (third ventricle 10.0 mm in diameter).

6. H.-P. B., 3½ years of age. Echo-EG No. 3338/67. 8.5 mm displacement of the midline echo 2 hours after a car accident. Questionable hematoma echo in the left temporal area. Death on the way to the operating room. Autopsy: massive cerebral edema with considerable displacement of the central fissure towards the right. Crush injury of the left temporal and occipital lobe to a depth of several centimeters. Numerous punctate hemorrhages in the brain stem. The questionable hematoma echo must have developed in the large temporal contusion area.

Earlier we had the impression that mass shifts due to contusions were greater in children than in adults. The present series, however, fails to show a definite difference in the degree of midline echo displacement between children and adults following severe head injuries. *The only difference is that midline displacements usually appear earlier in children* (see also Table 16).

In 59 patients, we carried out carotid angiography 30 times despite the absence of a shift because there was a deterioration in the clinical picture. Only in a single case was there no agreement between the echo-encephalographic and angiographic measurement (2%). 62 times a surgical procedure was performed usually with depressed fractures and open head injuries. In 32 patients we posses an autopsy finding. The amount of midline displacement could be well correlated with the echo-encephalographic findings in all cases.

If a badly battered temporal lobe is present together with a subgaleal hematoma, it is often impossible to find the midline echo from the side of the injury. Ultrasound is so strongly reflected and absorbed by tissues on the side of probe application that the structures responsible for the midline echo are no longer hit by sufficient quantities of ultrasound. The end echo is frequently not recognizable any more either. In such cases, an extremely prolonged initial echo consisting of multiple high reflections ("grass") is seen on the screen. In general it may be stated that the evaluation of an echo-encephalogram is more difficult if there is a contusion in the temporal lobe and that we must often depend on the record obtained from the opposite side alone. An illustration of this point is found in Fig. 128.

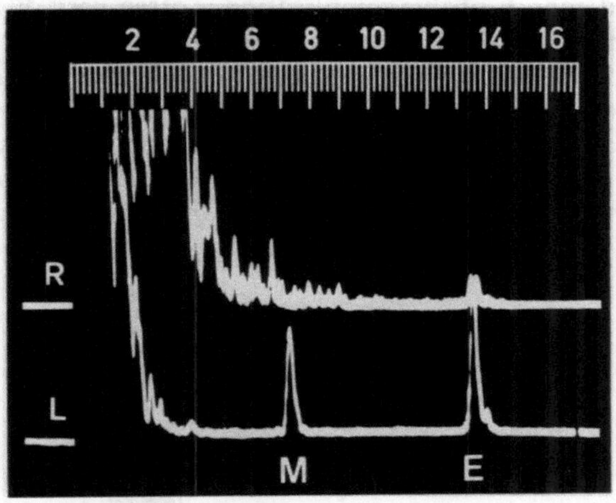

Fig. 128. Echogram of an 82-year-old man with an extensive right temporal subgaleal hematoma and a right temporal contusion. The echo-encephalogram can be evaluated only on sounding from the left because it is impossible to obtain an intelligible reflection from the right. Pat. J. G., Echo-EG No. 1730/65

The *width of the third ventricle* in cases of fresh contusions was measurable in only 208 of 302 cases (69%). In 22 patients we noted a dilatation of the third ventricle on the echo-encephalogram which had been observed in the acute stage already. Only in 4 elderly patients were there symptoms of a cerebrovascular disease so that the echo-encephalographically determined dilatation of the third ventricle could be related to this. However, the remaining 18 persons were almost all younger. Twice there must have been a tamponade of the ventricular system with blood. The widths of the third ventricle here were 9.0 and 11.0 mm. In the remaining 16 patients the abnormal width of the ventricles could not be definitively accounted for. Although the clinical findings pointed to a brain stem or cerebellar contusion in 11 patients, this suspicion could not be confirmed by operation or autopsy. Disturbances in the flow of CSF in the course of the aqueduct due to edema of the brain stem or the cerebellum may be discussed as causes for this phenomenon. In such cases, the width of the third ventricle lay between 7.5 and 9.0 mm. Many of these patients had a progressive dilatation of the third ventricle later on. Of special interest is the fact that the dilatation of the ventricular system ceased as soon as the patients became conscious. In patients who remained unconscious until death supervened, the dilatation continued to progress (Fig. 163).

3. Posttraumatic Intracranial Hematomas

In the last 5 years, echo-encephalography has proved to be a very valuable help in the diagnosis of space-occupying hemorrhages following traumatic head injuries. One special advantage of this method is that patients with fresh head injuries may be *observed continuously* by repeated ultrasound investigations thereby enabling us frequently to recognize freshly developing hematomas even before the appearance of alarming clinical signs.

Altogether a total of 245 patients with posttraumatic intracranial hematomas was observed and investigated by echo-encephalography at the Neurosurgical Clinic of the University of Erlangen in this period. The distribution of the different types of hemorrhages can be seen in the survey table (Table 15). In cases of combined hematomas, it was grouped under whatever form was more important from the viewpoint of ultrasound, e. g. all cases of epidural hematoma with coexistent subdural hematomas were grouped under epidural hematomas because the special reflection in the echo-encephalogram is due to the epidural hematoma in such a case. The chronic subdural hematoma group includes every case of hemorrhagic pachymeningitis with the exception of infantile subdural effusions which will be discussed in a separate section.

a) Epidural Hematoma

Since the introduction of echo-encephalography, we habe been able to observe 76 epidural hematomas. According to our own experience and reports in the literature, the clinical diagnosis of this complication is often very difficult and this is one of the reasons for the high mortality (see Wojahn, 1964; Hemmer, 1965; E. Müller, 1965). The echo-encephalographic findings with this special form of hematoma, however, are so significant that a detailed discussion was deemed necessary. If the midline structures are displaced by more than 2 mm after a fresh head injury, a space-occupying lesion may be assumed to exist in the corresponding cerebral hemisphere. Should the presence of a subgaleal hematoma which is impenetrable to ultrasound, make it necessary to undertake the measurement of the midline echo from one side only, then this measurement will have to suffice. The ultrasonic investigation is continued from the side opposite to the suspected space-occupying lesion. Every effort is now made to obtain reflections from the hematoma interfaces from an application point di-

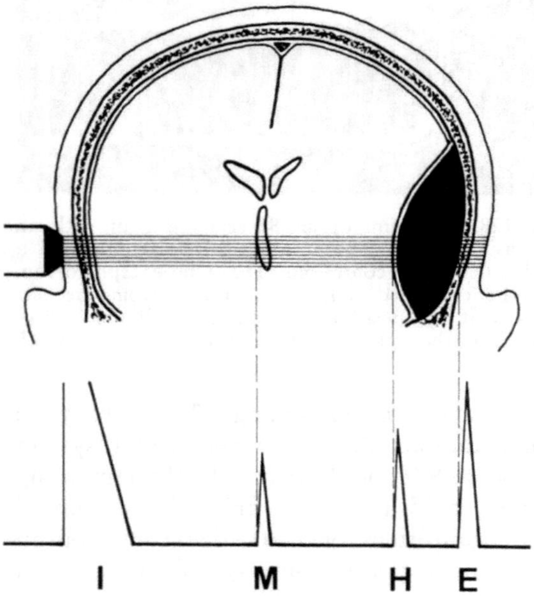

Fig. 129

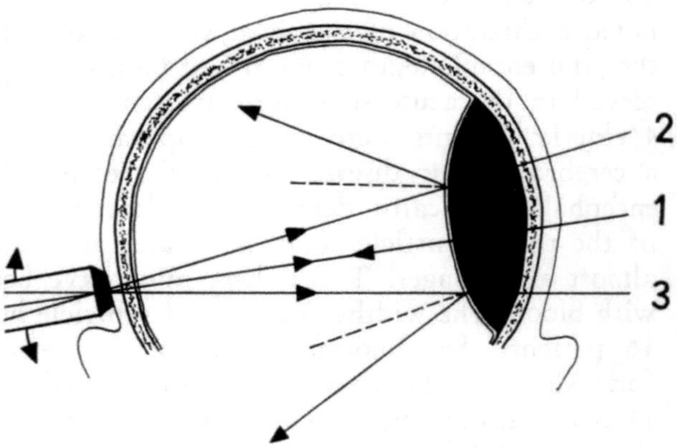

Fig. 130

Fig. 129. Diagrammatic illustration of the development of the echogram with a temporal epidural hematoma. A high reflection (H) can be recorded from the interface brain/dura/blood; this is the hematoma echo

Fig. 130. Effect of the direction of sounding on the demonstrability of a hematoma echo. Only if ultrasound hits the dura almost vertically, does enough energy return to the probe (1). Slight deviations from the perpendicular are enough to prevent the development of a hematoma echo (2.3)

rectly above the root of the ear (see p. 39); these appear directly ahead of the end echo on the screen (see diagram in Fig. 129). To do this, a large area of the opposite hemisphere is scanned with the ultrasonic beam by means of rotating and tilting movements of the probe, keeping good contact with the scalp throughout the procedure, however. By using this technique it is possible to detect epidural hematomas in a relatively large area by distinct hematoma echoes. The frontal, occipital, some of the temporo-basal hematomas, as well as those lying above the superior sagittal sinus, the so-called coronary hematomas are exceptions because they cannot be hit vertically by the ultrasonic beam when this technique is used. The primary requisite for the development of a hematoma echo is that the ultrasonic beam impinges vertically on the dura which has been forced off the inner table of the skull (Fig. 76). *The exact position and the centre of the hematoma can be ascertained from the direction in which the hematoma reflection has the greatest amplitude. The distance between the hematoma echo and the end echo represents the thickness of the hematoma.*

Because of the basic physical principles, the dura echo can be obtained from *one* certain sounding direction only, (but from different application points of the probe) namely, when

a
b
c

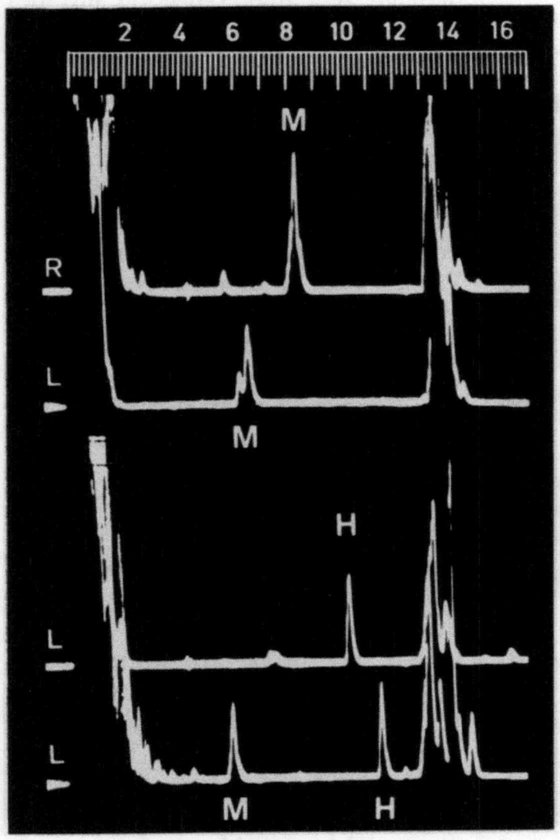

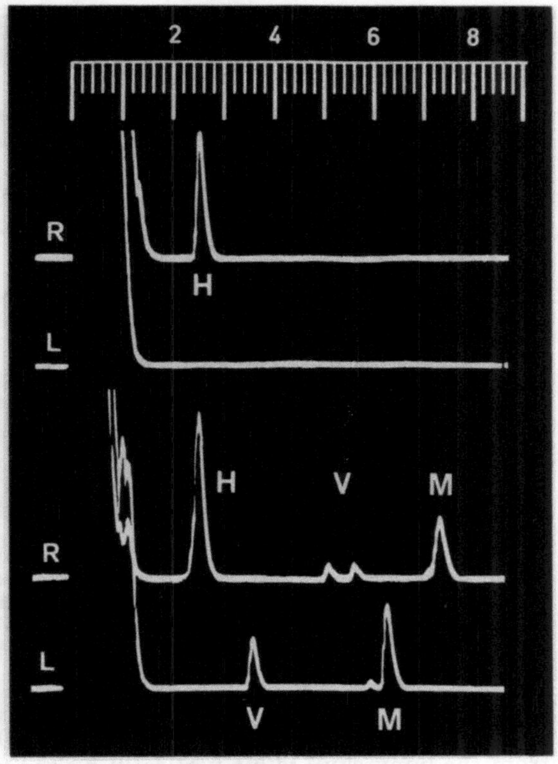

a
b

Fig. 131 Fig. 132

Fig. 131. Echo-encephalogram of a 55-year-old man with an epidural hematoma. a) Midline echo shift 9.0 mm towards the left. b) Hematoma echo (H) 32 mm ahead of the end echo on directing the ultrasonic beam towards the right parietal area. c) Another hematoma echo (H) 18 mm ahead of the end echo when the ultrasonic beam is directed towards the temporo-basal area. Further details see text. Pat. A. St., Echo-EG No. 2516/66

Fig.. 132. Echo-encephalography of an epidural hematoma using an elongated time axis and probes of higher frequency. a) Echogram using a 4 MHz probe 10 mm in diameter. With the exception of the hematoma echo, no further reflections can be recorded on sounding from the right. b) same technique of investigation as in a, this time with a 2 MHz probe 10 mm in diameter. In addition to the hematoma echo (H), the midline echo (M) and echoes from the lateral ventricle walls (V) become visible. Pat. E. H., 8 years of age. Echo-EG No. 686/63

the ultrasonic beam impinges vertically on the dura. The hematoma echo does not approach the end echo when the probe is slowly tilted upwards or downwards, rather it disappears immediately (see Fig. 130). Until now we have been able to record two different dura echoes in one and the same patient from one side only once. One dura echo appeared 18 mm ahead of the end echo when the ultrasonic beam was directed towards the temporal region, the other 32 mm ahead of the end echo when the beam pointed towards the parietal area (Fig. 131). At the operation which was performed without other investigations, the explanation turned out to be two adjacent epidural hematomas, a smaller one in the temporal area and a larger one in the parietal region.

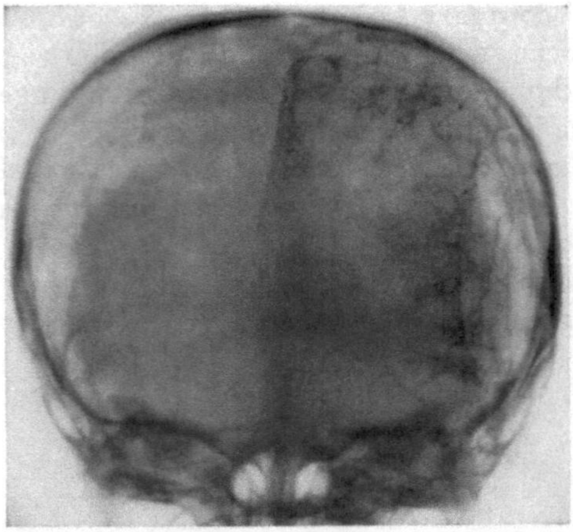

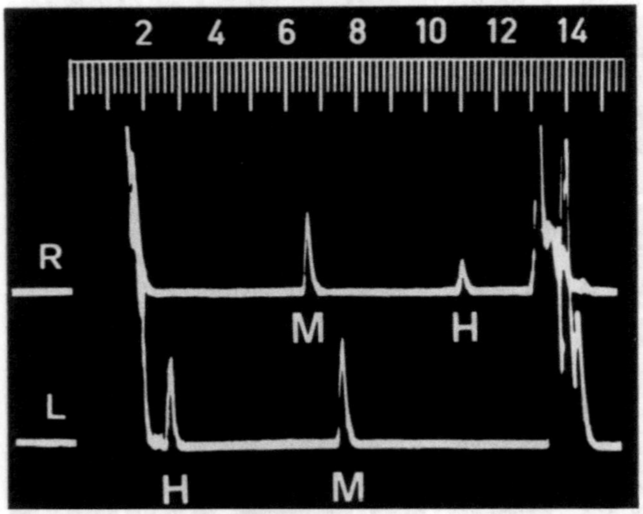

Fig. 133. Left temporo-parietal epidural hematoma. In addition to the midline echo shift, there is a hematoma echo (H) recordable from both sides. The distance between the hematoma echo and the end echo exactly equals the thickness of the hematoma in the angiogram. Pat. S.M., 14 years of age, Echo-EG No. 570/63

When the hematoma echo appears on horizontal sounding, the dura echo can also arise on the side of the hemorrhage as a high reflection just behind the initial echo complex (Fig. 132, Fig. 133, Fig. 137 top and 139). To see this, it is necessary either to switch on the sensitivity time control whereby the reflections which normally appear in the near field of the echo-encephalogram can be suppressed or to use probes of higher frequency. Probes with a diameter of 10 and 15 mm and a frequency of 4 Mc/s have proved particularly useful in this connection. In addition the time axis (longitudinal axis) can be elongated in order to have a better survey (Fig. 132 a and b).

In cases of temporo-frontal hematomas, the hematoma echo is highest if a more occipital application point for the probe is chosen on the healthy side of the head; for temporo-occipital lesions, the probe application point should be somewhat ahead of the ear. In cases of frontal pole epidural hematomas, a dura echo can be demonstrated in the near field of the echogram if the probe is applied in the frontal area and the dura which has been forced away from the bony skull is hit vertically by the ultrasonic beam. We were able to observe such a case just a short time ago (see case 19 and Fig. 151). A practiced person needs an average of 3 to 5 minutes for one examination, photograph included. Under adverse circumstances (very restless patient, subgaleal hematoma in the temporal area, temporal bruises or depressed fractures) the registration of the echo-encephalogram can last up to 10 minutes (see also Jefferson and Hill, 1966). Since this method generally offers a rapid and sure possibility of detecting posttraumatic complications without stressing the patient unduly, *the time loss is justified in every case*, especially since it is often possible to avoid angiographic examination, commonly employed when the suspicion of a hematoma arises, on the basis of the echo-encephalographic findings. In most cases, echo-encephalography does not mean a delay at all, because it can be performed while blood is being withdrawn, blood pressure is measured and preparations for the anaesthetic are made.

Differentiating the Dura Echo from Other Reflections

Besides the echo caused by the dura, there is a whole series of other reflections in the area between the midline and the end echo which can be mistaken for the former. However, with increasing experience a number of differentiating characteristics becomes evident which make an adequate differentiation possible.

1. Reflections in healthy patients barely 1 cm ahead of the end echo when the ultrasonic beam is directed towards the temporo-occipito-basal area. This echo was observed in a number of our patients; usually it can be recorded from both sides and develops at the *bony prominences* of the middle fossa in our opinion. If direction is slightly changed, the echo wanders; in addition its ascent is not as steep as that of the dura echo. A further echo which is obtained in the same direction is only 6 to 8 mm from the end echo and probably originates from the sigmoid sinus.

2. From the *external wall of the temporal horn* a high marked echo may be obtained in many patients — especially if the ventricles are dilated. Normally it is halfway between the midline and the end echo, in the case of an internal hydrocephalus it is closer to the end echo. Since mass shifts are hardly ever recorded here, this source of error is nearly eliminated. Furthermore a slight pulsation of the temporal horn may occasionally be observed on the screen, something we have never seen with a dura echo.

3. In case the *temporal horn is displaced* towards the midline by a circumscribed zone of cerebral edema or an acute subdural hematoma, the temporal horn echo may be obtained more medially, thereby creating diagnostic difficulties (see also Dreese, Hayes and Kempe, 1966). Occasionally, especially in older persons where there are *calcifications in the glomus of the choroid plexus,* a particularly high echo is obtained which can be differentiated only with great difficulty or not at all from a dura echo if there is an additional

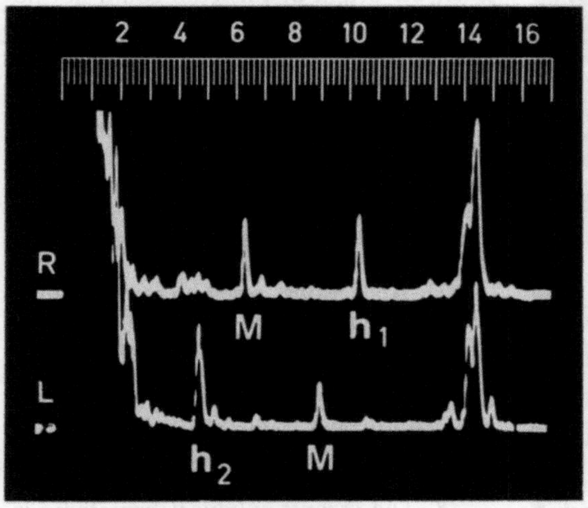

Fig. 134

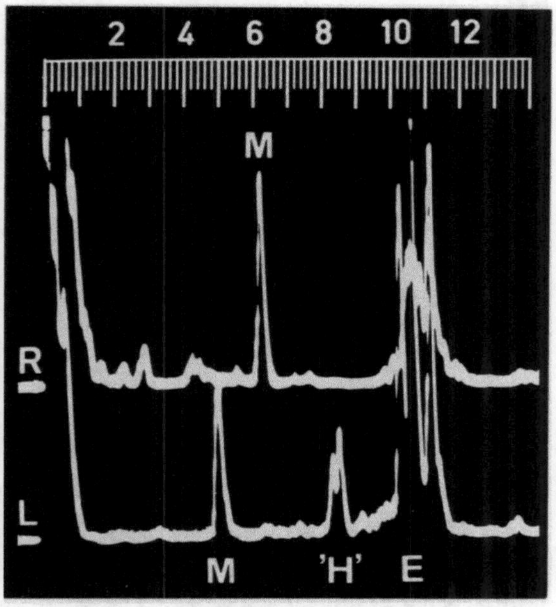

Fig. 135

Fig. 134. Erroneous interpretation of an echo-encephalogram in a 46-year-old man with an acute left-sided subdural hematoma. Besides the displaced M-echo, high reflections (h_1 h_2) appear bilaterally. These were initially regarded as hematomas echoes. Angiographic thickness of the hematoma, 1.5 cm. The additional reflection must have developed near the Sylvian fissure or from the lateral ventricles. Pat. A. Sch. Echo-EG No. 1288/64

Fig. 135. Reflection 'H' erroneously regarded as a hematoma echo, in a 6-month-old child with massive crushing of the brain in the right temporo-occipital region. Displacement of the M-echo by 6.0 mm. The reflection 'H' must have originated in the crushed area. Observe dicrotic notch on the supposed hematoma echo. The dura echo has only a single peak. Pat. M. T., Echo-EG No. 3051/67

midline displacement. We have observed such a reflection in combination with a mass shift only 3 times so far. Twice we were dealing with an acute subdural hematoma about 2 cm in thickness and once with an intracerebral hematoma above the Sylvian fissure. The reflection appeared about 4 cm ahead of the end echo and could be recognized distinctly behind the initial echo when sounding from the side of the hemorrhage (Fig. 134). Because of the exceptionally great distance of this reflection from the end echo — an epidural hematoma 4 cm in thickness was observed only once in an infant — angiography was performed and the correct diagnosis made.

Reflections from the region of the *Sylvian fissure* may also be mistaken for a dura echo. However, there are usually very marked *echo pulsations* originating from the ascending branches of the middle cerebral artery (see also p. 48), which may serve to clarify this finding, since the dura echo does not pulsate at all.

4. In cases of *acute subdural hematomas* one occasionally sees reflections from the surface of hematoma. As mentioned in the discussion on the hematoma echo, however, an echo of low amplitude develops here, whose diagnostic importance is much less than in cases of epidural hematomas. By observing the differences between the two forms of hemorrhage, we suspected that the dura is the source of the hematoma echo in cases of epidural hematomas. Fig. 153 and 154 are echograms of acute subdural hematomas.

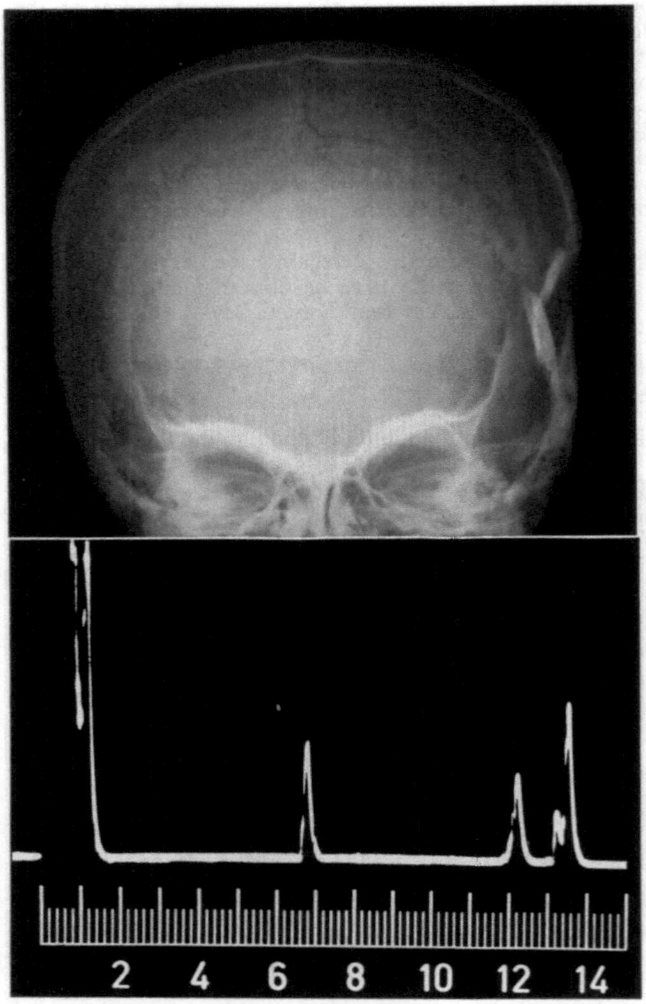

Fig. 136. Echo-encephalogram from the right in the case of a left temporal depressed fracture. High sharp echo ahead of the end echo originates from the depressed bone. Pat. R. M., 9 years of age, Echo-No. 1000/64

5. Reflections developing at the blood-brain interface may also be observed in the temporal area in cases of *intracerebral hematoma.* Usually there is a complex consisting of several reflections of different amplitudes which can hardly be mistaken for the very narrow dura echo. Fig. 156 demonstrates an echogram of a large intracerebral hematoma with a hematoma echo complex.

6. In cases of *contusions* of the temporal lobe, we have been able to observe high reflections ahead of the end echo on sounding from the healthy side in 4 cases. In two patients without a midline echo shift, the origin of this echo could not be ascertained. In the remaining 2 patients (children of the age of 1 and 4 years) there was a distinct shift of the M-echo (6.5 and 8.5 mm) on the echogram. According to the operation and the autopsy, the high additional reflection must have developed in the temporal lobe where the brain was badly crushed (Fig. 135). A similar observation has been reported by Barrows et al. (1968).

7. Reflections resembling the dura echo occur in cases of *depressed fractures* in the temporo-parietal area if the depressed bone fragment is largely perpendicular to the ultrasonic beam. It is possible to mistake it for a dura echo or an end echo. An x-ray in the sagittal direction clarifies the situation (see Fig. 135).

8. The *chronic subdural hematoma* is typically located in the parietal region. A distinct reflection frequently develops at the interface brain/hematoma membrane/blood if the probe is tilted upwards so that the ultrasonic beam hits the accumulation of blood vertically (Brückner, 1963; de Vlieger, 1964). However, the echo of the chronic subdural hematoma barely reaches the height of the dura echo so that a differentiation is always possible when the history is drawn into consideration.

9. In several cases of *tumor cysts,* we recorded an echo from the wall of the fluid-filled hollow space. However, this echo rarely has the steep rise and great amplitude of the dura echo. Sometimes it consists of several peaks. Actually differentiation is possible on the basis of the history alone.

Case Histories

Our experience with echo-encephalography with epidural hematomas is based on a total of *76 cases*, 66 being of traumatic origin and 10 epidural hematomas after surgery. The following histories will illustrate the importance of echo-encephalography in this clinical syndrome.

Case 10: Roland S., Echo-EG No. 1280/64. 7-year-old boy fell from bicycle on August 11, 1964. Apparently no injuries. Several hours later another fall from the bicycle. Immediately unconscious. Admission to hospital. Next day progressive improvement in the level of consciousness. Neurological findings unremarkable. On the morning of August 13, secondary unconsciousness, right pupil becomes dilated and the left side of the body exhibits a hemiparesis. Immediate referral on the suspicion of an intracranial hematoma. On admission around 10 a.m. on August 13, right-sided mydriasis and left-sided hemiparesis as well as bilateral pyramidal signs. Child reacted only slightly to strong pain stimuli. X-ray of the skull: thin fissure in the right temporal bone. The echo-encephalogram (Fig. 138) revealed a midline echo displacement of 10 mm and a hematoma echo 28 mm away from the end echo when sounding from the left. From the side of the hematoma, dura echo could also be demonstrated without difficulty. Thus the diagnosis of an epidural hematoma in the right temporo-parietal area, approximately 3 cm in thickness, was made. Immediate osteoplastic craniotomy on the right (centre: the site where the dura echo had the greatest amplitude). The expected hematoma appeared in the exposed area (Fig. 137); in the centre its thickness reached about 3 cm. The child recovered rapidly after this surgical intervention. 23 days later discharge in good condition. Residual oculomotor paralysis on the right.

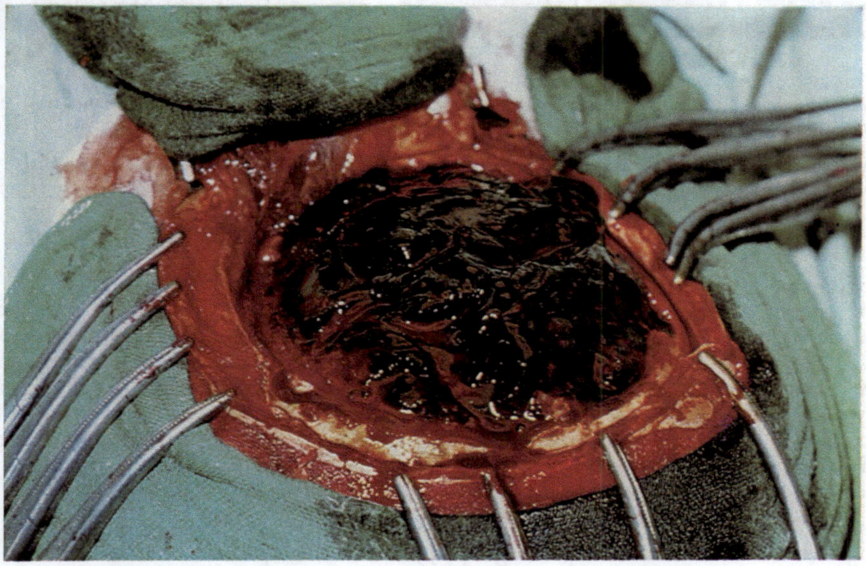

Fig. 137. Right temporal osteoplastic craniotomy over an epidural hematoma localized by means of echo-encephalography alone. Site of operation. See echo-encephalogram in Fig. 138. Pat. R. S., 7 years of age

The case just described here has the typical history and symptomatology of an epidural hematoma. The diagnosis would have been possible even without the echogram. However, the characteristic ultrasonic finding influenced the proceeding insofar that the osteoplastic craniotomy could be performed at the exact site of the hematoma. Angiography was not necessary, as it would not have been able to pinpoint the location any more exactly or determine the type of hematoma and would only have meant an unnecessary loss of time.

Case 11: Erich H., Echo-EG No. 686/63. 8-year-old boy. Fell from a cart on August 28, 1963 while playing. Unconscious for 5 minutes, then vomited. Admitted to hospital because of increasing headaches on August 30, 1968. Since left-sided pyramidal signs developed and short-lived disturbances in consciousness occurred, the child was referred to us. On admission on August 31, 1963, the patient was completely awake, pyramidal signs on the left could no longer be elicited, but the toes of the right foot spread in response to stimulation. Plain films of the skull: no signs of a fracture. Continuous observation of the child. On September 6, 1963 child was allowed to get up for the first time, but was dizzy, confused and remarkably disinterested. EEG: constant focus of delta waves in the right temporo-parieto-occipital region interpreted to be a brain contusion. Echo-encephalogram: surprising midline echo shift 6 mm to the left and high echo which could be recorded from both sides (Fig. 139).

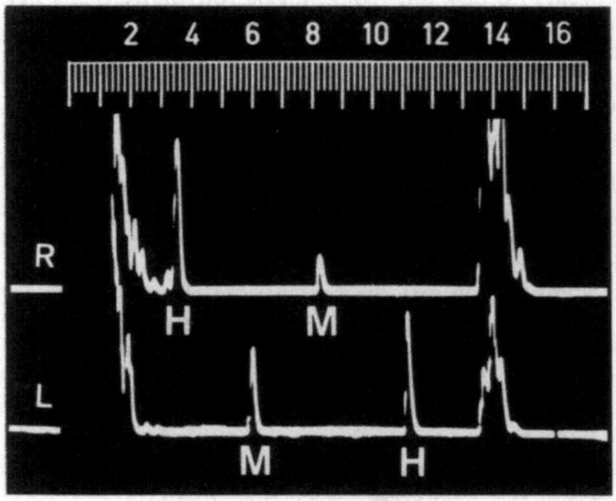

Fig. 138

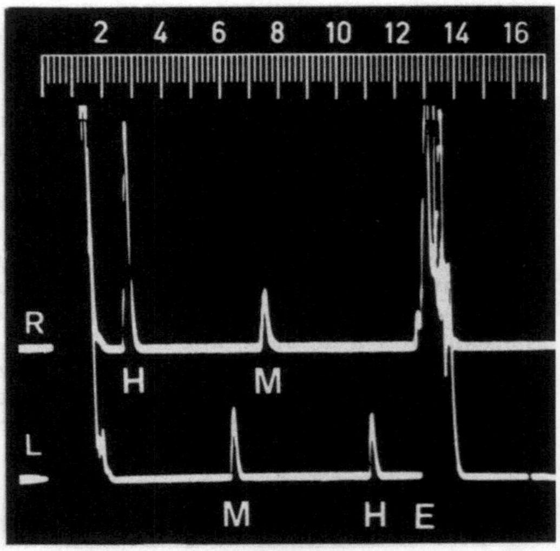

Fig. 139

Fig. 138. Echogram of a temporo-parietal epidural hematoma. Hematoma echo (H) from both sides indicates a 28 mm thick hematoma. Considerable M-echo shift. Pat. R. S., 7 years of age, Echo-EG No. 1280/64

Fig. 139. Echo-encephalogram of a temporo-occipital epidural hematoma. Midline echo shift of 6 mm and hematoma echo (H) which may be recorded from both sides. Further details see text. Pat. E. H., 8 years of age, Echo-EG No. 686/63

This reflection made us think of an epidural hematoma in the right temporo-occipital area with a thickness of 18 mm. In order to verify the presence of the hematoma, carotid angiography was performed on the right (Fig. 140). At first, the AP view of the angiogram did not seem to confirm the hematoma. The cerebral vessels were not forced away from the skull, only the phlebogram showed a small shift of the internal cerebral veins. Solely on account of the echo-encephalographic finding however, we made a series of x-rays with the head turned 30° towards the left (Fig. 141). Through this maneuver, the hematoma which had been diagnosed by echo-encephalography was portrayed on the film. Trepanation on Sept. 10, 1963: the epidural hematoma was partly organized already and surrounded by a hardening capsule. All complaints disappeared after surgery.

Through the use of echo-encephalography, a chronic epidural hematoma was diagnosed in the right temporo-occipital region, although the clinical findings gave no indication of a space-occupying hematoma. Without the echo-encephalogram, the hematoma probably would not have been recognized on the angiogram either. Among our 66 posttraumatic epidural hematomas, there is a total of 4 patients with a hematoma in this location. In all cases the course was subacute or chronic and failed to present secondary disturbances in consciousness.

Case 12: Werner S., Echo-EG No. 1296/64. 17 years of age. Had moped accident on August 18, 1964 around 9 p.m. Was able to go home without support, became confused somewhat later and vomited. Around 11 p.m. completely unconscious. Admission to out-of-town hospital. Because of suspected intracranial hematoma, immediate referral to us. Transport over 70 km. Admission at approximately 1 a.m. on August 19, 1964. Both pupils reacted weakly to light, right pupil a shade larger than the left one, bilateral pyramidal signs. In the echogram midline echo shift 6 mm towards the left (Fig. 142). Scanning from the left parietal region towards the right temporo-basal area (Fig. 142 bottom) revealed a reflection 25 mm ahead of the end echo which could be a hematoma echo.

Immediate craniotomy did not seem to be justified. Angiography on the right: temporo-basal epidural hematoma. During angiography the pupils became fully dilated and failed to react to light. Trepanation confirmed the echo-encephalographic and angiographic findings. The postoperative course was very stormy. After a 2-week phase in decerebrate rigidity, hyperthermia and non-reacting pupils, the state of consciousness cleared slowly but stupor was prolonged for weeks. Only after a 2-month interval was he able to speak a few words again. During this period the patient began to eat alone and to walk with support. Gradual improvement in the course of time, but psychic disturbances necessitated admission to a psychiatric clinic.

The above case demonstrated a very acute course of an epidural hematoma. The clinical findings alone were not enough to diagnose the location of the hematoma. Since we did not have much experience with echo-encephalography in cases of *temporo-basal* hematomas at this time, we did not regard the reflection ahead of the end echo as typical. The angiographic investigation undertaken for this reason meant a considerable loss of time, which proved to be most deleterious. Subsequently similar echo-encephalographic findings at such a late stage (bilateral pyramidal signs!) were sufficient indication for operating without the use of time-consuming dye studies.

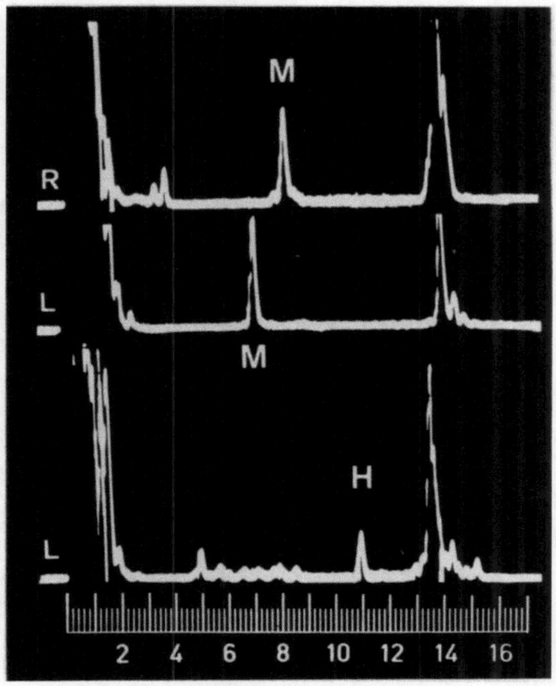

Fig. 142. Echo-encephalogram of a right temporo-basal epidural hematoma. Upper pair of traces: midline echo shift by 6 mm towards the left. Bottom: on sounding from a left parietal probe application point towards the right temporo-basal area, a possible hematoma echo (H) is demonstrated. Pat. W. S., 17 years of age, Echo-EG No. 1296/64

Case 13: Maria B., Echo-EG No. 676/63. The 53-year-old woman on a bicycle collided with a car around 6 a.m. on September 2, 1963. Immediately unconscious, admitted to clinic around 8 a.m. Ominous clinical picture: deeply unconscious, respiration considerably handicapped by marked bronchorrhea. Responded to pain stimuli with decerebrate rigidity, pupils somewhat wider than normal, reacted only slowly to light, equal. Monocle hematoma on the left. Aside from bruises in the shoulder region, no other signs of external injury. In addition, bilateral pyramidal signs. At first a brain stem contusion was suspected. Plain films of the skull: fracture line in the left temporal bone radiating towards the base. Therefore immediate echo-encephalography: M-echo displaced from the left to the right by 9.5 mm; sounding from the right temporal region towards the left parietal area revealed a reflection 28 mm ahead of the end echo which could be a hematoma echo (Fig. 143). Angiography and operation with osteoplastic craniotomy in the left parietal area confirmed the epidural hematoma suspected on echo-encephalography. Despite serious condition, patient recovered relatively quickly. 15 months later still headache and distinct emotional lability, probably due to the accompanying contusion.

Without the knowledge of the echo-encephalographic finding, this patient would not have undergone any further investigations at first, because primary unconsciousness, bilateral decerebrate rigidity and pyramidal tract signs were regarded as the sequelae of severe brain damage ("brain stem contusion"). Epidural hematomas which are masked by coexisting contusion symptoms and which do not run biphasic course, present great diagnostic difficulties. Here echo-encephalography has brought a decisive improvement in the detection and therefore in the prognosis of such multiple injuries.

Case 14: Josef N., Echo-EG No. 1309/64. 55-year-old male epileptic. Fell into a hole in the course of a seizure on the afternoon of September 11, 1964. Left parieto-occipital bruise. Awake shortly after, 2 hours later complained of headache and vomited. Admitted to out-of-town hospital because of further disturbances in consciousness, immediate referral because of a suspected epidural hematoma. Transport over 150 km. On admission patient deeply unconscious, no reflexes, no reaction;

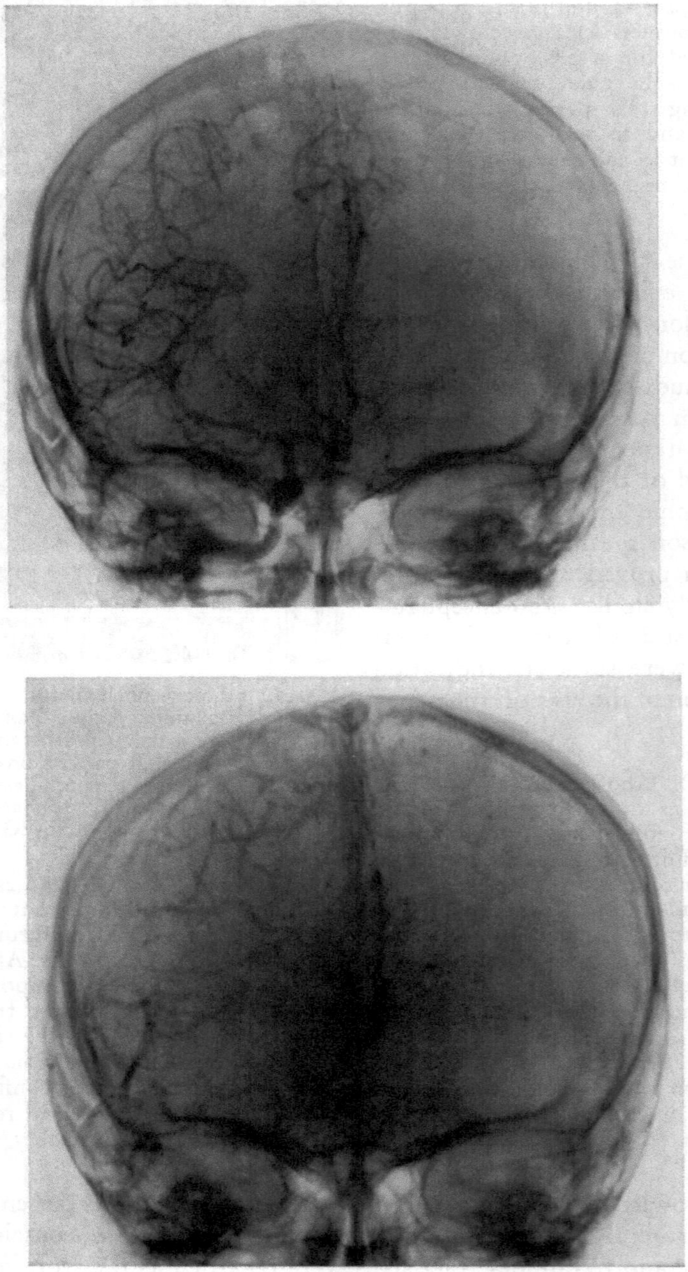

Fig. 140. Arteriogram and phlebogram in an AP view in a patient whose right temporo-occipital epidural hematoma was diagnosed by echo-encephalography. Only slight displacement of the internal veins, no avascular area in the temporal region. Pat. E. H., 8 years of age (see case 11)

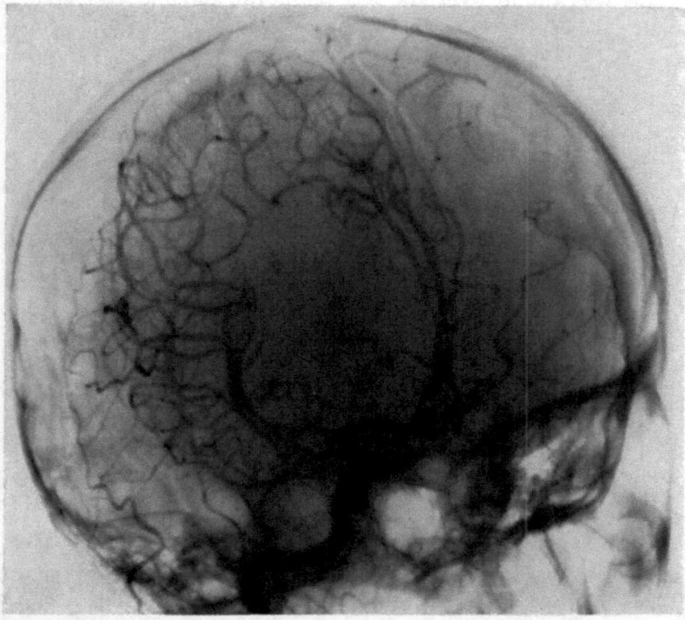

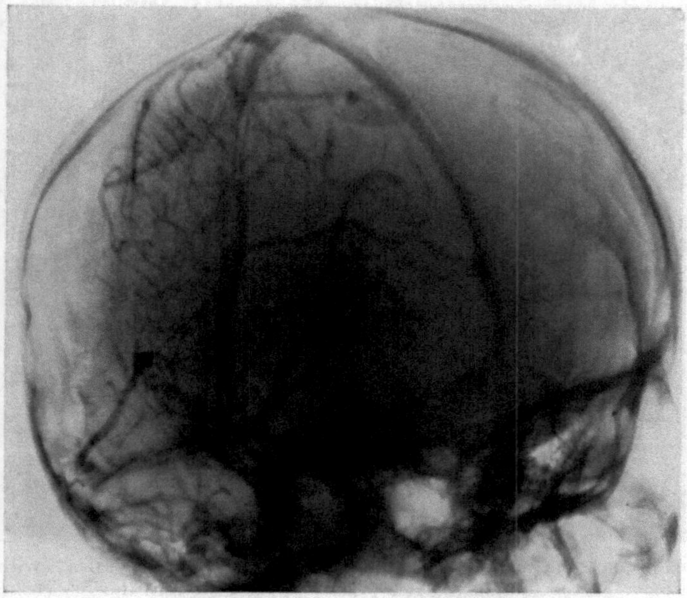

Fig. 141. Arteriogram and phlebogram with the head turned 30° towards the left in the same patient as in Fig. 140. Now one clearly sees the avascular space over the right temporo-occipital region. Confirmation of the echo-encephalographic findings in Fig. 139

pupils equal, narrow, no reaction to light. Deep regular respiration. Echogram: mass shift 7.5 mm from left to right and hematoma echo 27 mm ahead of the left end echo. Since this reflection was highest towards the parietal area (Fig. 144), trepanation of the left parietal bone was performed immediately — without preceding x-rays: a large epidural hematoma up to 3 cm in thickness, containing about 150 cm³ was found over the left temporal and parietal area. The hematoma was

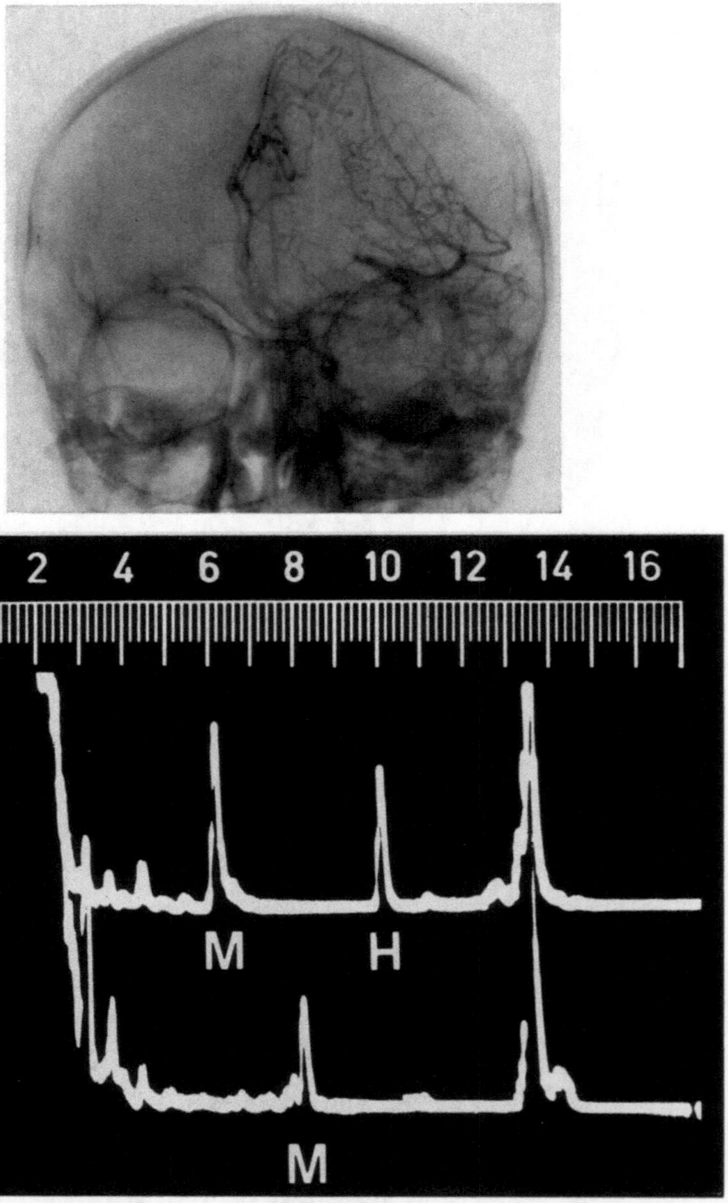

Fig. 143. Echo-encephalogram and carotid angiogram of a 53-year-old patient with a parietal epidural hematoma. Further details see text (Case 13). Pat. M. B., Echo-EG No. 676/63

evacuated within half an hour after admission to hospital. Nevertheless, the patient failed to recover and died on the 11th postoperative day. Autopsy: no signs of increased intracranial pressure, operative area unremarkable. We must assume that prolonged brain stem compression with tentorial herniation resulted in irreversible changes in a brain which was already damaged by epileptic seizures.

In the 55-year-old patient we found an epidural hematoma over the left hemisphere which ran a classical course. Despite immediate trepanation directly above the hematoma, on the basis of the echo-encephalographic finding, it was impossible to compensate the great time loss due to the long transport. Therefore it would be very worthwhile for small hospitals with surgical departments to become familiar with echo-encephalography in order to be able to trepanate immediately at the right place in emergency situations.

Case 15: Renate D., Echo-EG No. 866/63. 16-year-old girl had motorcycle accident on November 20, 1963 around 11 p.m. Said to have been briefly unconscious, but on admission to out-of-town hospital she responded and vomited. Plain films of skull: diastasis of coronal suture. Next morning unconsciousness and motor restlessness. Referral to our clinic. On admission (about 12 hours after the accident) patient deeply unconscious, very restless and responded to painful stimuli with decerebrate rigidity. Pupils initially equal, wider than normal and weakly responsive to light; a short timer later

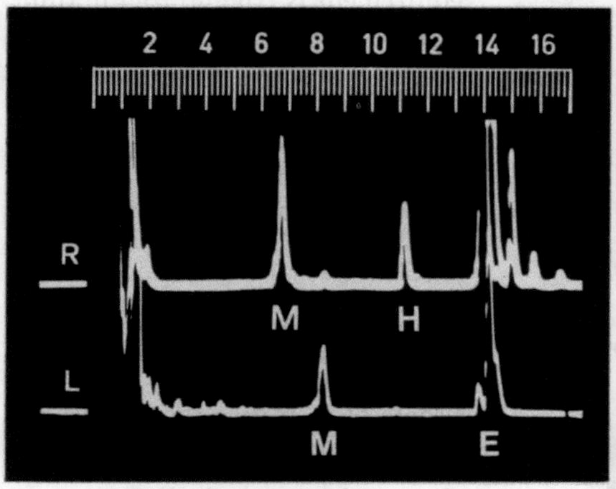

Fig. 144

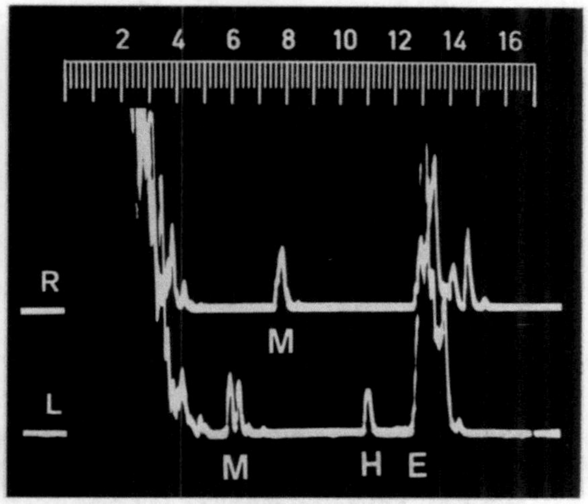

Fig. 145

Fig. 144. Echo-encephalogram of a left parietal epidural hematoma. The amplitude of the hematoma echo (H) was greatest when when the ultrasonic beam was directed towards the left parietal area. In this way the exact localization of the hemorrhage was successfully found. Pat. J. N., 55 years of age. Echo-EG No. 1309/64

Fig. 145. Echo-encephalogram of a right parasagittal epidural hematoma. For further details see text. Pat. R. D., Echo-EG No. 866/63

the *left pupil was completely dilated.* Babinski positive bilaterally, clonic reflexes of the lower extremities. The echogram revealed a *7.5 mm displacement from right to left.* On sounding from the left temporal region towards the right precentral area, a reflection resembling a hematoma echo was noted 18 mm away from the end echo (Fig. 145). Diagnosis: *right* precentral epidural hematoma despite a left-sided dilated pupil which caused the orderly to place the patient on the wrong side in the belief that the surgeon had made a mistake. On osteoplastic craniotomy in the right precentral area, a hematoma was evacuated whose thickness reached 2 cm. On the following day the patient was able to speak already. Discharge on the 16th postoperative day in good condition.

In this case echo-encephalography made possible the rapid localization of the hematoma which certainly contributed to the good course. The above case illustrates *the unreliabiliy of the clinical findings in the determination of the side on which the hematoma is located. Here echo-encephalography is undoubtedly superior.*

Case 16: Johann B., Echo-EG No. 1435/64. On the morning of November 29, 1964 around 3 a.m. the 45-year-old patient was found lying on the floor of his home by his wife. At first he could still babble, but within a few minutes he was completely unconscious. In the hospital, a monocle hematoma and increasing paresis of the right side of the body were noted.

No inequality of pupils. X-rays did not permit a definite evaluation (marked motor restlessness). In the next hour no special change; referral on the same day around 11 a.m. Neurological findings still unchanged (right-sided hemiparesis and deep unconsciousness). Pupils were the size of a pinhead, no reaction to light. Initially a hypertensive intracerebral hematoma was suspected since there was high blood pressure in the history. A space-occupying lesion was supported by the 6.5 mm displacement of the midline structures to the right in the echo-encephalogram. There was no hematoma echo. Immediate left-sided angiography: large space-occupying lesion in the left frontal area; in connection with a now distinctly visible fracture line in the left frontal bone, the diagnosis of an epidural hematoma could be made (Fig. 146). Osteoplastic craniotomy: epidural hematoma of almost goose-egg size over the left frontal pole extending as far as the superior sagittal sinus. Within a few days the patient recovered completely and was able to leave his bed on December 12, 1964.

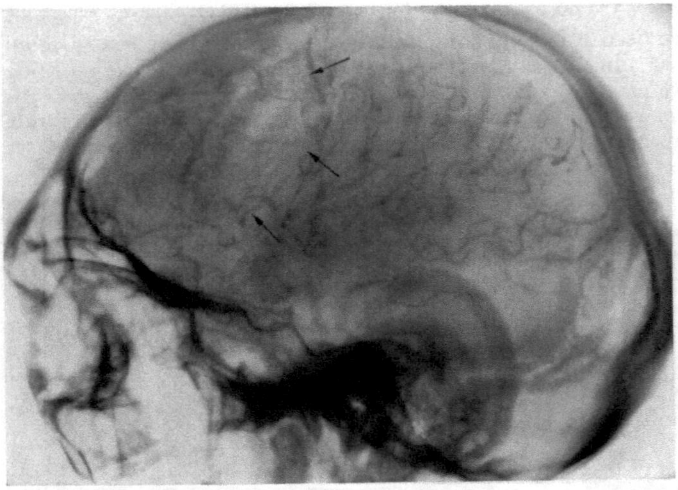

Fig. 146. Carotid angiogram (late arterial phase) in a frontal epidural hematoma. Surprising avascularity in the frontal area (arrows). Pat. J. B., 45 years of age

This history shows, that angiography is absolutely necessary for the exclusion of an atypically located space-occupying hematoma, if a hematoma echo is absent. This requirement is true up to a midline echo displacement of 7.5 mm. All space-occupying hematomas causing a shift of around 8 mm and more on the other hand, may be evacuated from a large temporoparietal craniotomy (also p. 161).

Case 17: Johann L., Echo-EG No. 962/64. 58-year-old man was thrown to the ground in the course of a fight on December 3, 1963. On December 10, 1963 admitted to hospital because of increasing somnolence and beginning paralysis on the right side. Variable course in the next few weeks. The consultant neurologist suspected a subdural hematoma over the left hemisphere. Referral to our clinic on January 3, 1964. Patient responded but was distinctly slowed and disoriented; the right pupil was mydriatic with a slow reaction to light. On the right side, generalized paresis with increased reflexes and positive pyramidal tract signs. Plain films of the skull: fracture line in the *right occipital bone*. Echogram: midline echo displacement 3 mm towards the right, only slight mass shift. At first no hematoma echo (Fig. 147 a). Left-sided carotid angiography showed a vessel running a convex course medially in the venous phase despite the absence of a shift in the anterior cerebral artery on the AP view; this was possibly the inner delineation of an occipital hematoma (Fig. 148 a). In order to verify the diagnosis, a further series of angiograms was made with the head rotated towards the right: distinct displacement of the vessels away from the inner table of the skull (Fig. 148 b). On the basis of the history, an epidural hematoma seemed the most likely diagnosis. Following angiography, an attempt was made to obtain a reflection from the displaced dura by using a more frontally situated probe application point (ultrasonic beam directed towards the occipital region). And we were actually successful in demonstrating a reflection 24 mm ahead of the end echo by sounding from the right fronto-temporal region, while sounding from the left in the

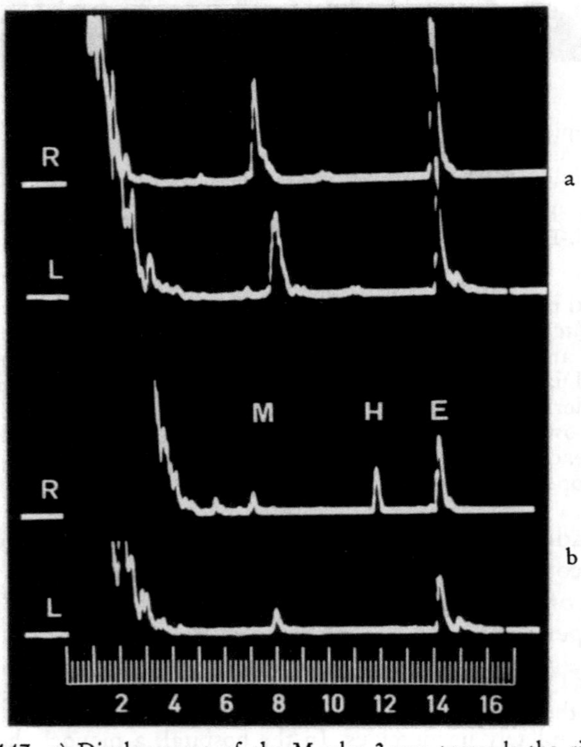

Fig. 147. a) Displacement of the M-echo 3 mm towards the right in an occipital epidural hematoma on the left. b) A hematoma echo (H) may be obtained only if the ultrasonic beam is directed towards the occipital area. Pat. J. L., 58 years of age, Echo-EG No. 962/64

same direction failed to show any additional reflections (Fig. 147 b). At operation, an encapsulated epidural hematoma, 2.5 cm in thickness, was found. The neurological deficits disappeared after surgery.

This last case shows that it is possible to make the diagnosis on the basis of the echo-encephalographic findings alone even with unilateral occipital epidural hematomas, if this localization is thought of and a suitable technique of investigation is used. In the meantime two further occipital hematomas have been detected by echo-encephalography in this manner.

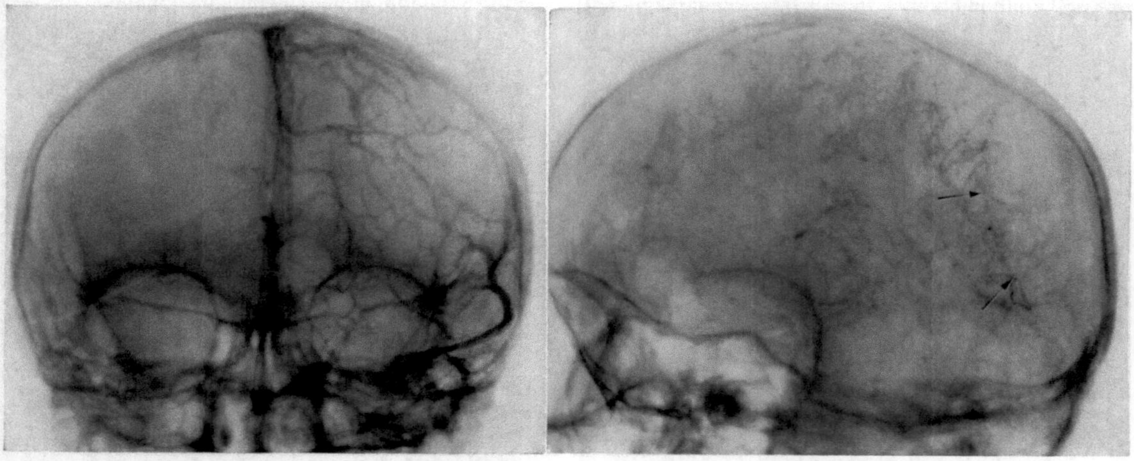

a b

Fig. 148. Angiograms of the same patient as in Fig. 147. a) AP view of the venous phase reveals only a vessel running medially in a convex arch, but no avascular zone. b) Only the diagonal views reveal an epidural hematoma (arrows). Pat. J. L., 58 years of age

Case 18: Ralf-Dieter F., Echo-EG No. 3334/67. On December 14, 1967 at about 11 a.m., the 5-month-old infant fell out of his basket on to the floor. Increasing pallor and apathy during the following hour. Referral to a pediatric clinic. On admission there at 2 p.m., the child was deathly pale (Fig. 149) and showed no obvious spontaneous respiration. Artificial respiration led to a certain

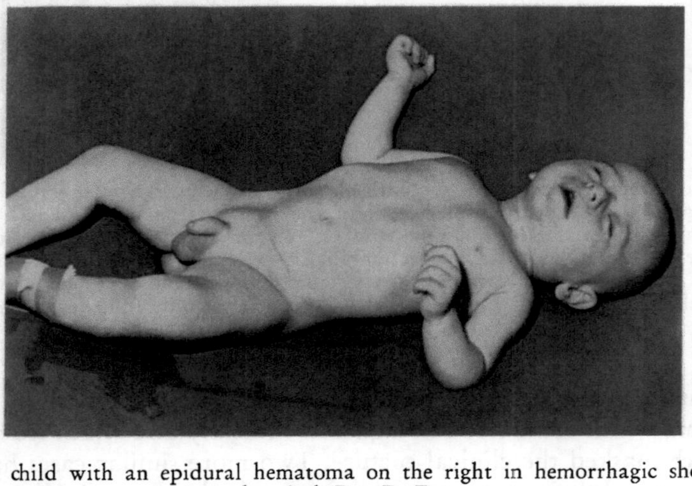

Fig. 149. 6-month-old child with an epidural hematoma on the right in hemorrhagic shock on admission to hospital. Pat. R. F.

improvement and the child moved arms and legs spontaneously. Neurological examination unremarkable. Clinical picture of a *hemorrhagic shock*. Peripheral pulse not palpable, hemoglobin 10 g%. Consultant examination by us: fontanelles under marked tension, slight swelling of the scalp behind the right ear. Echo-encephalogram: displacement of the midline echo 11.0 mm towards the left (Fig. 150 a). On directing the ultrasonic beam from the left temporal area towards the right parietal area, a high dura echo appeared 35 mm ahead of the end echo (Fig. 150 b). The same echo could be registered when the probe was applied in the right parietal area (Fig. 150 c). Immediate admission of the child to our service and osteoplastic craniotomy on the assumption of an epidural hematoma in the right temporo-parietal area. The hematoma thickness measured at operation was exactly the same as the ultrasonic finding (Fig. 150 d). Normalization of the midline echo within 3 days of surgery. Complete recovery.

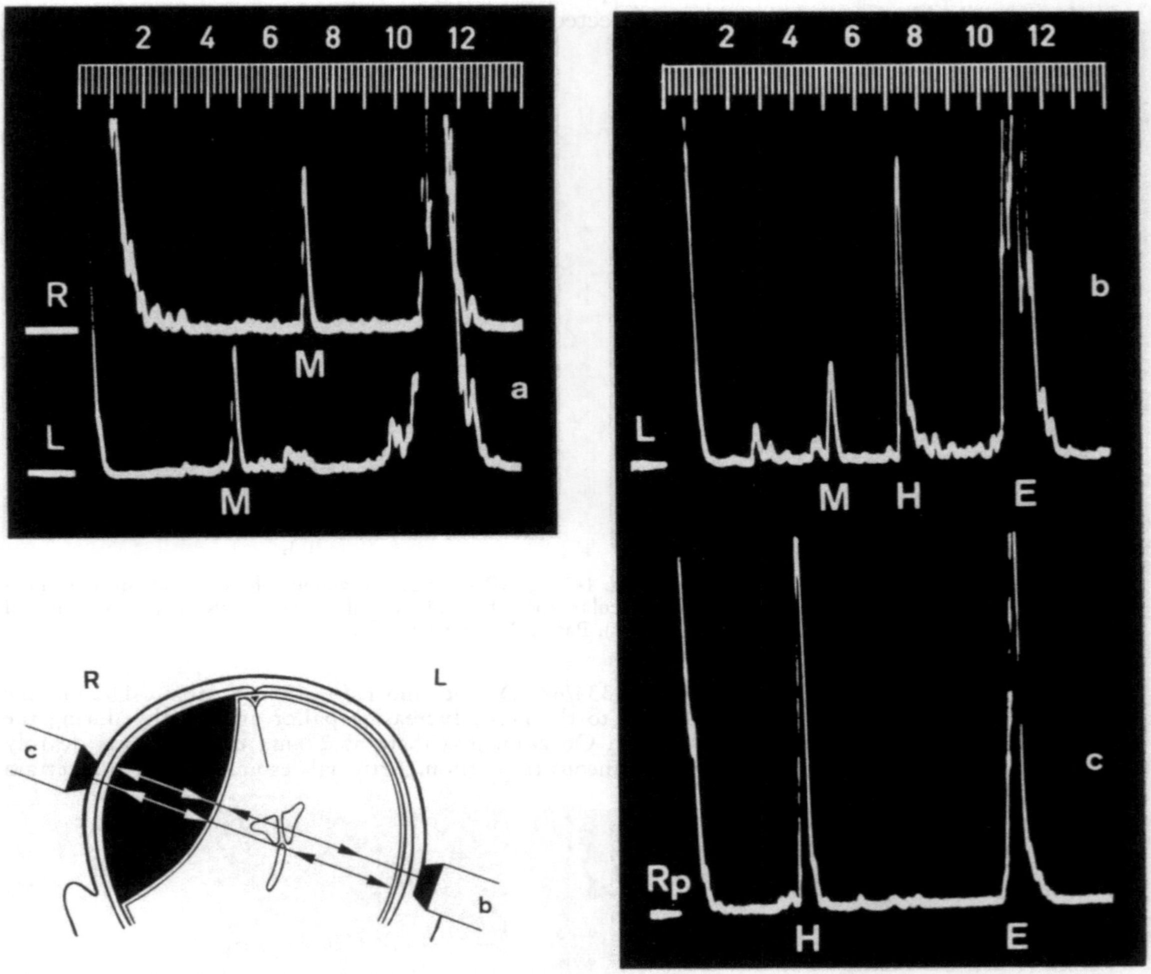

Fig. 150 a to d. Echo-encephalograms of a 6-month-old child in hemorrhagic shock following minor trauma (same case as in Fig. 149). a) 11 mm displacement of the M-echo from right to left. b) Record from left temporal to right parietal area. High hematoma echo (H) 35 mm ahead of the end echo. c) On applying the probe directly above the hematoma, a high reflection appears 40 mm from the probe. The diagram on the left shows the sounding technique in this case. Echo-encephalographic diagnosis: 3.5 cm thick epidural hematoma in the right parietal area; this was confirmed at the operation which followed immediately (Fig. 150 d). Pat. R. F., Echo-EG No. 1334/67

This case illustrates the value of echo-encephalography in an infant, who had a large epidural hematoma which could not be diagnosed clinically and which led to a severe hemorrhagic shock which masked the clinical picture. Two more similar cases have been observed in the last few years and their course was also favorable thanks to the immediate use of echo-encephalography. Pia and Geletneky (1968) also report such observations. In 1951 Campbell and Cohen pointed out that in infants with a history of trauma and subsequent anemia, an epidural hematoma must be assumed when other sources of bleeding have been eliminated. Even if the typical clinical symptoms are absent, echo-encephalography today enables us to trepanate at the proper site in all cases.

Case 19: Wolfgang G., Echo-EG No. 3509/68. On March 7, 1968 this 12-year-old boy fell during gymnastics and suffered a forearm fracture. The fracture was reduced under anaesthesia in an out-of-town clinic. Following anaesthesia, the boy failed to awaken fully. During the subsequent night, he had clonic seizures of the extremities, especially on the left. Further investigation gave rise to the suspicion of a hematoma in the left temporo-basal area and the patient was immediately transferred to our service. On admission, snoring respiration, bilateral non-reacting pupils, decerebrate rigidity, bilateral positive Babinski. Small subgaleal hematoma on the left forehead. The technically inadequate angiograms which were sent with the patient seemed rather to indicate a frontal space-occupying lesion. No fracture lines could be recognized. In the echo-encephalogram which was performed immediately, the midline echo was at the normal site (Fig. 151 a). On sounding from the front, a high echo was noted at a depth of 44 mm on the left; such an echo was absent on the right. We interpreted this reflection (Fig. 151 b) to be a dura echo and suspected a large left-sided epidural hematoma over the frontal pole. Immediate trepanation confirmed our suspicion. The hematoma had maximum thickness of 4 cm and extended far towards the base. After surgery, the boy recovered completely and could be discharged home on March 29, 1968 completely free of complaints.

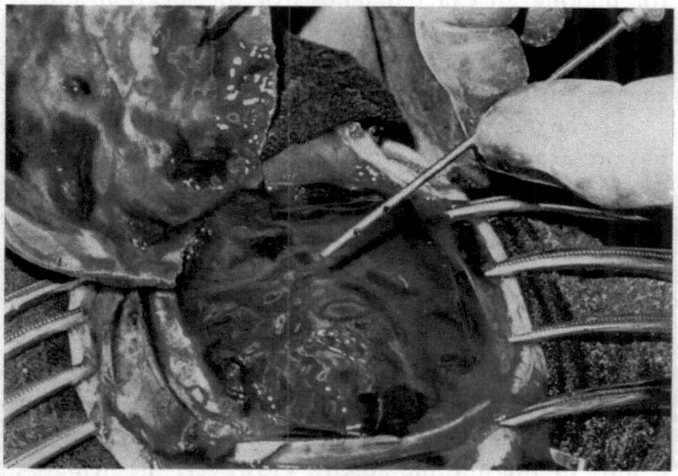

Fig. 150 d. Site of operation in the same child as in Fig. 149 and 150 a—c. The hematoma thickness measured at operation was 3.6 cm

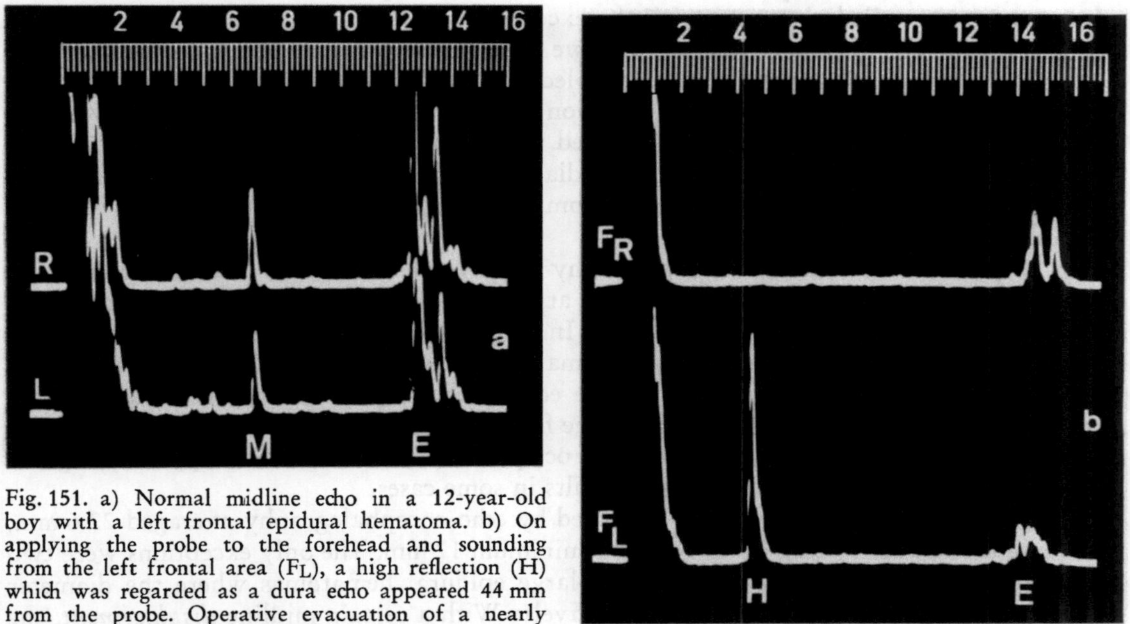

Fig. 151. a) Normal midline echo in a 12-year-old boy with a left frontal epidural hematoma. b) On applying the probe to the forehead and sounding from the left frontal area (F_L), a high reflection (H) which was regarded as a dura echo appeared 44 mm from the probe. Operative evacuation of a nearly 4 cm thick epidural hematoma over the left frontal pole. Pat. W. G., Echo-EG No. 3509/68

Some frontal epidural hematomas cause no displacement of the midline structures and therefore remain unrecognized if the investigation is limited to bitemporal sounding. On applying the probe to the forehead, a high echo can be recorded from the displaced dura and appears just behind the initial echo. The prerequisite for this is, however, that the skull bone is no thicker than 6 to 7 mm since one cannot obtain a useful reflection otherwise.

Discussion of the Echo-Encephalographic Results with Epidural Hematomas

In a large proportion of epidural hematomas, the echo-encephalogram alone is adequate for making an exact diagnosis of the location and type of the hematoma. The important criteria are:

1. the registration of a midline echo shift which indicates the side on which the hemorrhage is situated and

2. the demonstration of a so-called hematoma echo which originates from the dura which has been forced away from the inner table of the skull (see p. 73 f.).

In 72 of the 76 epidural hematomas, a *distinct displacement of the midline echo* (95%), averaging 7.2 mm was seen. JEPPSSON (1961) reports a mean value of 6.2 mm, LITHANDER (1961) 7 mm. The greatest average midline echo shifts occurred with acute epidural hematomas in the temporal and parietal region (9.0 and 7.7 mm). AMBROSE (1964)[1], GELETNEKY (1964)[1] and FISCHGOLD (1968) have reported that some frontal epidural hematomas fail to produce an echo-encephalographically recognizable mass shift and that they are missed for this reason. However, no less than 6 of our 7 patients with frontal epidural hematomas revealed a distinct midline echo shift whose average was 5.2 mm. In this way the side of the space-occupying lesion could be faultlessly determined in these cases. Only in one patient was a shift of the midline absent. However, the hematoma was diagnosed here from a frontal probe application point by an unusually high hematoma echo (dura echo) (Fig. 151 b). Chronic hemorrhages (10 patients, the interval between trauma and diagnosis being 8 days and longer) showed an average M-echo displacement of only 4.8 mm, in contrast to 7,9 mm in cases of acute and subacute hematomas.

In 59 of our 76 epidural hematomas, a so-called *hematoma echo* was obtained (78%). These were all the temporal, parietal, temporo-precentral and temporo-occipital hematomas as well as some of the temporo-basal ones. In frontal, occipital and coronary hematomas, a hematoma echo failed to appear with the exception of the cases just described. Twice the hematoma echo was questionable and twice we found the reflection only after angiography had been performed. Thus the dura echo enabled us to make the diagnosis in 55 cases (72%). Comparing this result with the differentiation which is possible with angiography, echo-encephalography proves to be very good indeed. According to TÖNNIS, FRIEDMANN, SCHMIDT-WITTKAMP and WALTER (1963) the correct diagnosis of the type of hematoma which was later confirmed by operation, was made from the angiogram in only 31 of 40 epidural hematomas (77.5%).

In 37 particularly acute cases, angiography was avoided and yet it was still possible to perform the osteoplastic craniotomy directly at the site of the hematoma on account of the echo-encephalographic findings alone (49%). In the absence of definite signs of a hematoma in the echogram (midline echo shift + hematoma echo), carotid angiography is indispensable, if the other findings in connection with the echographic shift do not permit an accurate localization of the site. This is particularly true for frontal and occipital epidural hematomas. For the demonstration of hematomas over the occipital lobe, a more frontal application point of the probe may improve the diagnostic results in some cases.

The *thickness of the hematoma* as measured by echo-encephalography averaged 22.6 mm, the maximum in adults being 32 mm and the minimum 12 mm. The only exceptions were two 5- and 6-month-old infants with unusually large epidural hematomas where the diameter of the hemorrhage was 36 and 40 mm respectively. With a certain midline *displacement*, the expected hematoma thickness in cases of epidural hematoma is equal to three times the thickness of the midline echo shift.

According to the literature, the *mortality* of the epidural hematoma until now has always been disproportionately high. This is due to concomitant brain injuries on the one hand, and on the time of operative evacuation on the other. The use of echo-encephalography markedly improves the prognosis of the epidural hematoma by accelerating the diagnostic procedure.

[1] personal communication

Table 17 shows the *complete change in the recognition of epidural hematomas* which is accompanied by a considerable lowering of the mortality in the course of the last few years since the introduction of echo-encephalography. Subtracting the 10 postoperative epidural hematomas, all of whom survived, only 11 of the remaining 66 posttraumatic cases which

Table 17. *Diagnostics and results of treatment with 103 extradural hematomas in the years 1958 to 1968* (Neurosurgical Clinic of the University of Erlangen)

Year	No. of Cases	Angiography	Diagnosed by Echo-Encephalography and Angiography	Echo-Encephalography	Died
1958—1962	27	21 [a]	—	—	12 (44%)
1963—1964	36	—	23	13	7 (19%)
1965—1968	40 [b]	2	12	24	4 (10%)
Total	103 [b]	23	35	37	23 (22.3%)

[a] The remaining 6 patients were trepanated on the basis of the clinical findings alone without angiography.
[b] The horizontal addition results in 101 patients only. Two cases were diagnosed at operation or autopsy. First case: Midline echo normal. On lifting a depressed fracture in the parietal region, an extradural hematoma 1.5 cm in thickness was found. Second case: Midline echo displaced by 5.0 mm. No hematoma echo. Further diagnostic procedures were avoided because of multiple serious injuries. Autopsy revealed a frontal extradural hematoma 3 cm in thickness.

were treated in our clinic since the use of echo-encephalography, have died. Even in patients with bilateral decerebrate rigidity, immediate directed trepanation led to a relatively high percentage of good results. Thus 6 of 8 patients with such complications have survived in the last two years; one even had a respiratory arrest just before operation and still recovered.

Of the 11 deceased patients, 6 had a free interval. However, some of them were transported as much as 100 kilometers after the diagnosis was suspected and thus had a very poor prognosis prior to operation (bilateral dilated pupils which failed to react to light, no reflexes, respiratory disturbances, decerebrate rigidity). 4 patients had such a serious additional brain injury that the fatal outcome must be attributed to it. One 68-year-old patient had bronchitis and emphysema as well. The two patients had suffered thoracic injuries in addition to the epidural hematoma which led to irreversible brain damage through hypoxia. Assuming ideal diagnostic and operative facilities as well as a well-organized staff, it should be possible to lower the mortality of the epidural hematoma to about 10% (see also Böhmer and Carlsson, 1964).

With the help of echo-encephalography a decrease in the midline displacement i. e. in the M-echo shift can be observed after the hematoma is evacuated. In cases of epidural hematomas the midline echo shift usually normalizes within a few days; in our experience the amount of shift is already less, shortly after the hematoma is evacuated. The complete normalization usually takes from 8 to 14 days. If brain contusion is also present however, the return of the M-echo to the midline can take up to 4 weeks. The data in the literature on this topic are very variable. Brückner (1965) reported on a nearly complete or complete normalization of the echo-encephalogram within 12 to 24 hours after surgery. Hotermans and Thiry (1967) on the other hand, have observed that the midline echo shift remains unchanged for the first 5 to 7 days and then slowly decreases within 12 to 14 days. Fischgold et al. (1968) report a time period of 15 to 20 days on the average, but in one case they even found a time period of 50 days before the midline echo returned to normal. According to Ito et al. (1968) the time interval between hematoma evacuation and return of the midline echo to the original site is 40 days.

The differences in the reports are probably due to the fact that the time interval in which the midline echo returns to the original site is dependent on many factors (age of the patient, extent of the preoperative midline echo shift, site and size of the hematoma, concomitant contusion). Therefore it is impossible to establish a firm rule.

b) Acute and Subacute Subdural Hematoma

The echo-encephalographic findings in cases of acute subdural hematomas are not as diagnostically conclusive as those of epidural hematomas, because a characteristic hematoma echo is missing (see p. 74). The reflection developing at the interface between the surface of the brain and the blood accumulation has a much smaller amplitude than the dura echo (see Fig. 73 b and 74). Of the 82 acute subdural hematomas which we examined by echo-encephalography, 80 had a pathological ultrasonic trace which enabled us to establish the diagnosis of an intracranial space-occupying lesion (see also Table 15). The average midline echo displacement was 7.2 mm which is the same as that produced by epidural hematomas. Cases in which the mass shift was less than 6.0 mm revealed pancake hematomas with a thickness of only 4.0 to 8.0 mm or were situated in the frontal or precentral region. Cases running a very acute course usually led to M-echo displacements of 8.0 to 11.0 mm. Only in one case was the existing tissue displacement not indicated in the echo-encephalogram. While the angiogram showed a displacement of the internal cerebral vein of barely 1 cm, the echogram was exactly midline even at the control examination following angiography. We could find no explanation for this error in measurement or more correctly, negative record. In one patient with an acute subdural hematoma, 1 cm in thickness, a midline echo shift was absent. The angiogram also did not reveal a marked displacement of the vessels towards the opposite side.

In 38 cases (46%), an additional reflection was found ahead of the end echo on sounding from the side opposite to the hematoma; this reflection had an amplitude of only a few millimeters and its distance from the end echo averaged 12.6 mm. A demonstration of the hematoma echo from the side of the lesion is not possible in the case of an acute subdural hematoma in our experience. Usually the initial echo is prolonged; in a few cases we can barely record the midline echo from the side of the hematoma (see also Dreese, Hayes, Kempe, and McHenry, 1966; Jefferson and Hill, 1966).

A hematoma echo is more frequently found in subacute forms than in acute ones. During operation it is apparent that the sulci and gyri are still distinctly visible in cases of acute subdural hematomas (see Fig. 152) while subacute forms result in progressive flattening of the brain surface. Naturally the reflection conditions for ultrasound are altered by this.

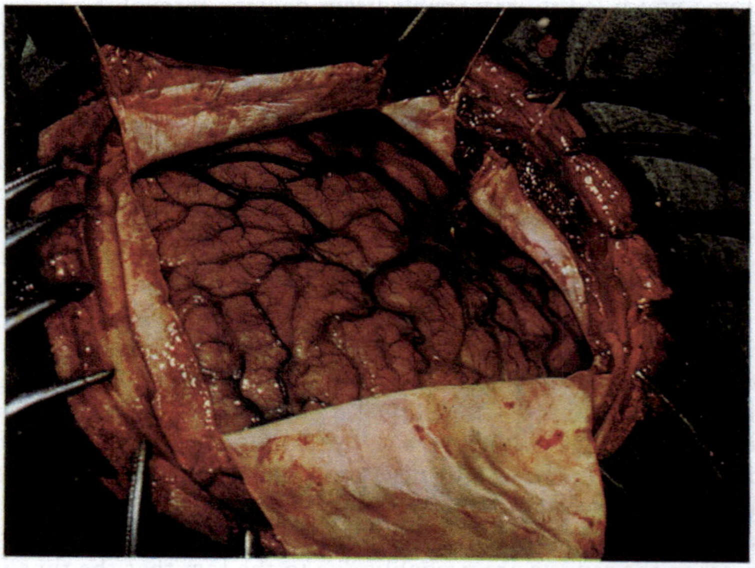

Fig. 152. Operative site after evacuation of a hematoma localized by echo-encephalography in the left parasagittal area (see echogram of the same patient in Fig. 154). Despite the fact that the brain was compressed by the hematoma the gyri and sulci are still distinctly visible. This results in poor reflection conditions at the hematoma interface. Pat. H. B., 60 years of age

Increasing experience enabled us to determine better the type of hematoma from the echo-encephalogram directly. However, a longer period of experience is necessary for as we have repeatedly indicated, the hematoma echo is not nearly as characteristic as that of the epidural hematoma. Nevertheless in 29 cases of acute hematoma, the echo-encephalographic findings were so unequivocal that we were able to do without dye studies of any kind (35%). In the remaining 53 patients, carotid angiography was performed to verify the diagnosis.

In acute subdural hematomas, the echo-encephalographic follow-up study proved to be very valuable. Again and again the echo-encephalogram was found to be normal when the traumatized patient was admitted to hospital: half an hour later, a displacement of 4.0 to 6.0 mm appeared and in very acute cases, we have even seen a midline displacement of 10 mm develop within an hour. *A midline echogram must never mislead us to stop checking the patient!*

Case Reports

Two histories of patients with acute subdural hematomas will illustrate the use of echo-encephalography in this type of hematoma.

Case 20: Luise H., Echo-EG No. 1932/65. The 37-year-old patient was admitted to an out-of-town hospital on the morning of October 10, 1965 in an unconscious state. No history of trauma; after a night spent carousing the patient could not be awakened the following morning. After a coma due to a metabolic disorder or poisoning was excluded and a left-sided hemiparesis developed, the patient was referred to us on November 1, 1965 with the tentative diagnosis of *right-sided cerebral hemorrhage*. Findings on admission: no external signs of injury. Deeply unconscious, snoring respiration; defensive movements with the right arm and leg in response to painful stimuli, beginning left-sided convulsions. Pupils slightly unequal, a shade wider on the left than on the right, diverging bulbi. Reflexes brisk, equal on both sides, Babinski bilaterally positive. Some neck stiffness, CSF slightly hemorrhagic. Echo-encephalogram: 7.0 mm displacement of the midline echo from left to right (Fig. 153 b). Directing the ultrasonic beam towards the left parietal area revealed a small reflection, possibly indicative of a subdural hematoma, 11 to 12 mm ahead of the end echo (Fig 153 a). Because of the unclear history and the homolateral paresis, carotid angiography was performed on the left in order to be certain; this confirmed the presence of an *acute subdural hematoma over the left hemisphere* (see Fig. 153 c). Evacuation of the hematoma through osteoplastic craniotomy. It originated from a small contusion in the left parietal area where a cortical artery had been severed. Slow postoperative convalescence and disappearance of left-sided hemiparesis. Rehabilitation treatment in a neurological clinic because of Parkinsonism.

While the clinical finding of a left-sided hemiparesis seemed to speak for a space-occupying lesion in the right cerebral hemisphere, echo-encephalography made it clear immediately that this was a homolateral paresis. An acute subdural hematoma was suspected right away because of the small reflection which was probably a hematoma echo. In view of the unclear history however, left-sided carotid angiography was performed to be safe.

Case 21: Hans B., Echo-EG No. 1965/65. The 60-year-old man was found unconscious in the hallway of his home around 6 a.m. in the morning on November 16, 1965 and was subsequently brought to us. He had probably fallen around 11 p.m. the previous evening already. Findings on admission: deeply unconscious, snoring respiration, considerable hypothermia (rectal temperature 28° C). Weak defensive movements of the right-sided extremities in reponse to painful stimuli. Left pupil dilated, does not react to light, right pupil of normal size, also fails to react to light. No muscular reflexes, no pyramidal tract signs. No external signs of injury. Echo-encephalogram: massive displacement from left to right by 10.0 mm (Fig. 154). On directing the ultrasonic beam from the right towards the left precentral area, a low amplitude reflection appears 23 mm ahead of the end echo; in connection with the clinical findings, this made us suspect an extensive acute subdural hematoma in the left precentral area (Fig. 154 top). Osteoplastic craniotomy performed in the left temporo-precentral area without further diagnostic procedures. Evacuation of an acute subdural hematoma maximally 2.5 cm in thickness, which originated from a torn ascendant vein in the area of the Sylvian fissure (see Fig. 152). Postoperatively the level of consciousness improved within the first week. After a variable course, death due to circulatory failure supervened on December 12, 1965.

The echo-encephalographic finding in this case enabled us to make the full diagnosis so that dye studies could be avoided. Osteoplastic craniotomy was performed at the site of the hematoma echo. In this manner it is usually possible to gain about half an hour which can be of life and death importance in a patient with an acute intracranial hematoma. However, the 60-year-old patient was no longer in a position to compensate the damage due to brain stem compression and hypothermia.

Postoperative examinations show that the midline echo shift remains unchanged for some time after the evacuation of the hematoma and decreases only very slowly. In our experience at least 4 to 6 weeks are necessary before the echogram becomes normal (see also Fig. 155).

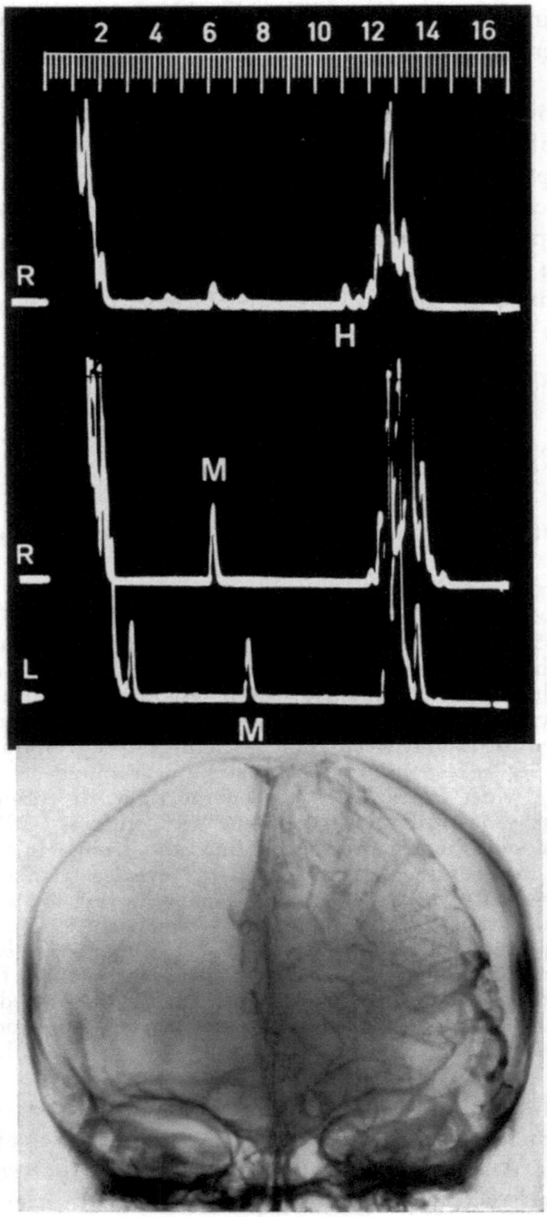

a

b

c

Fig. 153

Fig. 153 a to c. Echo-encephalograms and angiogram of a 37-year-old patient with an acute subdural hematoma over the left hemisphere. For further details see text. Pat. L. H., Echo-EG No. 1932/65

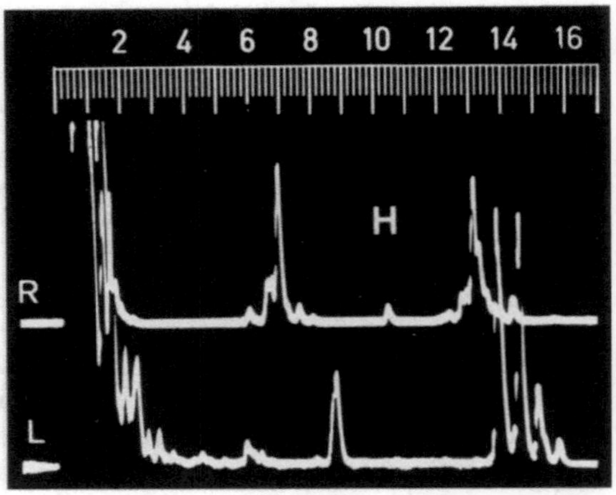

Fig. 154

Fig. 154. Echo-encephalogram of a 60-year-old man with a left precentral acute subdural hematoma. Pat. H. B., Echo-EG No. 1965/65

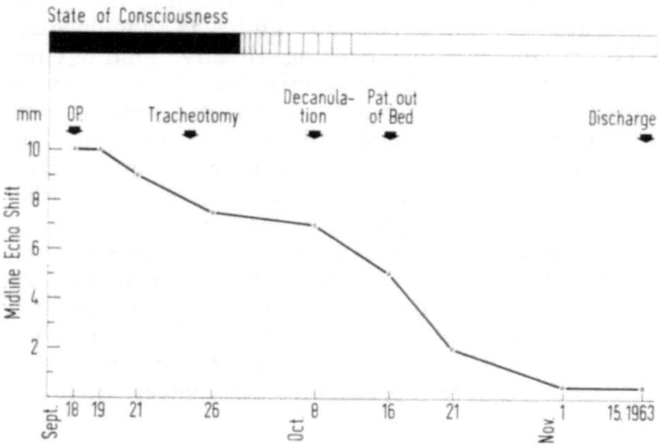

Fig. 155. Echo-encephalographic follow-up study following the evacuation of a subacute subdural hematoma in a 55-year-old patient. Pat. J. R., Echo-EG No. 758/1—9/63

c) Intracerebral Hematoma

Intracerebral hematomas also lead to a *distinct mass diplacement* which can be recognized by a shift of the midline echo in the echo-encephalogram. The 28 unilateral hematomas produced an average midline echo shift of 6.4 mm. This relatively small value is due to the fact that 14 (50%) of the unilateral intracerebral hematomas were located in the frontal area (see also p. 51). The average midline displacement in cases of frontal hematomas was 5.5 mm. All were thereby unequivocally recognized as space-occupying lesions. It is remarkable that this average displacement considerably exceeds the mean value for frontal gliomas (3 mm). The intracerebral hematomas located in the temporal or precentral area produced an average midline echo shift of 7.2 mm. Hematomas in the occipital lobe were not observed; this corresponds to reports in the literature. The degree of midline displacement does not aid in clarifying the differential diagnosis.

Four of the frontal intracerebral hematomas were bilateral. Three of these had an unremarkable echo-encephalogram, in one case a midline echo displacement of 2.5 mm was observed; this was due to the difference in the size of the two hematomas.

Pathological echo complexes produced by the hematomas themselves were recorded only if the hematoma was located in the temporal or precentral region because frontal hematomas are not picked up by the ultrasonic beam when our technique of investigation is used. Hematoma echo complexes were found in 9 of 14 hematomas (64%) located in the temporal lobe or somewhat above the Sylvian fissure (see Fig. 156); these made a more exact localization of the lesion possible.

Angiography was avoided in 7 of the patients because of a particularly fulminant clinical course. An example of this is seen in the following history:

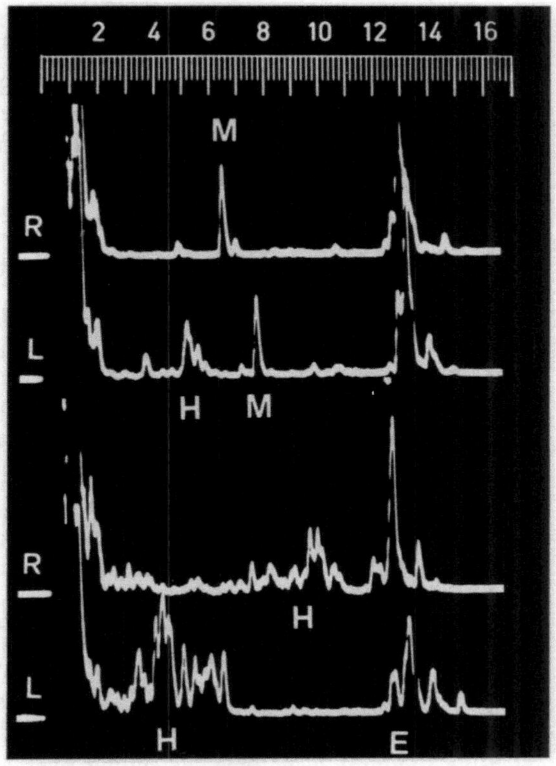

Fig. 156. Echo-encephalogram of a 45-year-old patient with an intracerebral hematoma in the left temporal lobe. Upper pair of traces: displacement of the midline echo (M) by 5.5 mm from left to right, additional reflections (H) resembling a hematoma echo ahead of the midline echo when sounding from the left. Lower pair of traces: when the ultrasound beam is directed towards the temporo-basal area, a distinct complex of pathological echoes (H) can be seen which corresponds to the hematoma. Pat. R. M., Echo-EG No. 1074/64

Case 22: Dieter B., Echo-EG No. 1864/65. The 17-year-old boy had a moped accident shortly after 7 p.m. on September 9, 1965. On admission to the clinic, he was able to speak at first; motor restlessness was noted. Plain films of the skull showed a depressed fracture of the frontal bone bilaterally, more so on the left than on the right. Within a few minutes, a *left-sided* hemiparesis developed, the patient became deeply unconscious, the left pupil dilated completely and failed to respond to light. The right pupil also became wide and soon failed to react to light. The echo-encephalogram which was performed immediately, revealed a 6.5 mm displacement of the midline structures from the left to the right side. This indicated a space-occupying lesion on the *left*. Obviously the paresis was homolateral. Because of the clinical fiindings, left frontal osteoplastic craniotomy was performed immediately (45 minutes after the accident). After slitting the dura which was stretched tightly and shimmered blue-black, a thick acute subdural hematoma was found extending from the frontal to the temporal area. Following the evacuation of this hematoma, an intracerebral hematoma approximately 60 cm³ in size was found in the left frontal pole; it originated from a larger artery which had been injured by a bone sliver.

The patient recovered slowly over a period of 4 weeks after surgery. A year later there were still marked cerebral deficiencies and the vision of the left eye was almost completely gone. The echogram showed a distinct dilatation of the ventricular system.

Before the introduction of echo-encephalography carotid angiography would have been carried out because clinical findings alone seldom reveal whether the paresis is homolateral. Here echo-encephalography has proved to be a very valuable help in the diagnosis of complications following head injuries. Especially in frontal or precentral hematomas, a homolateral paresis occurs quite frequently in our experience. The pressure produced by the hematoma is transmitted diagonally through the head leading to a compression of the midbrain at the contralateral edge of the tentorium (see KERNOHAN and WOLTMAN, 1929). The first patient suspected of having a hematoma, who was therefore investigated by echo-encephalography in 1962 by us, had this symptomatology. Since our ultrasonic studies were just beginning, the result of the echo-encephalographic examination was regarded as erroneous and angiography was performed on the side opposite to the hemiparesis. The anterior cerebral artery was found to be displaced towards the same side. Angiography on the other side then revealed an intracerebral hematoma in the frontal area. *Echo-encephalography is considerably more reliable in determining the side of a space-occupying lesion than the clinical findings.*

d) Chronic Subdural Hematoma, Unilateral and Bilateral

Among the 55 chronic subdural hematomas which we have observed, 30 were unilateral and 16 bilateral (Table 15). Nearly all *unilateral* hemorrhages showed a *distinct displacement of the midline echo* averaging 8.8 mm (JEPPSSON 8.6 mm, 1961; PIA and GELETNEKY 8 mm, 1968). This average is somewhat higher than that due to acute intracranial hematomas. If one considers only the older hematomas (which are lens-shaped in the angiogram) in the calculation of the mean value, the average rises to 9.3 mm. Only one unilateral hematoma located high up in the parietal area which was 1.5 cm in thickness had a normal midline echo. Although the anterior cerebral artery was displaced in the angiogram by 4 mm here, there was no shift of the internal cerebral vein.

Among the *bilateral subdural hematomas* 7 had a shift of the midline echo — due to the difference in the amounts of fluid accumulated on both sides; naturally this shift was not very great (one each 2.0, 3.0, 5.0, 6.0 and 7.0, twice 4.0 mm). In the remaining 9 patients the midline echo was at the normal place.

In one patient an error in measurement occurred. This was a 12-year-old boy with an extensive subdural hematoma, extending over almost the entire right hemisphere. Simultaneously he had a right temporal arachnoidal cyst which made the determination of the midline echo very difficult. Since there was also a pathological focus on the left side in the EEG as well, we were unable to determine the side of the lesion from the echogram. The diagnosis was finally established by carotid angiography.

In many cases of chronic subdural hematoma, an additional reflection may be obtained from the brain/hematoma capsule/hematoma interface. Using the normal sounding technique however, only the midline displacement can be determined on the echogram because the hematoma itself usually lies higher up, where the ultrasonic beam does not reach it.

By means of a special technique of investigation (BRÜCKNER and KAZNER, 1963; DE VLIEGER, 1964) where the probe is applied in the temporal area and is then tilted upwards pointing towards the opposite parietal region, is it possible to demonstrate the *hematoma echo* (see diagram in Fig. 157). The reflection appears only if the ultrasonic beam hits the interface almost completely vertically. The amplitude of the hematoma echo in these cases is approximately one quarter to one half the amplitude of the end echo. Only four unilateral chronic subdural hematomas which we examined and the error in measurement mentioned above, failed to show this reflection in the echogram, since we had not thought of this special method of investigation.

Even in cases of *bilateral subdural hematomas*, we were able to detect *hematoma echoes* in 12 out of 16 cases by diagonal sounding (10 times bilateral, twice only unilateral from the larger hematoma). Although the diagnostic reliability of the hematoma echo in cases of unilateral hematomas is very great in our experience and although the exact diagnosis could really be made on the basis of ultrasound alone in most cases, a reflection resembling a hematoma echo and appearing ahead of the end echo on both sides cannot be regarded as evidence of a bilateral subdural hematoma. Similar reflections are also found in a number of patients with cerebrovascular disease. These echoes originate from the dilated subarachnoid spaces. Nevertheless, echo-encephalography can be of great use in some cases of bilateral chronic subdural hematoma also, as will be discussed below.

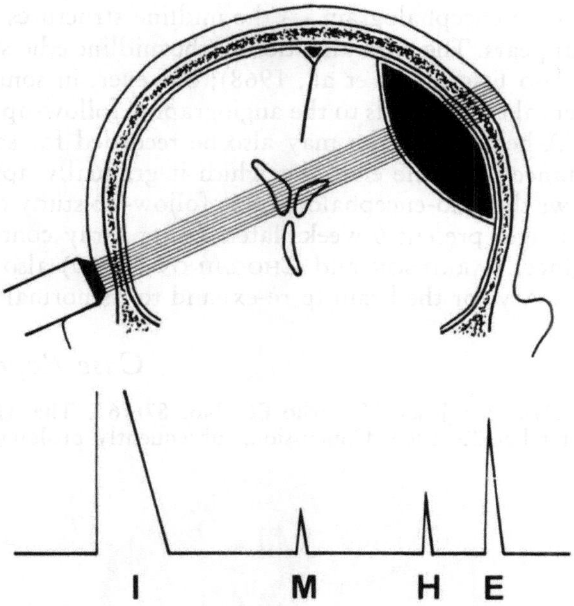

Fig. 157. Sounding technique in cases of chronic subdural hematoma

In cases of unilateral hematomas, the average distance of the hematoma echo from the end echo was 20.0 mm (10.0 to a maximum of 37.0 mm), in bilateral hematomas 16.0 mm (8.0 to a maximum of 30.0 mm). By means of carotid angiography which was performed in all cases and at operation, the thickness of the hematoma as measured by ultrasound was confirmed. While we always carry out a preoperative contrast medium study right after echo-encephalography, a control angiogram is unnecessary after the hematoma is evacuated because the success of the operation can be observed on

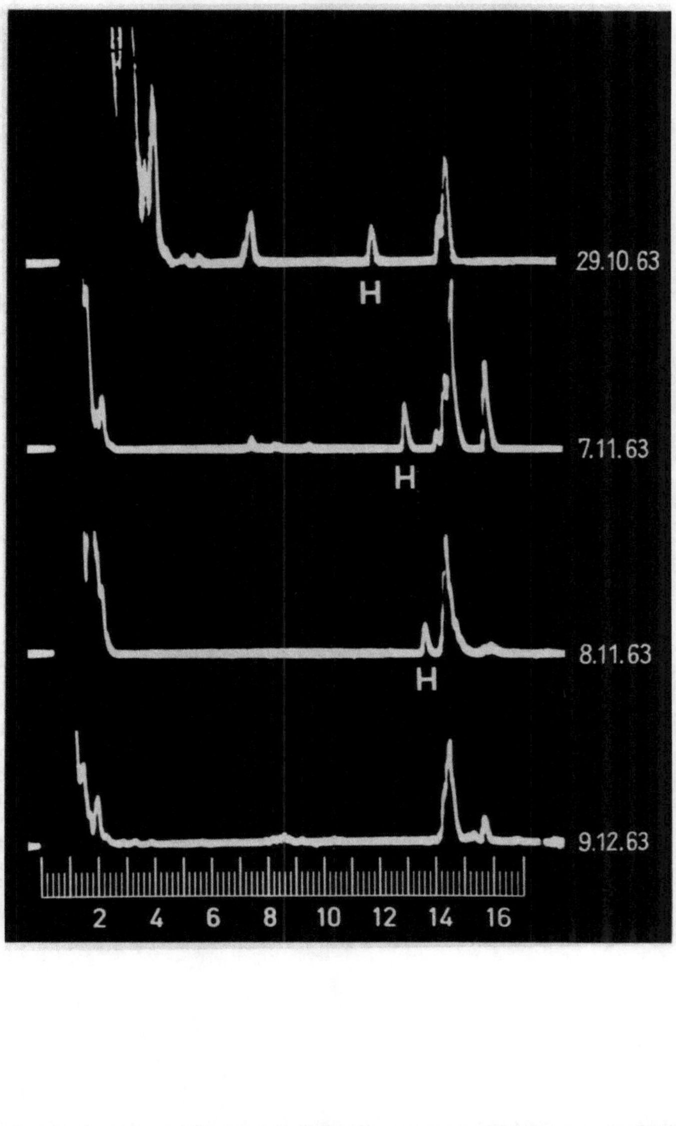

Fig. 158. Echo-encephalographic follow-up study of normalization of the ultrasonic trace following operative evacuation of a chronic subdural hematoma. The hematoma echo (H) slowly approaches the end echo. 6 weeks later the hematoma echo is gone. Pat. B. Sch., 59 years of age. Echo-EG No. 821/63

the echo-encephalogram — the midline structures return to normal and the hematoma echo disappears. The normalization of the midline echo shift generally takes 4 to 6 weeks (Fig. 158) see also FISCHGOLD et al., 1968); however, in some cases it can take even longer. This time interval corresponds to the angiographic follow-up studies of MCLAURIN and HELMER (1966).

A hematoma echo may also be recorded for some time yet, although at a much smaller distance from the end echo which it gradually approaches. The echogram series in Fig. 158 shows the echo-encephalographic follow-up study of a hematoma where the hematoma echo is no longer present 6 weeks later. Using x-ray control of silverclips on the dura and cortical surface, PARKINSON and CHOCHINOW (1960) also found that a period of 30 to 60 days is necessary for the brain to re-expand to its normal volume and position.

Case Reports

Case 23: Josef M., Echo-EG No. 876/63. The 41-year-old patient suffered a head injury on September 20, 1963. Concussion, subsequently prolonged head ache. Towards the end of November 1963 increasing confusion. Admission to our clinic on November 26, 1963. Patient is markedly slowed in his responses and not well-oriented, neurological findings unremarkable. Bilateral papilledema. EEG without any pathological signs. Echo-encephalogram: 11.0 mm displacement of the midline echo from the left to the right (Fig. 159 bottom), on sounding diagonally from the right to the left parietal area, a hematoma reflection appeared 26 mm ahead of the end echo (Fig. 159 top). Diagnosis: chronic subdural hematoma over the left cerebral hemisphere, 26 mm in thickness. Confirmation of the echo-encephalographic findings by carotid angiography (Fig. 159) and subsequent operation.

The location of a unilateral hematoma can be determined immediately by echo-encephalography even in the absence of neurological or EEG abnormalities. By means of diagonal sounding both the type of hematoma and its thickness can be determined.

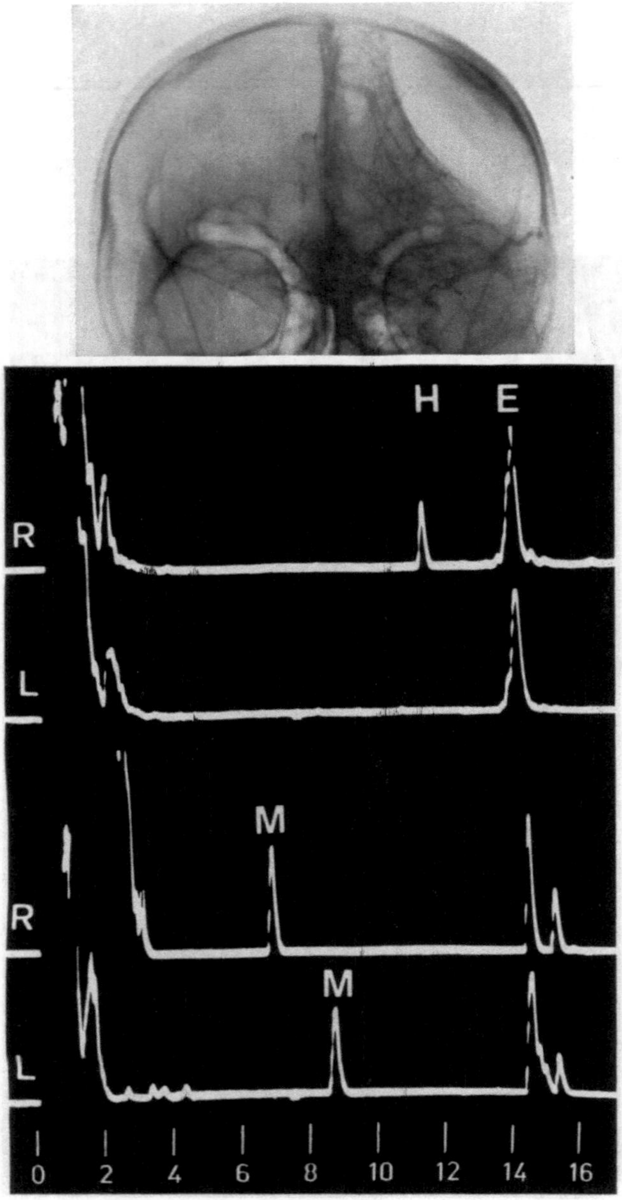

Fig. 159. Echo-encephalogram of a 41-year-old man with a chronic subdural hematoma over the left hemisphere. Lower pair of traces: displacement of the midline echo (M) by 11.0 mm towards the right. Upper pair of traces: a hematoma echo (H) 26.0 mm away from the end echo (E) which indicates the presence of a thick chronic subdural hematoma in the left partietal area, appears only when the diagonal sounding technique is used. Top: confirmation of the echo-encephalographic diagnosis by means of carotid angiography. Pat. J. M., Echo-EG No. 876/63

Case 24: Frieda H., Echo-EG No. 2177/66. The 67-year-old woman was referred to us from an out-of-town neurological clinic because of a chronic subdural hematoma over the right hemisphere which was found at angiography (Fig. 160). The angiographic finding of an avascular space in the parietal area and and unequivocal displacement of the anterior cerebral artery seemed to speak for a unilateral localization. In the echo-encephalogram the midline structures were shifted by only 6.0 mm with a right parietal hematoma echo 24.0 mm ahead of the end echo (Fig. 160). Such a slight

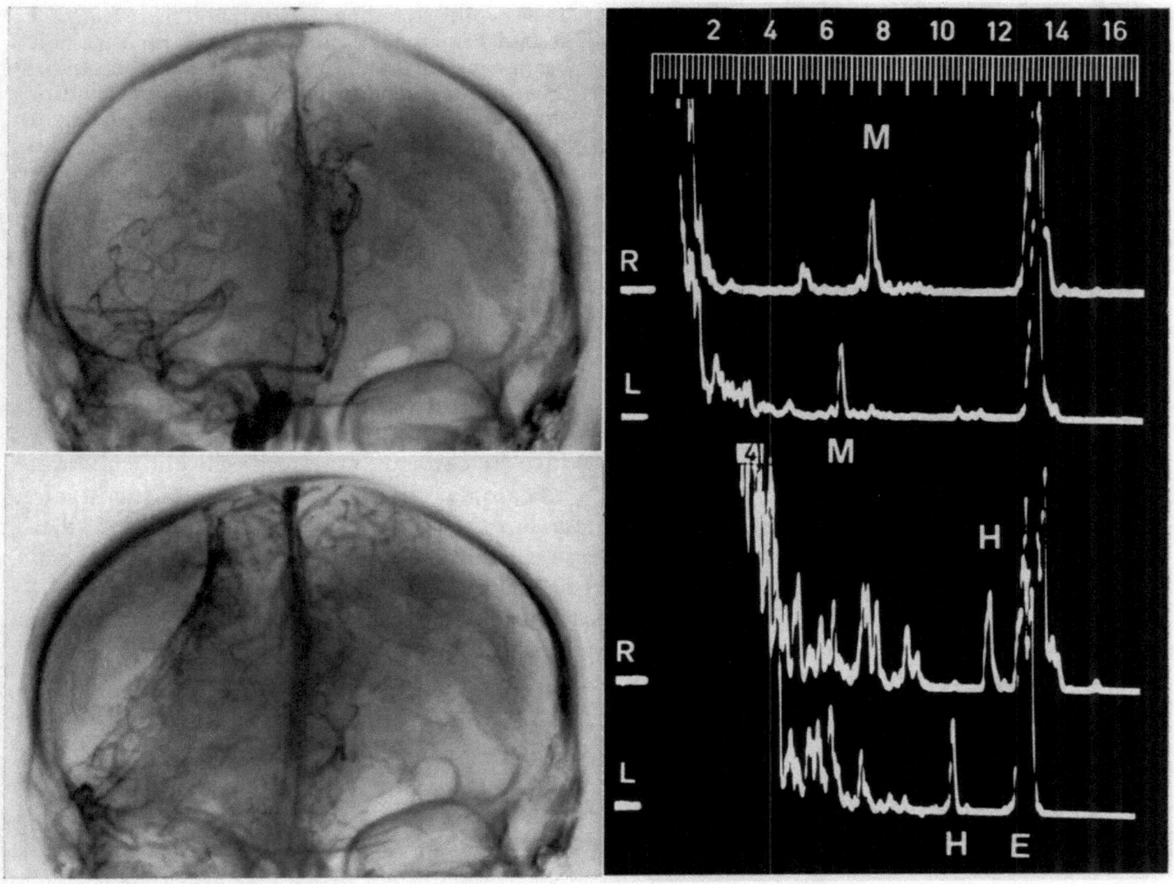

Fig. 160. Echo-encephalographic diagnosis of a bilateral chronic subdural hematoma. Despite a clear displacement of the anterior cerebral artery and internal cerebral veins, this is a case of bilateral hematoma since diagonal sounding demonstrates hematoma echoes (H) on both sides (lower pair of traces). Thickness of the hematoma on the right is 24 mm, on the left, 12 mm. Confirmation by operation. Pat. F. H., 67 years of age, Echo-EG No. 2177/66

M-echo shift in the presence of such an extensive hematoma led us to suspect a second hematoma over the left hemisphere. Diagonal sounding from right to left revealed a further probable hematoma echo from the parieto-occipital area 12.0 mm ahead of the end echo (see Fig. 160 bottom). After evacuating the right-sided hematoma through a parietal burr hole, a second trepanation was performed in the left parieto-occipital area. The second hematoma, suspected at the echo-encephalographic examination, was found to have a maximum thickness of 1.5 cm.

The above case shows the value of *echo-encephalography as a complementary examination to angiography.* Since the thickness of a hematoma on one side is usually only 2.3 times the midline echo displacement in our experience, a second hematoma had to be suspected despite the apparently unequivocal angiographic findings, since the value here was 4 times the distance of the M-echo shift. And we found an unequivocal hematoma echo on the side which had not been examined by angiography. The ultrasonic findings were confirmed at operation.

Case 25: Josef R., Echo-EG No. 1034/64. The 65-year-old man was admitted because of a chronic subdural hematoma over the left hemisphere which had been diagnosed by angiography already. Since the anterior cerebral artery was almost midline, angiography was also carried out on the right but it failed to show any avascular zone. The echo-encephalogram showed a minimal displacement of the midline echo (1.5 mm towards the left). On diagonal sounding from the right a hematoma echo was visible immediately (Fig. 161) while sounding from the left revealed a second hematoma echo only when the ultrasonic beam was directed towards the right occipito-parietal area. The echo-encephalographic finding thus indicated a second hematoma on the right side, probably more in the occipital area. This assumption was confirmed at operation. Bilateral burr-holes allowed us to evacuate the hematomas which measured 2 cm in diameter.

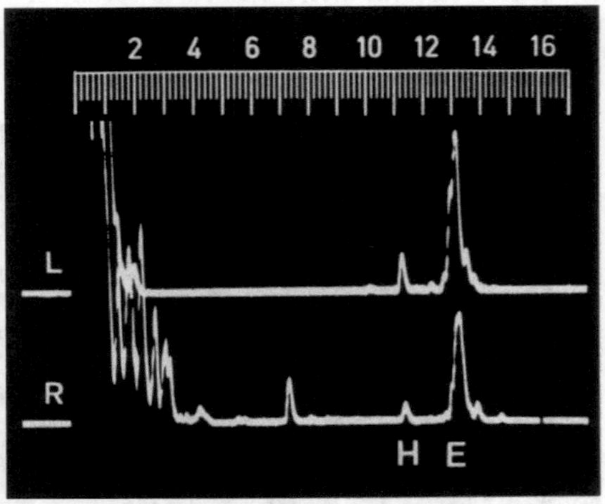

Fig. 161. Echo-encephalogram of a 65-year-old man with a bilateral chronic subdural hematoma. Hematoma echo (H) on both sides preceding the end echo (E). Further details see text. Pat. J. R., Echo-EG No. 1034/64

In this patient with a bilateral subdural hematoma, echo-encephalography was also a valuable adjunct to angiography.

In our experience, it is always possible to obtain a reflection from the hematoma interface in cases of unilateral chronic subdural hematoma, so that there is no difficulty in distinguishing it from a tumor-dependent mass shift. This is true on the condition that diagonal sounding is actually carried out because the echo-encephalographic possibilities can be exhausted completely only in this way. After some practice, the hematoma echo can be successfully demonstrated in nearly every case. In cases of bilateral "lateral peaks" the danger of *mistaking an atrophic brain lesion for a bilateral hematoma* must always be drawn into consideration. This is why an angiographic study must be carried out as a matter of course in all such cases before surgery.

e) Differential Diagnosis of Posttraumatic Intracranial Hematomas Using Echo-Encephalography

The degree of midline echo shift does not permit *a differentiation between the individual types of hemorrhage,* since all unilateral posttraumatic hematomas lead to similar high M-echo shifts (an average of 7.2 mm with epidural hematomas, of 7.2 mm with acute subdural hematomas, of 6.4 mm with intracerebral hematomas and of 8.8 mm with chronic subdural hematomas). The only way of *distinguishing the types of hematomas is the hematoma echo, which varies in form and amplitude with the different hematomas* (Fig. 162) (see also Ito et al., 1966; Tanaka, 1968).

Hematoma reflections were registered in 151 of 245 intracranial hematomas following head injuries. This results in a percentage of 61.6. *Intracerebral hematomas* were correctly identified 9 times by a special hematoma echo complex (28%). The quota is so low here primarily because no less than 18 hemorrhages were located in the frontal region and only 14 in the temporal lobe. Considerably better results are achieved with *acute subdural hematomas* where a hematoma echo was found 38 times (46%) in 82 cases. This reflection usually has a height of only a few millimeters which distinguishes it clearly from epidural hematomas. The high, needle-sharp echo originating from the dura is the characteristic finding in a large percentage of epidural hematomas making a clear differentiation from other forms of intracranial hematoma possible. 59 of 76 epidural blood accumulations showed a special reflection which enabled us to make the unequivocal diagnosis of an epidural hematoma 55 times (72%).

Here the accuracy of echo-encephalography in diffeffrential diagnosis *approaches that of angiography.* Atypically located epidural hematomas are practically the only ones that escape direct echo-encephalographic demonstration. The hematoma echo is at least one-third as high as the end echo in cases of epidural hemorrhage and if the ultrasound hits the dura at right angles exactly the amplitude of the HE even exceeds that of the end echo. Ways of differentiating the dura echo from other lateral reflections have been discussed in detail already (see p. 135 f.).

In cases of chronic subdural hematomas, the results are also very favorable. In our series, the presence of the hematoma echo in 33 of 39 cases (85%) of unilateral hematoma permitted a more precise differentiation of the lesion, whereas bilateral hematomas showed a much lower percentage with 12 in 16 cases (75%). However, the hematoma echo can be obtained only if the diagonal sounding technique is used. It has an amplitude of a few millimeters, rarely becoming higher.

We hope that these results will be improved upon yet, since our survey also includes the first patients with posttraumatic hematomas, examined by this method. By using more suitable probes and contact media, a reflection from the interface between the brain and the hematoma could be detected more and more frequently in the last two years especially in cases of acute subdural hematoma (20 of 32 hematomas, corresponding to 63%).

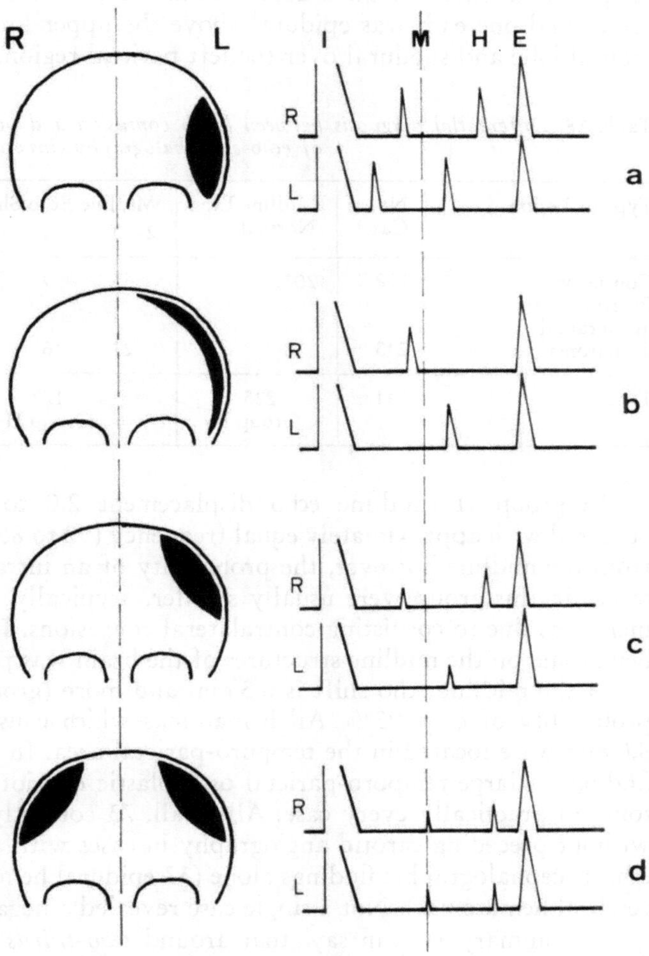

Fig. 162. Diagrammatic representation of the echograms with various forms of extracerebral hemorrhages. a) epidural hematoma; aside from the midline echo shift, hematoma echoes (H) may be registered from both sides. b) Acute subdural hematoma: midline echo shift but no unequivocal hematoma echo. c) Unilateral chronic subdural hematoma; midline echo shift and hematoma echo on diagonal sounding. d) Bilateral chronic subdural hematoma: no midline echo shift. Diagonal sounding reveals hematoma echoes on both sides. (From SCHIEFER, KAZNER and BRÜCKNER: Fortschr. Neurol. Psychiat. 31, 457 [1963])

f) Differential Diagnosis between Postcontusional Edema and Posttraumatic Intracranial Hematoma Using Echo-Encephalography

Of great interest is the question whether echo-encephalography permits *a differentiation between postcontusional cerebral edema and a space-occupying intracranial hematoma.* Our investigation of this special problem is based on the echo-encephalographic findings of a total of 541 patients with contusions and hematomas (see Table 18).

If a normal echo-encephalogram (group I in Table 18) is recorded in the acute phase of a head injury, the probability of finding a space-occupying intracranial hematoma is very low, *however, a normal echogram cannot exclude such a complication with absolute certainty.* Among the 18 hematomas without a midline echo displacement, 9 were bilateral chronic subdural hematomas, one was a unilateral chronic subdural hematoma located high up in the parietal area, 1.5 cm in thickness, and 2 were subacute epidural hematomas above the upper longitudinal sinus. Of the 6 acute hematomas, 3 were intracerebral (bilateral frontal) hematomas and one each was epidural above the upper longitudinal sinus, epidural over the left frontal lobe and subdural over the left parietal region.

Table 18. *Differential diagnosis between brain contusion and posttraumatic intracranial hematoma by means of echo-encephalography (investigation of 541 cases)*

Type of Lesion	No. of Cases	Midline Echo Normal	Midline Echo Shifted (mm)							
			2—3	—4	—6	—8	—10	—12	—14	more
Contusion	298	207	63	9	13	4	2			
Posttraumatic intracranial hematoma	243	18	21	16	55	62	46	18	6	1
Total	541	225 Group I	177 Group II			139 Group III				

In group II (midline echo displacement 2.0 to 6.0 mm) hematomas and contusions occurred with approximately equal frequency (92 to 85). If the M-echo deviated 4.5 to 6.0 mm from the midline however, the probability of an intracranial hematoma is 80%. The hematomas in this group were usually smaller, atypically located or the slight midline displacement was due to coexisting contralateral contusions. In this way the pressure exerted by the hematoma on the midline structures of the brain was partly compensated.

If the midline echo shift is 6.5 mm and more (group III), a hematoma is present with a probability of over 95%. All hematomas which caused a midline echo shift of more than 8.0 mm were located in the temporo-parietal area. In the case of such echo-encephalographic findings, a large temporo-parietal osteoplastic craniotomy enables us to evacuate the hematoma in practically every case. All in all, 73 correctly placed craniotomies were performed without preceding carotid angiography in cases with a very acute course on the basis of the echo-encephalographic findings alone (37 epidural hematomas, 29 acute subdural and 7 intracerebral hematomas). Not a single case revealed a negative result at operation.

In summary we can say, that around *two-thirds of all intracranial hematomas can be differentiated from contusions by the extent of the midline echo shift.* Tönnis and Frowein (1966) arrived at a similar conclusion according to a series of 191 patients with suspected hematomas (Table 19). Not considered here is the extraordinarily important *hematoma echo*

Table 19. *Echo-Encephalography in suspected hematomas* (Tönnis and Frowein, 1966)

	Midline Echo Displacement													
	0	1	2	3	4	5	6	7	8	9	10	11	< mm	
Hematomas		4	8	9	13	16	9	7	7	5	2	1	2	
Brain injuries *without* hematoma	31	29	20	14	8	5	1							
	31	33	28 (2–4: 36%)	23	21	21	10	7	7	5	2	1	2	191

which leads to further differentiation. Furthermore the *time* at which a midline echo shift occurs following the head injury, is of considerable significance. While all except 6 cases of acute posttraumatic hematoma showed a midline echo displacement within the first 24 hours, only 41 of 298 cases (= 13.8%) of contusion demonstrated an echo-encephalographically evident midline shift on the first day. These were primarily children up to the age of 10, in whom an accurate differentiation is most difficult of all (Table 16, see also GELETNEKY, 1965).

Considering all the echo-encephalographic criteria (degree of midline echo shift, hematoma echo, onset of midline displacement) it is possible to differentiate between a hemorrhage and a contusion in 9 out of 10 patients with a severe head injury. The consequences of this fact for the indication for carotid angiography will be described in more detail in the following section.

g) Echo-Encephalographic Findings with Head Injuries and Indications for Carotid Angiography

At the beginning of our echo-encephalographic investigations every finding was naturally checked by an angiogram in order to delineate the possibilities and limitations of the pulse-echo technique. For about 4 years, the following indications for carotid angiography in cases of head injury have been used when the ultrasonic finding is considered.

Carotid angiography is necessary:

1. if the midline echo shift is between 2.0 and 6.0 mm, no hematoma echo can be obtained and the clinical findings point to a hematoma;
2. if the midline echo shift reaches a value of 6.5 to 8.0 mm which is typical for a hematoma, a hematoma echo is absent and the site of the hemorrhage cannot be determined exactly on the basis of the clinical finding;
3. *if there is suspicion of a hematoma on account of the clinical symptomatology despite a normal echo-encephalogram.*

Carotid angiography can generally be *avoided* if the displacement of the midline echo is greater than 8.0 mm even if there is no hematoma echo to be found, because every hematoma which produces such a great midline displacement may be evacuated from a large osteoplastic craniotomy in the temporo-parietal area. A contusion is improbable here.

Carotid angiography is no longer permissible:

1. if the echogram localizes the hematoma exactly and the patient has already lost consciousness;
2. if hematoma reflections are absent in an echo-encephalogram which shows a typical midline echo shift, but the hematoma can be well localized by the clinical findings and plain x-rays of the skull and signs of midbrain herniation are already apparent.

These aspects have evolved from a critical evaluation of the echo-encephalographic findings together with the clinical findings and therapeutic results of around 190 acute intracranial hematomas following head injury. One limiting factor is that such a course of investigation can be followed only by those who are familiar with echo-encephalography. *In case of doubt the alternative is always an angiogram.*

4. Sequelae of Head Injuries

The fact that the width of certain sections of the cerebral ventricles, especially that of the third ventricle, can be determined echo-encephalographically, may be used for the follow-up study of head injuries. Our series of investigations includes 204 patients who had suffered head injuries between 4 weeks and 32 years before the echo-encephalographic examination (see Table 20).

A *displacement of the midline echo* was recorded only 5 times. In all cases it was 2.0 mm which corresponded to the pneumoencephalographic finding; 3 of these patients had had open brain lacerations and 2 patients had severe closed head injuries. The displacement of the intracranial structures always occurred towards the side of the trauma due to unilateral atrophic processes. In 26 patients *who had suffered concussions, not a single case of posttraumatic dilatation of the third ventricle was found.* One patient whose third ventricle measured 8.0 mm in diameter (see Table 20) was found to have a cerebrovascular lesion.

Table 20. *Width of the third ventricle in the echo-encephalogram with sequelae of head injuries*

State after	No. of Cases	ME Normal	ME Shifted	Width of the Third Ventricle in the Echo-EG (mm) Normal		Enlarged					
				—6.0	—7.0	—8.0	—9.0	—10.0	—11.0	—12.0	More
Concussion	26	26	—	20	5	1					
Contusion with unconsciousness up to 1 week	27	27	—	16	10						
Contusion with unconsciousness exceeding 1 week	44	42	2	1	4	17	5	8	4	2	3
Open head injury with circumscribed contusion	50	47	3	21	14	8	1	2			
Intracerebral hematoma	7	7	—	1	2	1	3				
Epidural hematoma	21	21	—	10	2	3	3	2			1
Acute subdural hematoma	13	13	—	2	2	3	2	1	3		
Chronic subdural hematoma	16	16	—	3	2	5	2	2	1		
Total	204	199	5	74	41	38	16	15	8	2	5
				115		84					

Brain contusions which caused periods of unconsciousness lasting up to one week also did not demonstrate an unequivocal dilatation of the third ventricle. On the other hand, of 44 patients with severe closed head injuries and unconsciousness for over 8 days, 39 (= 88.6%) revealed *a dilatation of the third ventricle* which indicated obvious *dependence on the duration of the unconscious period.* We regarded it as the end of the unconscious period when the patient first opened his eyes on being addressed. It is remarkable that there is no further increase in posttraumatic hydrocephalus after that unless there are disturbances in the CSF circulation due to leptomeningeal adhesions. The 5 patients with prolonged unconsciousness in whom no ventricular dilatation could be measured, were children and young people up to the age of 26 years.

The greatest diameter of the third ventricle where the patient had suffered a serve *closed head injury* and was subsequently able to work perfectly normally without any personality changes was 11.0 mm. We have observed survival in patients with third ventricular widths up to 13.0 mm; however, these patients continued to need care and presented the clinical picture of posttraumatic stupor over a period of months. All injuries which resulted in even more marked ventricular dilatations died subsequently, the greatest width of the third ventricle being 21.0 mm. *Until now it has been necessary to perform multiple pneumoencephalograms in order to detect a progressive posttraumatic hydrocephalus, today echo-encephalography can do this rapidly without any stress to the patient.* Fig. 163 shows two echo-encephalograms obtained 4 and 6 weeks after a brain contusion; they indicate a distinct increase in ventricular dilatation. The first measurement was controlled by pneumo-encephalography (Fig. 163 top), later we were able to do without this investigation. The width of the third ventricle continued to increase up to 19.0 mm until the time of death; the finding was confirmed by autopsy.

Open brain lacerations led to an increase in the width of the third ventricle only if there was a considerable loss of brain tissue. Smaller, circumscribed contusions with depressed fractures and fronto-basal injuries where the dura was open, were not sufficient to produce a measurable increase in third ventricular width. Only if the concomitant brain injury resulted in prolonged unconsciousness, did these cases evolve an enlargement of the third ventricle.

Posttraumatic dilatation of the third ventricle often begins at the end of the first week after the injury already, according to our observations. Fig. 164 shows the follow-up study of the width of the third ventricle in a 23-year-old woman, who had suffered a severe closed head injury. We have seen normalization of a posttraumatic hydrocephalus only once so far; that was in a 16-year-old boy, who had had a fronto-basal depressed fracture with subsequent meningitis where a transient obstruction of the CSF flow was probably the cause of the ventricular dilatation. Ulbricht and de Seixas (1966) have reported a number of such cases.

In a series of *57 posttraumatic intracranial hematomas* observed by echo-encephalography for a prolonged period of time after operative evacuation, only 24 (42%) were found to have ventricles of normal width. The proportion of epidural hematomas was highest here (12 of 21 cases). A later dilatation of the third ventricle occurred in this form of hematoma only if there was decerebrate rigidity before the operation and prolonged unconsciousness after surgery. In contrast, more than two-thirds of the patients who survived an acute subdural hematoma showed ventricular dilatation without indicating any dependence on the preoperative findings but rather on the coexistent contusion. The enlargement of the third ventricle after evacuation of a chronic subdural hematoma, is probably due to a separate cerebrovascular lesion, because these were chiefly older men without a severe cranio-cerebral trauma.

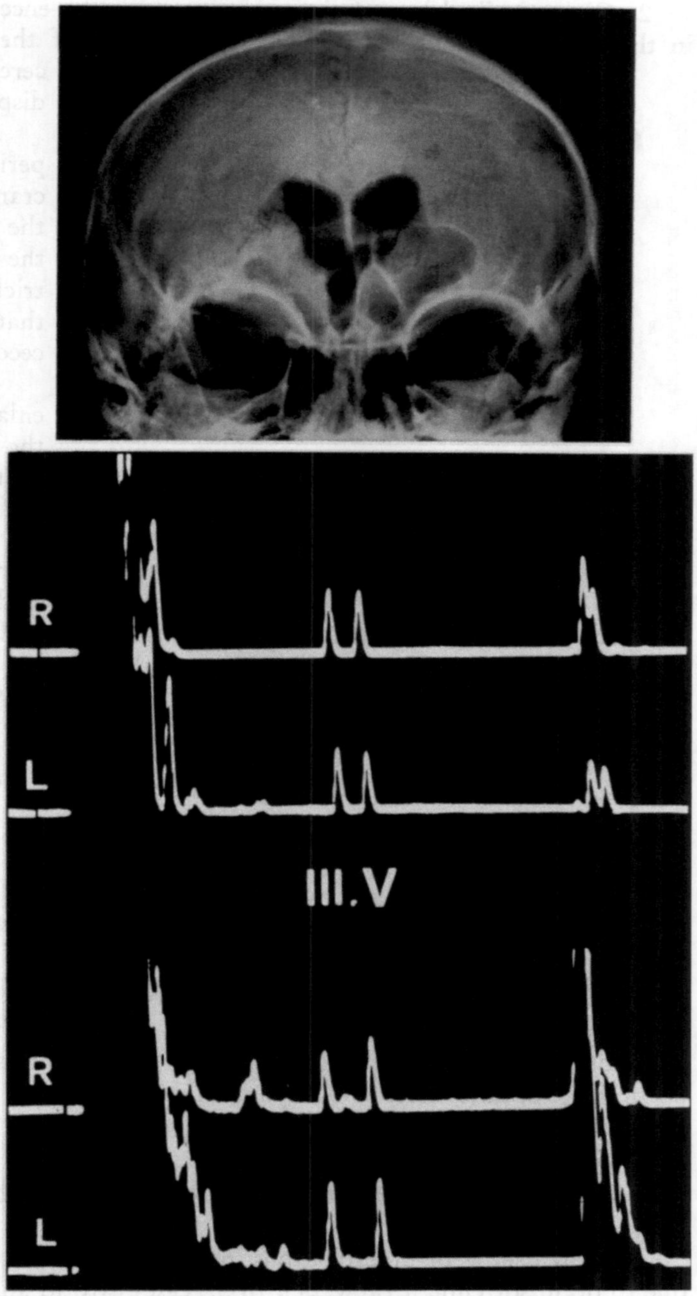

Fig. 163. Increasing dilatation of the third ventricle following severe closed head injury. Upper set of traces: 4 weeks after the injury the width of the third ventricle is 9.0 mm. Confirmation through pneumoencephalography (top). Lower set of traces: 6 weeks after the injury the width of the third ventricle has increased to 14.0 mm. Pat. G. W., 22 years of age, Echo-EG No. 896/63

Of these facts, the following aspects are of importance for the evaluation of the sequelae of head injuries:

1. Sequelae of head trauma in the ventricular system can generally not be demonstrated by echo-encephalography in uncomplicated concussions and contusions where the period of unconsciousness does not exceed 7 days.

2. Circumscribed loss of tissue also escapes echo-encephalography. A midline echo shift in the echo-encephalogram can be expected only if there is considerable atrophy of one cerebral hemisphere which results in a displacement of the ventricular system.

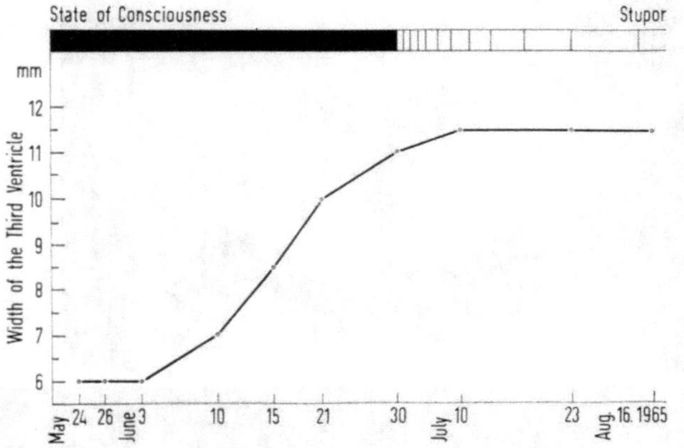

Fig. 164. Echo-encephalographic follow-up study of posttraumatic ventricular dilatation in a 23-year-old patient. R. U., Echo-EG No. 1639/1—10/65

3. In patients with a prolonged period of unconsciousness following craniocerebral trauma, a dilatation of the third ventricle may be recorded on the echo-encephalogram. A third ventricle of normal width makes it unlikely that the period of unconsciousness exceeded 10 days in adults.

4. Frequently one recognizes an enlargement of the third ventricle in the echo-encephalogram in cases of posttraumatic intracranial hematomas which run a very stormy course.

Until now, our remarks have referred to the determination of the width of the third ventricle. Just as with occlusive hydrocephalus, however, the position of the temporal horn echo can also be determined by echo-encephalography so that further deductions can be made regarding the degree of dilatation of the lateral ventricles. The registration of the echo from the lateral wall of the temporal horn assumes importance only in cases of advanced posttraumatic hydrocephalus. The echo-encephalographic brain mantle index (BMI) (p. 61) is usually increased only if the width of the third ventricle reaches 9 to 10 mm or more.

D. The Echo-Encephalogram in Cases of "Stroke"

By the term "stroke" we mean all neurological deficits which develop spontaneously either acutely or within a few hours and days and which are not due to an inflammatory process or a tumor. In other words this is a group of divers vascular lesions where the diagnostic possibilities of echo-encephalography will be discussed.

1. Encephalomalacia (Occlusive Cerebrovascular Disease)

Our own series of investigations includes 72 patients in whom a stroke was found to be due to brain softening (Table 21). In 37 cases carotid angiography was performed and once vertebral angiography. 25 patients were found to have *an occlusion of a larger cerebral vessel,* usually of the middle cerebral artery (Fig. 165). The echo-encephalographic findings obtained in this way are presented in Fig. 168. In 9 patients *the midline echo was shifted an average of 3.7 mm.* All these cases were confirmed by angiography.

More marked shifts of the midline structures are rare in cases of brain softening. Only 2 patients were observed in whom one entire hemisphere underwent softening. In one case here, the midline displacement was 7.0 mm and in a second case even 10.0 mm. The echogram and the autopsy findings of the last case are seen in Fig. 166. All the main cerebral arteries on the left side were blocked by an ascending thrombosis of the carotid artery in this 49-year-old patient.

Usually, the midline echo shift develops only after a period of 1 to 2 days following vascular occlusion in our experience until now, to that a displacement of the midline echo in the acute stage of a case of encephalomalacia is a rare occurrence.

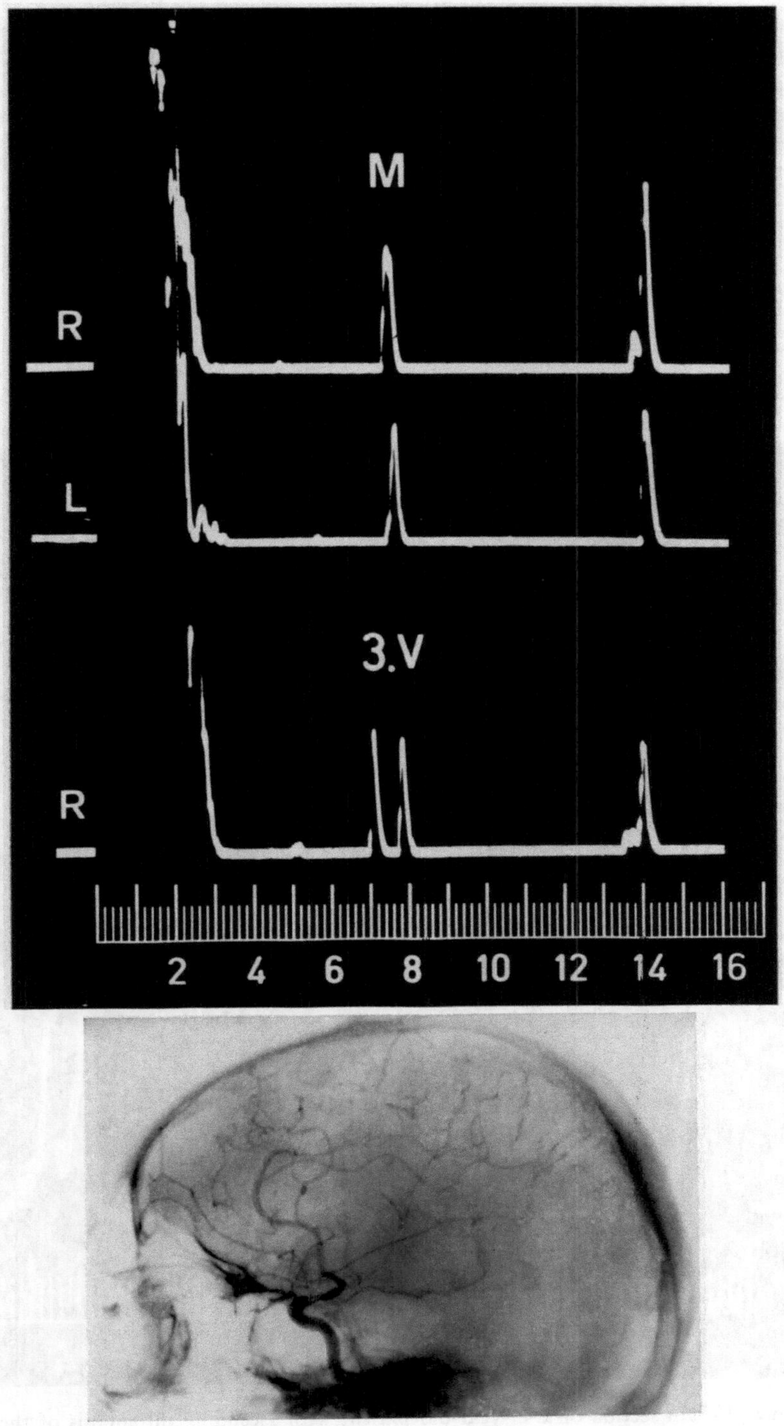

Fig. 165. Normal echo-encephalogram in a case of fresh occlusion of the middle cerebral artery (see angiogram). Pat. A. F., 56 years of age, Echo-EG No. 2721/67

In the 47 patients with softening of unknown origin, angiography was performed 13 times without finding the cause of the stroke. We avoided angiography of the carotid system in the other 34 cases because the patients were either too old or in some cases, because the

Table 21. *Echo-encephalographic findings in 154 patients with "stroke"*

Lesion	No. of Cases	ME Normal	ME Shifted	Width of Third Ventricle Exceeding 7.0 mm	Hematoma Echo
Intracerebral hematoma	32	1	31	3	7
Intracerebellar hematoma	2	2	—	2	—
Ventricular hemorrhage	6	6	—	5	—
Thrombo-embolic vascular occlusion	25	16	9	4	—
Other forms of encephalomalacia	47	37	10	17	—
Spontaneous subarachnoid hemorrhage	42	40	2	18	—
Total	154	102	52	49	7

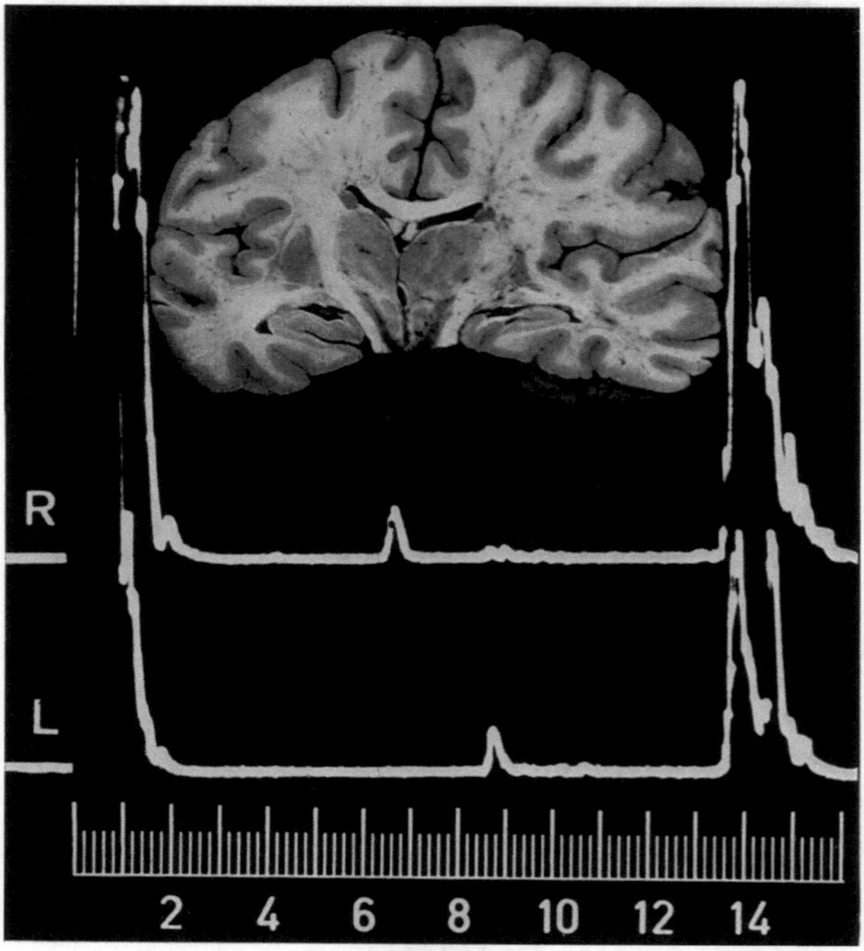

Fig. 166. Echogram and brain section of a 49-year-old patient with ascending thrombosis of the carotid artery. 10 mm midline echo displacement corresponds to the autopsy finding. Pat. M. S., 49 years of age, Echo-EG No. 111/62

neurological deficits subsided within a short time. The indication for contrast medium studies was set chiefly by the echo-encephalographic findings. In cases of midline shifts, angiography was usually performed immediately. Among the 47 patients only 10 showed a slight displacement of the M-echo, the average being 2.9 mm (Fig. 168). All other echograms were unremarkable as far as the midline echo was concerned. However, the fact that 17 patients had a *dilatation of the third ventricle* in the echogram during the acute stage of the "stroke" is especially remarkable. This enlargement ranged from 7.5 to 13.00 mm with a mean of 9.4 mm. In other words, an atrophic lesion was present in these patients even before the acute event. Such a finding was obtained only 4 times in 25 cases of vascular occlusion on the other

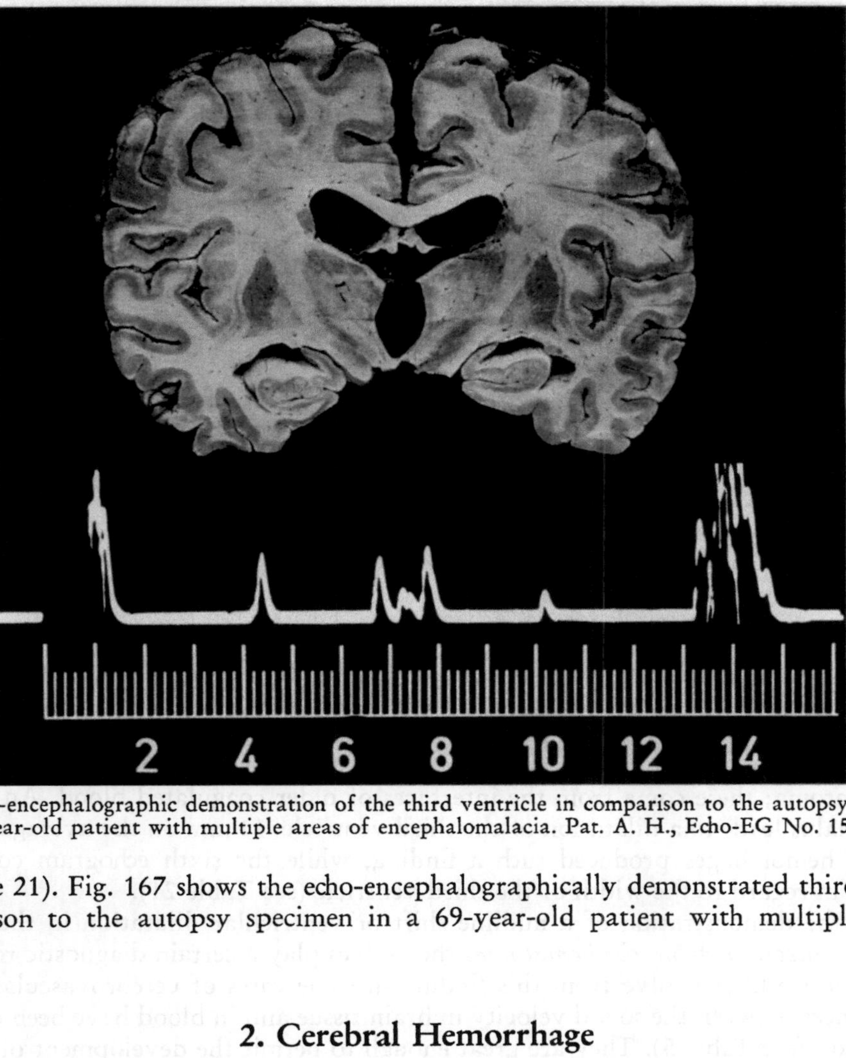

Fig. 167. Echo-encephalographic demonstration of the third ventricle in comparison to the autopsy specimen in a 69-year-old patient with multiple areas of encephalomalacia. Pat. A. H., Echo-EG No. 1579/65

hand (Table 21). Fig. 167 shows the echo-encephalographically demonstrated third ventricle in comparison to the autopsy specimen in a 69-year-old patient with multiple areas of softening.

2. Cerebral Hemorrhage

Echo-encephalography was performed in 40 cases of cerebral hemorrhage. All cases were confirmed either by angiography, operation or autopsy. The different causes for the hematomas were as follows:

19 intracerebral hematomas due to angiomas or aneurysms
17 so-called spontaneous intracerebral hemorrhages
4 hypertensive hemorrhages

32 hematomas were located in the cerebral hemispheres, 4 times rupture into the ventricular system occurred. Two hematomas were located in the cerebellum and 6 primarily

in the ventricular system (Table 21). Of the *32 cases of intracerebral hematoma,* 31 showed *a displacement of the midline echo* which averaged 5.2 mm (Fig. 168). No differences were found between the different groups concerning the extent of midline displacement which would have made a more accurate differentiation possible. In the case of an older frontal intracerebral hematoma, a displacement of the M-echo was absent.

Cerebellar hemorrhages led to a *dilatation of the third ventricle* by blocking the flow of CSF; this was recognized in the echo-encephalogram. The widths of the third ventricle in the two cases we observed, were 11 and 12 mm respectively. Similar conditions pertain to hematomas leading to a tamponade of the entire ventricular system. Here too, an abnormal double echo can be obtained from the walls of the dilated third ventricle (Fig. 169). Occasionally several smaller reflections appear between the ventricular wall echoes on the echogram of such a hematoma; these come from the interfaces of older coagulated blood. An acute total hematocephalus leads to a dilatation of the third ventricle, 11 mm on the average. 5 out of 6 ventricular hemorrhages produced such a finding, while the sixth echogram could not be interpreted in regard to the width of the third ventricle (see Table 21).

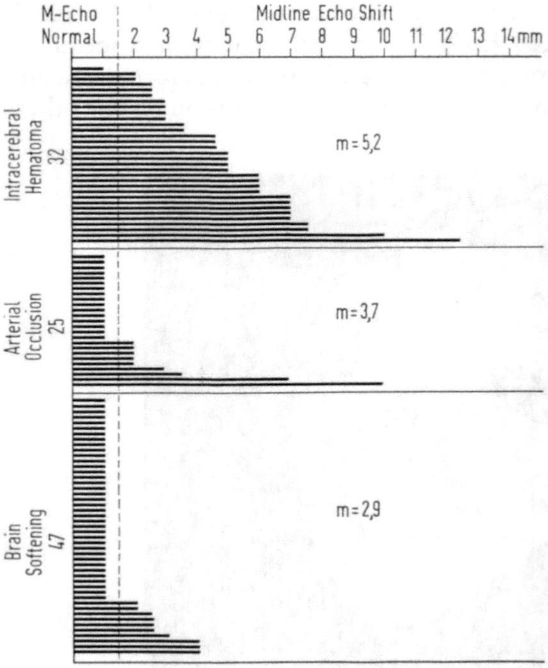

Fig. 168

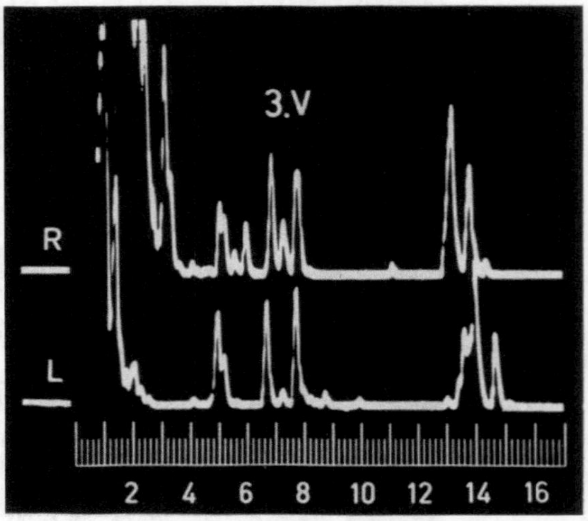

Fig. 169

Fig. 168. Behavior of the midline echo following a "stroke" (Results in 104 patients)

Fig. 169. Echo-encephalogram of a ventricular hemorrhage. Abnormal double echo from the walls of the third ventricle. Pat. K. U., 47 years of age, Echo-EG No. 1702/65

Besides the demonstration of a midline shift or ventricular dilatation in the echo-encephalogram, *reflections from the hematomas* themselves play a certain diagnostic role. Further diagnostic possibilities evolve from this finding in some cases of cerebrovascular accidents. The differences between the sound velocity in brain tissue and in blood have been discussed in detail already (see Table 5). They are great enough to permit the development of recordable reflections between blood and brain tissue. However, if the usual sounding technique is used, a hematoma can be picked up directly only if it is located in the temporal lobe or in the area of the basal ganglia. We were able to pick up pathological reflections originating from hematoma interfaces only 7 times in a series of 32 intracerebral hematomas.

Fig. 170 a shows the echo-encephalogram of a 58-year-old male with hypertensive hemorrhage and rupture into the ventricles. Besides the midline echo shift, a large complex of pathological echoes corresponding to the extent of the hematoma can be seen. A somewhat higher reflection develops at the interface between brain tissue and blood on each side; in between there are multiple smaller echoes from the interfaces of the onion-skin arrangement of the hematoma which consists of

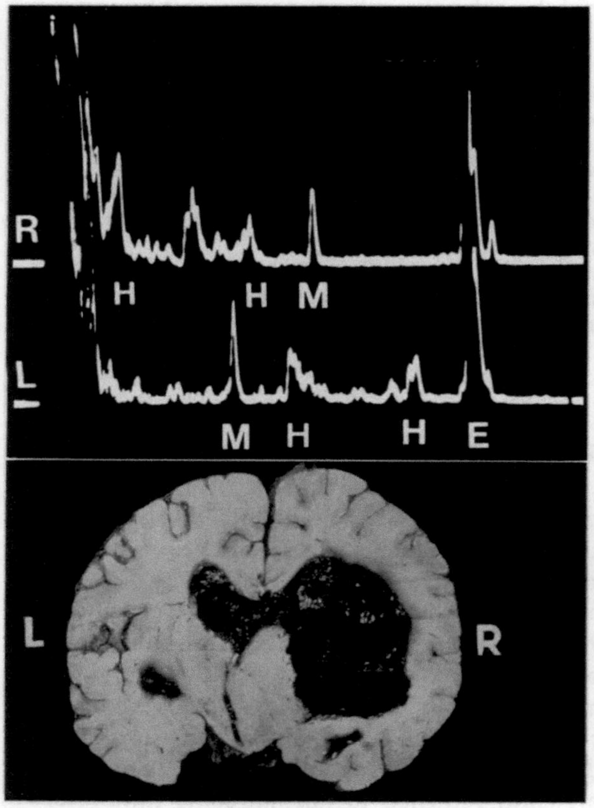

Fig. 170

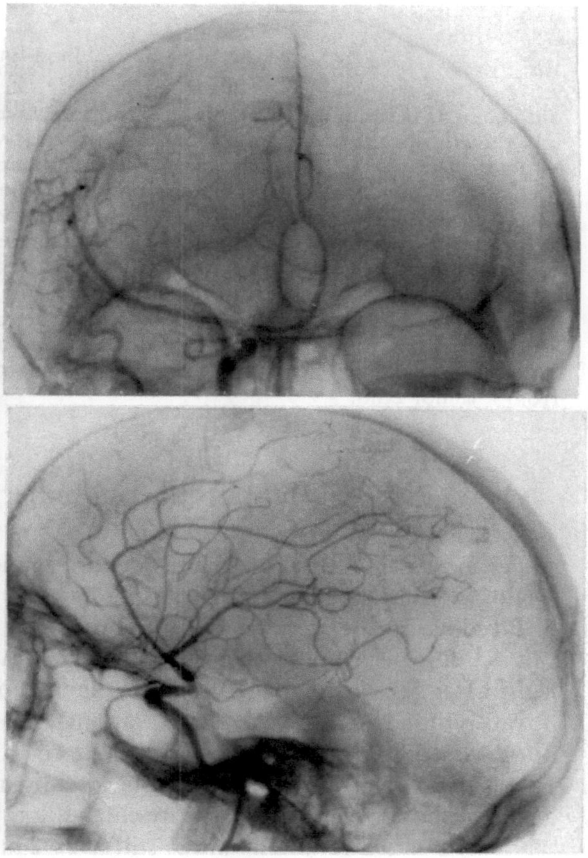

Fig. 171

Fig. 170 a and b. Echogram and autopsy finding in a 58-year-old man with hypertensive hemorrhage and penetration into the ventricle. 12.5 mm midline echo shift. Complex of pathological echoes 45 mm wide H—H, which corresponds to the hemorrhage in the brain section exactly. Pat. G. K., Echo-EG No. 1262/64

Fig. 171. Angiogram of the same patient as in Fig. 170. In contrast to the echogram the signs of displacement are slight

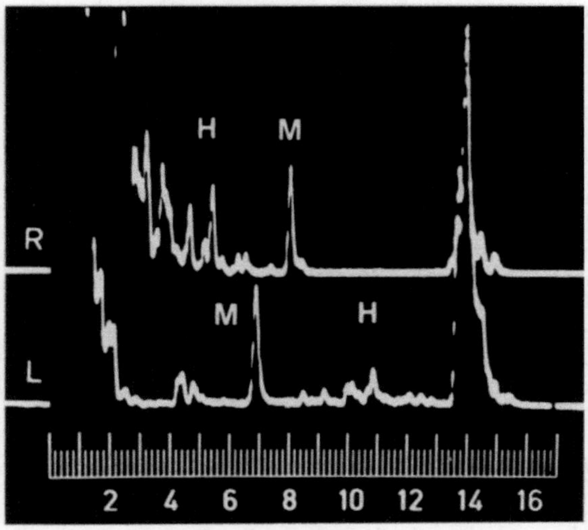

Fig. 172. Echo-encephalogram of a 39-year-old woman with a microangioma of the branches of the middle meningeal artery (see enlarged section of the angiogram) and an intracerebral hematoma. The complex of pathological echoes from the hemorrhage (H) is seen in both traces. Pat. G. R., Echo-EG No. 1003/64

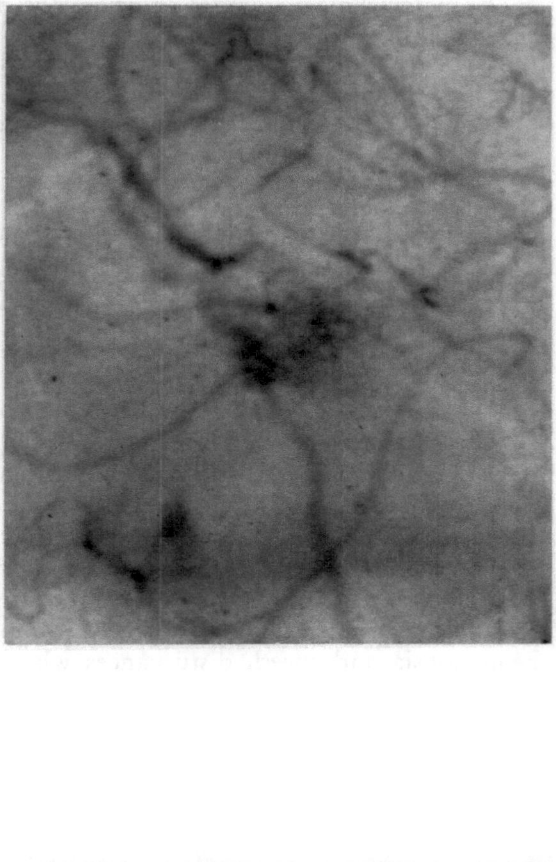

old coagula. Despite the large blood/brain interface which is at right angle to the direction of sounding, the echo peaks originating from this are relatively small and have an unusually broad base. The patient died within an hour after the examination; therefore we can assume that the autopsy specimen of the brain largely corresponds to the relationships within the brain during the echo-encephalographic examination (Fig. 170 b). The good correlation between echogram and brain section at the level of sounding is distinctly recognizable. The diameter of the hematoma in the echo-encephalogram was 45 mm; the same value was found when the autopsy specimen was measured. The carotid angiogram (Fig. 171) revealed a surprisingly small shift.

Fig. 172 shows a further echogram with a hematoma echo complex in a 39-year-old woman with a microangioma of the middle cerebral artery and an intracerebral hematoma. Our own investigations failed to reveal a characteristic shape of the hematoma echo whereby it could be clearly distinguished from a tumor echo.

3. Spontaneous Subarachnoid Hemorrhage

42 patients with spontaneous subarachnoid hemorrhages were examined by echo-encephalography. Of these, 22 patients had aneurysms, 4 had angiomas and 26 showed no source of bleeding on the angiogram. In the majority of the patients echo-encephalography was not carried out until one to two weeks after the acute event. Those patients, who were sounded right after the hemorrhage occurred, had a *normal echogram* if there was not an additional intracerebral hematoma present. Patients, who were examined later often showed a *slight dilatation of the third ventricle* up to a maximum of 10 mm. This finding was obtained in 18 patients (43%). It seems probable that disturbances in the CSF flow due to adhesions, play a role in the development of ventricular dilatations. In 40 cases the midline echo was normal, while two cases revealed initial shifts of 2 mm which returned to normal some time later (see Table 21). No unequivocal midline displacement was found in the angiogram here.

4. Echo-Encephalographic Differentiation between Hemorrhage and Brain Softening

Even for the experienced clinician, the differentiation between massive cerebral hemorrhage and brain softening can be very difficult, if not impossible in some cases. Dalsgaard-Nielsen (1956) for example, reported a series of 239 cases diagnosed as cerebral hemorrhage clinically, of which only 155 = 65% were confirmed by autopsy; in 139 cases with the clinical diagnosis of brain softening, only 81 cases = 58% could be thus confirmed.

Occasionally one encounters the statement that patients with hypertensive hemorrhage are younger than those with arteriosclerotic brain softening. However, it is impossible to make an unequivocal differentiation by this criterion. In addition, hemorrhages are said to occur during work while softening occurs during a period of rest primarily at night. But this symptom is not very useful in a specific case either. The speed at which neurological deficits occur, is also said to differ. According to Scheid's investigation (1963) however, this is not true either. No doubt severe clinical neurological deficits, especially disturbances in consciousness occur more frequently in cases of cerebral hemorrhage. But one-third of all cases of encephalomalacia also lead to deep unconsciousness. Even the appearance of blood in the CSF is not limited to cases of cerebral hemorrhage.

Of the complementary investigations, the EEG is certainly of great importance. However, the EEG can contribute little to the indication for an eventual operation. An isotope investigation in order to establish the diagnosis is out of the question during the acute stage.

Without doubt, the most accurate method for determining the cause of a stroke is *cerebral angiography. However, the indication for this investigation must be examined very carefully because patients with cerebrovascular disease are at a greater risk.* Bull, Marshall and Shaw (1965) for example, observed complications of angiography such as hemiparesis, hemianopsia and speech disturbances which normalized rather slowly, 9 times in a series of 80 patients, i. e. 11%. TÖNNIS and SCHIEFER have pointed out the special dangers of angiography for patients with diffuse cerebrovascular disease especially those with arteriosclerosis (see also SCHIEFER and STEINMANN, 1958).

Therefore angiography during the acute stage is justified only if it has therapeutic consequences. The best examination to precede this investigation is echo-encephalography in our opinion.

In their investigations TAYLOR, NEWELL and KARVOUNIS (1961) arrived at the conclusion that echo-encephalography makes it possible to differentiate between hemorrhage and softening in every case. JEPPSSON (1961) found a distinct displacement of the midline echo in all 14 cases of intracerebral hematoma, whereas 56 cases of encephalomalacia produced a midline displacement, only 8 times which was statistically significantly less than with hemorrhages. Later investigators largely confirmed these statements. Table 22 summarizes 239 cases of

Table 22. *Echo-encephalographic findings with "stroke" (according to the literature)*

Author	Year	Intracerebral Hematoma		Brain Softening	
		No. of Cases	Midline Echo Shifted	No. of Cases	Midline Echo Shifted
JEPPSSON	1961	14	14	56	8
SUGAR and UEMATSU	1964	3	3	7	—
NAGAI et al.	1964	7	6	20	1
TANAKA et al.	1964	41	41	67	—
ACHAR et al.	1966	13	10	71	2
MITSUNO et al.	1966	40	37	10	1
WIDÉN et al.	1966	15	13	50	3
DE VLIEGER	1967	49	33	383	68
SCHIEFER and KAZNER	1967	32	31	72	19
PIA and GELETNEKY	1968	25	18	7	5
Total		239	206	743	107
Percentage of Midline Echo Shifts			86.2		14.4

hematoma and 743 cases of brain softening found in the literature. A midline echo shift was found in 206 (86.2%) of the hematomas, in contrast to only 107 (14.4%) with encephalomalacias. However, these summarized results necessitate further discussion.

Our own case material shows that it is not so simple, however. If a midline echo shift exists on the first day or within the first hours, the probability of finding a hematoma is great. If echo-encephalography is carried out somewhat later, the decision is considerably more difficult since vascular occlusion and other forms of encephalomalacia also lead to mass shifts due to cerebral edema in 20 to 25% of cases. Even though this shift is generally not as great as with hematomas, it is still not possible to differentiate between them with certainty through echo-encephalography when faced with a specific case. *This proves that the decisive echo-encephalographic criterion is the time at which the midline echo shift appears.* We perform angiography immediately in all patients with signs of a mass shift in the echogram and prefer to wait in patients with a stroke and normal ultrasonic findings.

The demonstration of a *complex of pathological echoes* in the region corresponding to the area of the basal ganglia or the temporal lobe can support the diagnosis of an intracerebral hematoma providing the clinical symptoms point to such a hematoma. A definite distinction between these and tumor echo complexes on the basis of the ultrasonic trace alone is not always possible in our experience. Echo-encephalography, however, is a valuable adjunct to angiography in these cases because the angiograms of patients with lesions in the basal ganglia often reveal only very slight shifts of the cerebral vessels. PIA and GELETNEKY (1968) are of the opinion that operative evacuation of a hematoma can be performed even without preceding angiography if hematoma echo complexes are demonstrated.

Some authors go so far as to state that echo-encephalography is more reliable for the detection of intracerebral hemorrhages than carotid angiography (MITSUNO, KANAYA, SHIRIKATA, OSHAWA and ISHIKAWA, 1966). They believe that although angiography reveals displacements of the anterior cerebral artery and middle cerebral artery, these signs are not absolute proof of massive intracerebral hemorrhage, since edema and other space-occupying lesions can also produce such vascular displacements. With echo-encephalography on the other hand, a clear decision can be made from the midline echo shift (92,5%) together with the characteristic hematoma reflections (95%).

KANAYA, YAMASAKI, SAIKI und FURUKAWA (1968) were able to confirm these observations on a large series of patients. A midline echo shift was found in 43 of 52 cases (83%) of intracerebral hemorrhage. A hematoma echo described as a "characteristic multiple spike-like pattern" was found in 49 of 52 of these cases (94%) thereby making it possible to differentiate them from cases of brain softening. 7 of the 12 patients, who failed to reveal a displacement of the anterior cerebral artery in the angiogram, had a midline echo shift in the ultrasonic trace. The authors therefore believe that the diagnostic accuracy of the ME shift is superior to that based on a displacement of the anterior cerebral artery.

Especially in older patients with arteriosclerotic hemorrhages at the typical site in the basal ganglia, a midline echo shift of about 4.0 mm may be observed, while the angiogram does not show signs of displacement. Occasionally one sees a flattening of the arch formed by the anterior and middle cerebral artery (RICHARDSON, 1962) or a displacement of the vessels of the basal ganglia (lenticulostriate and choroid arteries). However, it seems entirely possible that smaller hemorrhages in the region of the internal and external capsules in the presence of an atrophic brain lesion fail to produce any signs of displacement in the echogram either. Further investigations are necessary here.

A complete etiological diagnosis of stroke and the localization of a massive cerebral hemorrhage is possible only in individual cases. However, the method simplifies the selection of patients, who are possible candidates for instrumental diagnostic investigation and eventual operation. A distinct mass shift can be detected in cases of intracerebral hematoma on the day of the acute incident already, while cases of brain softening do not cause a shift at first. Some time must pass before displacements of the midline structures can be detected in even a portion of the vascular occlusions and other forms of brain softening. Only in exceptional cases — for example when softening affects an entire hemisphere — is it possible to record a more extensive displacement of the midline echo. Hemorrhages in the internal or external capsule often produce a complex of abnormal reflections at the site of the bleeding. Fresh encephalomalacias in older patients not infrequently reveal a moderately dilated third ventricle in the echo-encephalogram indicating previously existing brain atrophy. The echo-encephalographic picture of ventricular dilatation may be found in cases of cerebellar hematoma and hemorrhage into the ventricular system as well. The extent of third ventricular dilatation exceeds the values measured in cases of softening, by about 2 mm. All in all, echo-encephalography represents an important adjunct to the differential diagnosis of stroke; however, a differentiation is not possible in every case.

5. Postapoplectic States, Atrophic Processes

63 patients, who were examined by echo-encephalography had a history of an earlier apoplectic insult or were investigated for a space-occupying lesion because of unclear

Table 23. *Width of the third ventricle in the echo-encephalogram in 63 patients with cerebrovascular disease*

Width of 3rd Ventricle in the Echo-EG (mm)	—7.0	—8.0	—9.0	—10.0	—11.0	—12.0	More
No. of Cases	4	14	13	22	7	1	2

neurological symptoms. In 59 cases the echo-encephalogram showed a *dilatation of the third ventricle* (more than 7.0 mm in width). In 4 patients the width of the third ventricle reached the borderline value of 7.0 mm. Since these were exclusively younger patients, a dilatation of the third ventricle had to be assumend here, too. The width of the third ventricle in the 63 patients with cerebrovascular disease is presented in table 23. The majority of these patients had third ventricles whose width ranged from 8 to 10 mm with a mean of 9.3 mm, corresponding to a ventricular diameter of 11.5 to 13.0 mm in the pneumoencephalogram.

In 27 patients the echo-encephalographic findings were checked by pneumoencephalography. When the enlargement due to the x-ray technique was drawn into consideration, complete agreement was found in all cases. Fig. 173 shows the echo-encephalogram of a 58-year-old woman with an atrophic brain lesion where the abnormal double echo of the third ventricle is 10 mm apart. The pneumoencephalogram which was performed later revealed the same width of the third ventricle (Fig. 173 top).

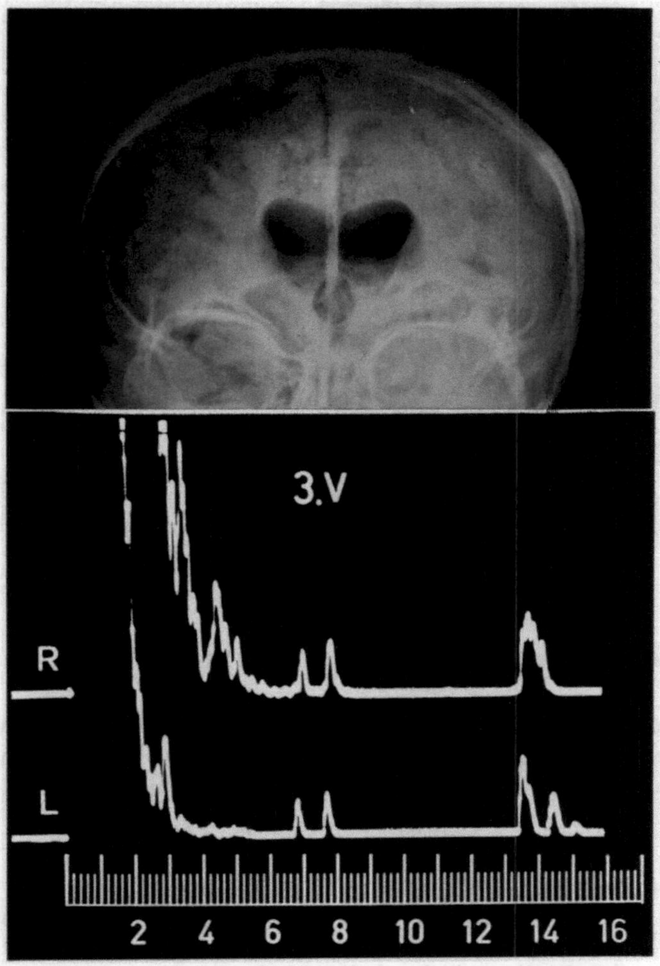

Fig. 173. Echo-encephalographically demonstrated dilatation of the third ventricle to 10 mm in the case of an atrophic brain lesion. The pneumoencephalogram which was performed later is on top. Pat. E. F., 58 years of age. Echo-EG No. 17/62

Displacements of the midline echo are generally absent in cases of atrophic brain lesion in older patients. Only once did we see a 2.0 mm shift of the M-echo which could be confirmed by pneumoencephalography and which was found to be the result of a slight asymmetry.

Occasionally the echo-encephalographic investigation can contribute to the exclusion of a tumor of the cerebral hemispheres by revealing a distinct ventricular dilatation; the following case history is an example in point:

Case 24: Hermann B., 59 years of age. Echo-EG No. 938/63. The patient was referred to us for operation from an out-of-town neurological clinic with the diagnosis of a right-sided brain tumor in the temporal lobe. The angiograms which were sent along showed an elevation of the middle cerebral artery while the anterior cerebral artery was midline. Clinically, the patient had a left-sided hemiparesis which had worsened at intervals. In the echo-encephalogram a considerable dilatation of the third ventricle up to a width of 12.5 mm was seen (Fig. 174 bottom). This was already enough to exclude a space-occupying cerebral lesion. Pneumoencephalography revealed a marked dilatation of the right lateral ventricle which explains the elevation of the middle cerebral artery; the width of the third ventricle was confirmed (Fig. 174 top). Therefore this was a case of cerebrovascular disease with intermittent encephalomalacias of the right cerebral hemisphere.

LE MAY and OJEMANN pointed to such difficulties in the angiographic differentiation between temporal lobe tumors and circumscribed dilatations of the temporal horn as early as 1963. With the help of echo-encephalography, confusion can be largely avoided.

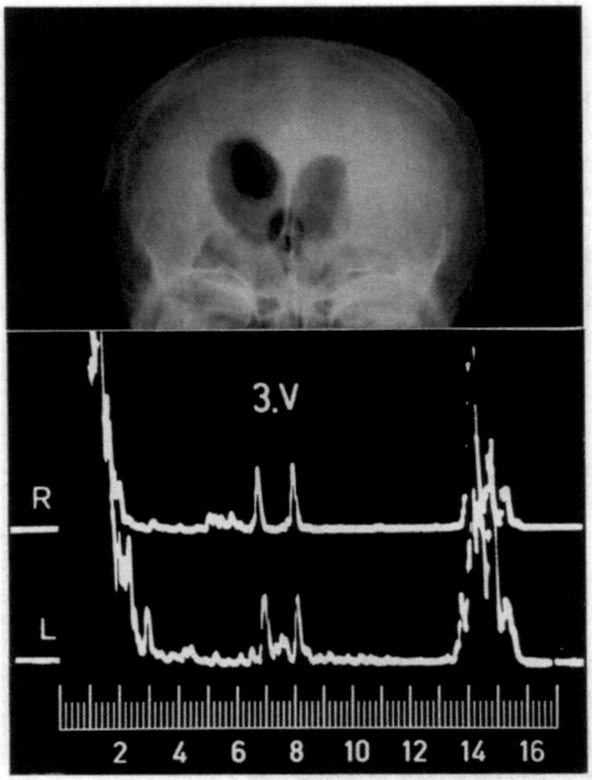

Fig. 174. Exclusion of a brain tumor by echo-encephalography. Further details in text. Pat. H. B., 59 years of age. Echo-EG No. 938/63

E. Echo-Encephalography with Infantile Hydrocephalus

The syndrome of infantile hydrocephalus is one of the most worthwhile areas for the use of echo-encephalography. LITHANDER (1961) was the first to report on the measurements of the width of certain sections of the ventricular system (third ventricle, lateral ventricles) in children with the help of the ultrasound pulse echo method. Control pneumoencephalographic investigations indicated general correlation between the findings. 10 cases of ventricular dilatation in the echo-encephalogram were reported.

SCHIEFER, KAZNER and BRÜCKNER (1963) also used echo-encephalography for the diagnosis of infantile hydrocephalus and were able to confirm LITHANDER's results. In addition to the determination of the width of the third ventricle and the anterior horns, they emphasized the *measurement of the brain mantle thickness around the temporal horn.* Besides this, they found that echo-encephalography was very useful for follow-up studies in cases where ventricular drainage systems had been inserted. In 10 children, a slight postoperative decrease in ventricular dilatation was diagnosed by ultrasound.

GELETNEKY (1964) and HEMMER (1964) also confirmed the usefulness of echo-encephalography in the diagnosis and follow-up of infantile hydrocephalus. HATA, MARUTA, YOSHIDA, ISHIDA and INABA (1964) performed post-operative ultrasonic investigations in children with hydrocephalus in order to evaluate the sucess of the operation. They used not only the temporal sounding method but also application points in the frontal area and at the fontanelles. In 1965 UMBACH and KLEY published a report on echo-encephalographic investigations for the diagnosis and follow-up of infantile hydrocephalus in 75 children, which concludes that echo-encephalography is a useful procedure for the detection of normal and dilated lateral and third ventricles. Comparisons with helium encephalograms showed an exact correlation between the two methods. In 28 children post-operative follow-up studies were carried out which made it possible to state the size of the ventricles and the thickness of the brain mantle at any given time. The autors arrive at the conclusion that echo-encephalography represents a considerable simplification in diagnostics especially after shunt operation. UMBACH and KLEY suggest 4 different measurements in children with hydrocephalus; 1. third ventricle, 2. width of the lateral ventricles so that the SCHIERSMANN's index can be calculated, 3. thickness of the brain mantle in the occipital area, 4. brain mantle thickness in the frontal area. On the basis of echo-encephalographic and pneumoencephalographic comparative investigations in children with hydrocephalus of various etiologies, SJÖGREN (1964, 1965, 1967, 1968) also arrived at the conclusion that "echo-ventriculography" is a simple, useful and safe technique for early diagnosis and follow-up study of infantile hydrocephalus. WEST (1967) is of the same opinion.

The ultrasonic trace of infantile hydrocephalus is characterized by reflections of exceptionally high amplitude. This is partly due to the fact that the reflected ultrasound must penetrate only a relatively thin bony lamella (0.5 to 4.0 mm, GELETNEKY, 1966) and partly because the enlarged ventricular walls reflect more ultrasound. In most cases, the thickness of the brain mantle in the temporal horn area can be measured echo-encephalographically without any difficulty (Fig. 175 and 176). For this it is necessary to apply the probe somewhat higher up and to direct the ultrasonic beam slightly towards the base of the skull. Comparisons between pneumoencephalograms on an anterior-posterior view with the child lying on its side and the echo-encephalographic results show a nearly complete agreement of the findings (see also WEST, 1967). In cases of slight to moderate ventricular dilatation, *the third ventricle can also be demonstrated by means of an abnormal double echo* on the screen; in more marked cases of internal hydrocephalus, it is often impossible to obtain satisfactory reflections from the lateral walls of the third ventricle. The borderline value for the width of the third ventricle to be recorded by echo-encephalography appears to be about 25 to 26 mm (compare Fig. 177) in our experience. Ventricular dilatations exceeding this diameter apparently push the structures bordering the third ventricle so far apart that it is no longer possible to hit vertically both lateral walls with the ultrasonic beam simultaneously.

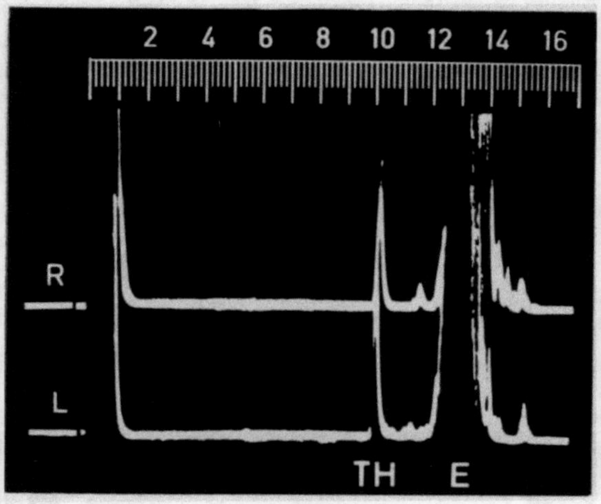

Fig. 175

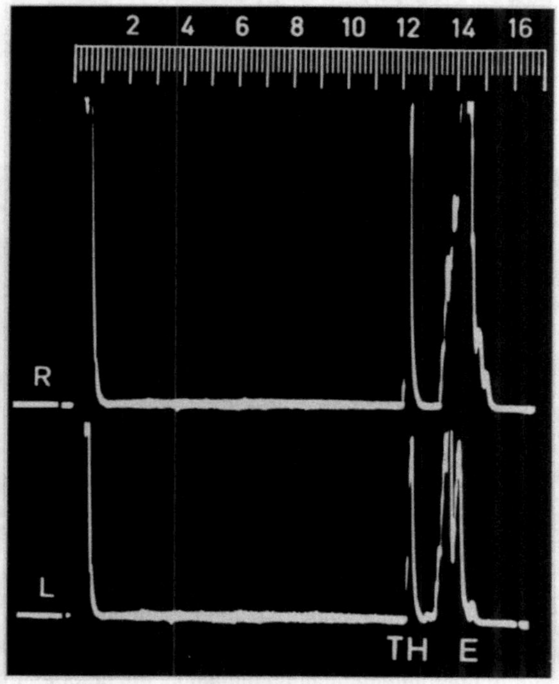

Fig. 176

Fig. 175. Measuring the brain mantle thickness in the temporal horn area (TH) with the help of echo-encephalography in infantile hydrocephalus. Brain mantle index 2.7 Echo-EG No. 1565/65

Fig. 176. Echo-encephalogram showing the external wall of the temporal horn in a severe hydrocephalus. Brain mantle thickness 10 to 12 mm. Brain mantle index 5.5. Pat. S. G., 7 months of age, Echo-EG No. 1995/65

The *echo-encephalographic brain mantle index* (p. 61 f.) is often extremely high in cases of infantile hydrocephalus. Values between 3 and 5 are no rarity; the maximum value in our investigation series was 9.0 with a brain mantle thickness of 6 mm. In cases of severe internal hydrocephalus (from a BMI of 3.5 upwards) the thickness of the brain mantle in the frontal and occipital area can also be measured by echo-encephalography (see Fig. 178).

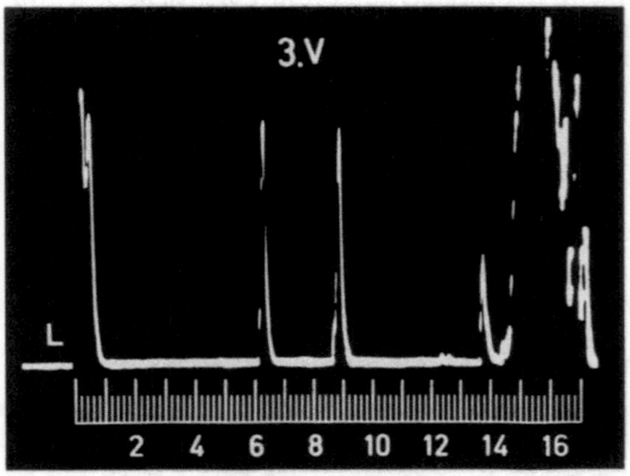

Fig. 177. Markedly dilated third ventricle in a case of infantile hydrocephalus. Abnormal double echo 26 mm apart at the centre of the ultrasonic trace. Pat. M. S., 18 months of age, Echo-EG No. 2576/66

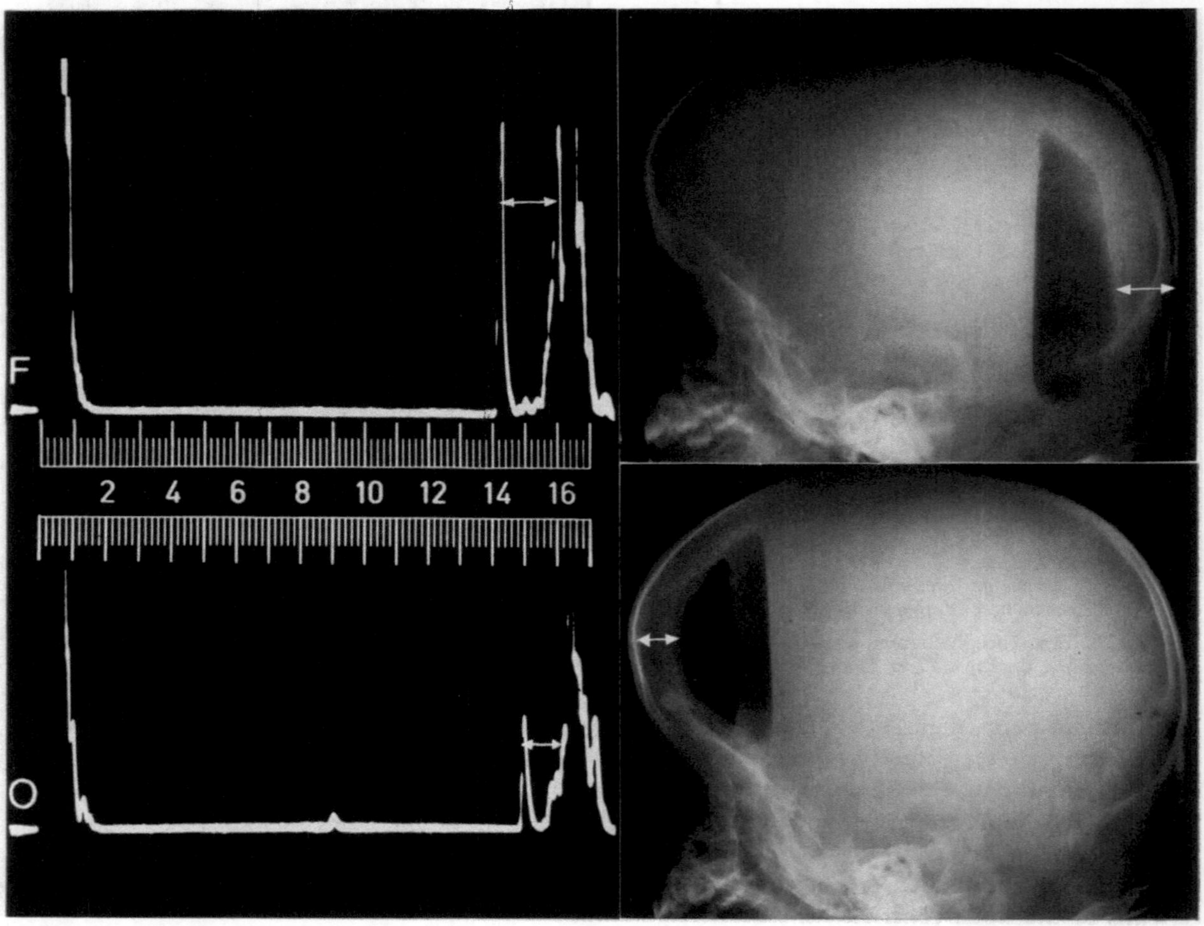

Fig. 178. Determining the thickness of the brain cortex in the frontal and occipital area by means of ultrasound. F: sounding from the frontal area towards the occipital region. High echo from the ventricular wall ahead of the occipital end echo. O: Sounding from the occipital area towards the frontal region. Echo from the ventricular wall ahead of the frontal end echo. Complete agreement with the pneumoencephalogram. Pat. M. S., 7 months of age, Echo-EG No. 2518/66

The technique of investigation used for *determining the width of the lateral ventricle in the body region* is shown diagrammatically in the lower part of Fig. 179. Above it, is an original echogram of a 3-month-old infant with moderate ventricular dilatation which shows the 3 characteristic reflections of this trace: high midline echo from the cerebral falx or the septum as well as the two echoes from the outer walls of the lateral ventricles which appear in front of and behind the midline echo. From these values the SCHIERSMANN ventricular index can be calculated. A further example of this is seen in Fig. 180. In most cases the amount of hydrocephalus on the two sides is equal. In a series of 103 children, we found genuine asymmetry resulting in a displacement of the third ventricle up to a maximum of 3 mm and a unilateral thinner brain mantle remnant only twice. An example of this is presented in Fig. 181. In 3 patients the midline echo was entirely absent (agenesis of the corpus callosum?). The results of echo-encephalographic examination of 103 children with hydrocephalus of various causes are summarized in Table 24.

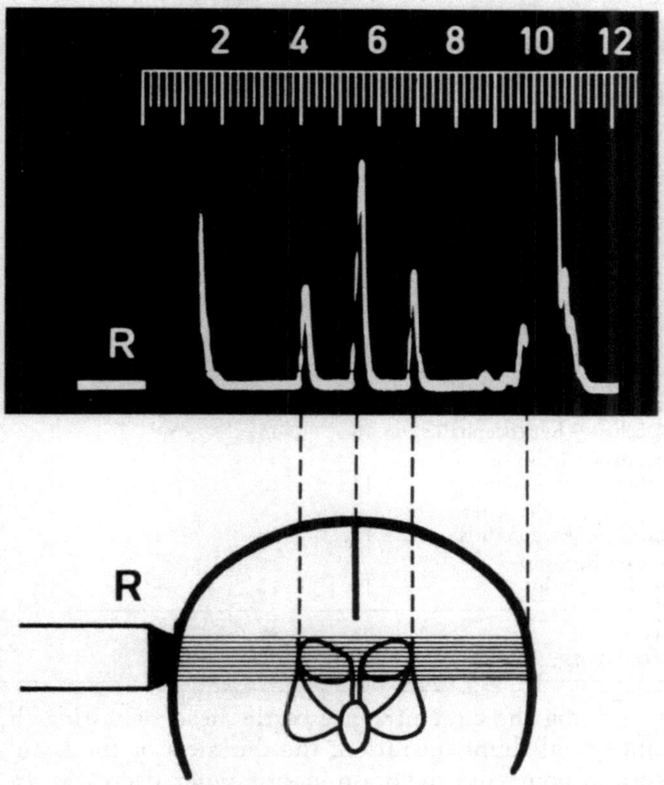

Fig. 179. Echogram with echoes from both lateral ventricles and diagrammatic illustration of ventricular sounding. Pat. R. G., 3 months of age, Echo-EG No. 1402/64

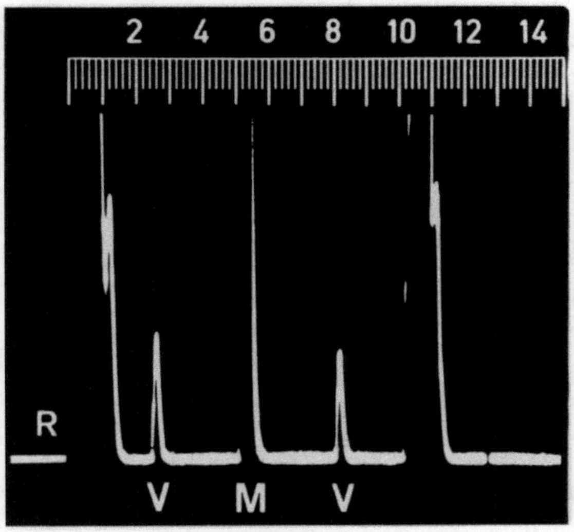

Fig. 180

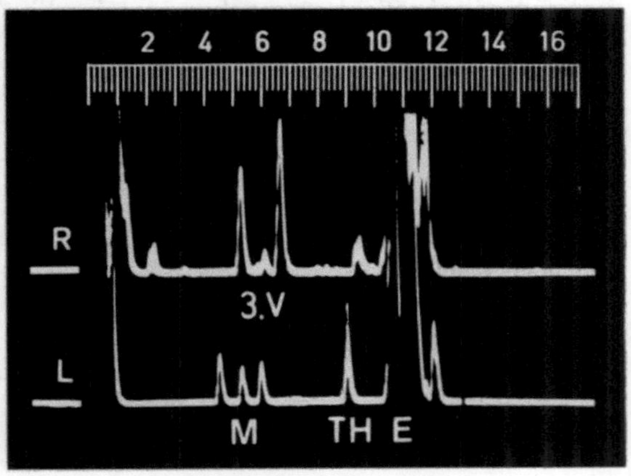

Fig. 181

Fig. 180. Measuring the width of both lateral ventricles (V—V) in a severe hydrocephalus. Pat. R. Z., 9 months of age, Echo-EG No. 2093/66

Fig. 181. Asymmetrical ventricular dilatation. The double echo from the third ventricle is shifted 3 mm towards the left. The brain mantle on the left is considerably thinner than on the right (10 mm to 13 mm). Pat. D. B., 6 months of age, Echo-EG No. 1263/64

Up to the present, echo-encephalography is the only method of demonstrating the size of the ventricles precisely following ventricular shunt operation. The success of such operations can be followed accurately in this manner and the method is much more reliable than

Table 24. *Results of echo-encephalography in 103 children with hydrocephalus of various causes*

Lesion	No. of Cases	Midline Echo Normal	Midline Echo Shifted	Midline Echo not Detectable	Width of the 3rd Ventricle Normal	Width of the 3rd Ventricle Enlarged	Width of the 3rd Ventricle not Measurable	Brain Mantle Index (BMI) Exceeding 2.3
Congenital communicating hydrocephalus	20	19	—	1	—	17	3	20
Hydrocephalus with/after myelocele	25	24	1	—	—	25	—	22
Occlusive hydrocephalus	16	15	—	1	—	13	3	16
Hydrocephalus due to infantile brain damage	13	13	—	—	—	13	—	6
Hydrocephalus after meningo-encephalitis	10	9	1	—	—	10	—	6
Other forms of hydrocephalus	19	18	—	1	—	16	3	8
Total	103	98	2	3	—	94	9	78
Percentage	100	95	2	3	—	91	9	76

measuring the circumference of the head or testing the function of the valve. After ventriculo-atrial shunt operation, the thickness of the brain mantle increases continually up to a certain point and the brain mantle index decreases. In the same way, the width of the third ventricle decreases if no complications occur. Occasionally if the operation is performed during an early stage, complete normalization may occur. A relapse due to improper functioning of the drainage system or complete blockage may be detected in the echo-encephalogram immediately if the hydrocephalus remains stationary or begins to increase.

In 35 children, *postoperative follow-up studies* were carried out following ventriculo-atrial shunt operations (Holter valve). The results of these follow-ups are presented individually in Table 26 and summarized in Table 25. In 28 children (80%), an improvement

Table 25. *Echo-encephalographic findings after ventriculo-atrial drainage* (Holter system) *with infantile hydrocephalus*

Lesion	No. of Cases	Echo-Encephalographic Finding Postoperatively		
		Improved	Unchanged	Deteriorated
Communicating hydrocephalus	10	8	2	—
Hydrocephalus with/after myelocele	12	11	1	—
Occlusive hydrocephalus	10	8	—	2
Other forms of hydrocephalus	3	1	2	—
Total	35	28	5	2
Percentage	100	80	14	6

was noted, i. e. the thickness of the brain mantle had increased. Correspondingly a decrease in the echo-encephalographic brain mantle index (BMI) was calculated and it was found that the width of the third ventricle had also decreased in most of these patients. 5 children (14%) remained practically unchanged, partly due to failing or absent valvular function caused by plexus tissue which had grown into the ventricular catheter, partly because atrophic brain lesions contributed to the development of the hydrocephalus. In 2 children we had to register an increase in the hydrocephalus. In a single case, the flow through the valve was not enough although the valve was completely patent, in the other child the entire drainage system had to be removed because of recurrent E. coli meningitis. In addition, there was an decrease in ventricular dilatation at first which was followed by a subsequent increase in the hydrocephalus in 3 children due to a block in the ventricular catheter; we were able to diagnose this immediately by echo-encephalography. After replacing the ventricular catheter, the hydrocephalus again decreased. If the valve fills only very slowly after manual emptying, then there is nearly always a stagnation in the ventricular width or even a renewed increase in the hydrocephalus (see Table 26).

Our investigations have always shown that the brain mantle increases in width relatively quickly at first following a CSF drainage operation, then more and more slowly. This is especially so in cases of hydrocephalus accompanying or following meningomyeloceles. In older children, the tendency of the internal hydrocephalus to return to normal is distinctly less.

The degree of ventricular dilatation and the decrease in the hydrocephalus can be determined best by measuring the thickness of the brain mantle in the temporal area by echo-encephalography and calculating the echo-encephalographic brain mantle index from this. We prefer this evaluation of the echo-encephalogram to other calculations because the natural enlargement of the head over a period of time never results in a falsification of the measurements. The third ventricle especially is not a very good standard of controlling cases of infantile hydrocephalus because it must always be regarded in the relation to the diameter of the head and this distance increases rapidly in the first years of life (see Table 26). Lateral ventricular indices are also a good standard (UMBACH and KLEY, 1965; SJÖGREN, 1965—1968).

The results obtained during echo-encephalographic follow-up studies permit us to conclude that a general normalization in ventricular width can be expected in infants with a brain mantle index up to 4.0. But even in very severe cases of hydrocephalus (BMI up to 7.0 or more) do not appear to be completely hopeless, an operation which will decrease the amount of CSF pressure should be attempted even if the brain mantle is only about 1 cm thick, especially since little experience has been accumulated in this area.

Of more importance however, is the *early diagnosis of hydrocephalus which is greatly simplified by echo-encephalography*. In recent years we have routinely carried out echo-encephalographic investigations at regular intervals in children with meningomyeloceles before and after the malformation was operated. If the brain mantle index is 3.0 to 3.5, we perform the operation without previous pneumoencephalography, solely on the basis of the ultrasonic findings and the clinical symptomatology. Such a procedure is certainly justified since there are no doubts about the cause of the hydrocephalus in these cases. A similar course is followed in cases of occlusive hydrocephalus as a sequel of meningitis. However, in all other cases of hydrocephalus, the cause must be elucidated by means of pneumoencephalography or ventriculography, so as not to overlook a cerebellar tumor for example.

The undeniable advantages of an early operation and echo-encephalographic follow-up studies in cases of infantile hydrocephalus are illustrated by the following history:

Case 27: Bettina L., born July 20, 1965. Echo-EG No. 1759/65. Admission at the age of four days because of a lumbar myelocele. The width of the third ventricle before the malformation was operatively removed, was 6.0 mm, the brain mantle index 3.0. Diagnosis: moderate ventricular dilatation. Repeated postoperative control investigations revealed an increase in the width of the third ventricle up to a diameter of 9.0 mm and the BMI was calculated to be 3.5. On August 17, 1965 a CSF-drainage-operation was performed withouth previous pneumoencephalography. The width of the third ventricle had become almost normal within 6 weeks of the operation already (third ventricle 7.0 mm in width, BMI 2.25).

9 months later, sudden rapid increase in the circumference of the head. The valve can be emptied but refills only very slowly. Echo-encephalogram: third ventricle 18.0 mm, BMI 4.5. After replacing the ventricular catheter which had been plugged by choroid plexus, the hydrocephalus decreases to a BMI of 2.4 and the third ventricle to 10.5 mm within 4 months!

Table 26. *Echo-encephalographic follow-up studies of 35 children with operated hydrocephalus*

Echogram No.	Name	Age at Operation	Age at Control	Echo-Encephalographic Finding: Before Operation — Third Ventricle (mm)	Before Operation — BMI	At Control — Third Ventricle (mm)	At Control — BMI	Remarks	Type of Lesion
52/63	O. F.	3 mo.	3 y. 3 mo.	14.5	2.8	9.5	2.5		Hydrocephalus with myelocele
296/62	R. H.	16 mo.	2 y.	—	4.6	—	3.6		Hydrocephalus with myelocele
619/63	W. L.	4 mo.	3 y. 4 mo.	16.0	3.4	18.0	4.0	Slight increase in hydroc. despite good valve function	Occlusive hydrocephalus
834/63	R. P.	3 mo.	3 y. 5 mo.	10.0	—	—	6.7	Removal of valve because of meningitis	Occlusive hydrocephalus after E. coli meningitis
1051/64	P. L.	6 y.	8 y.	22.0	4.3	17.0	2.7		Occlusive hydroc.
1220/64	I. M.	$2^1/_2$ mo.	1 y.	12.0	4.0	13.5	3.6	Valve fills only slowly	Hydrocephalus with myelocele
1263/64	D. B.	4 mo.	1 y. 3 mo.	20.0	4.4	15.0	3.1	Transient deterioration, insertion of a new ventricular catheter	Occlusive hydrocephalus
1295/64	J. H.	1 mo.	2 mo.	14.0	4.8	11.5	3.8		Communicating hydrocephalus
1402/64	R. G.	2 mo.	2 y. 2 mo.	9.5	2.7	6.0	2.3	Transient increase of BMI up to 3.4; insertion of a new ventricular catheter	Communicating hydrocephalus
1439/64	D. K.	2 mo.	9 mo.	—	—	7.0	2.5	Distinct improvement in comparison to operative finding	Hydrocephalus with myelocele
1441/64	O. M.	3 mo.	2 y.	10.0	3.0	8.0	2.3		Hydrocephalus with meningocele
1501/64	K. K.	9 mo.	1 y. 10 mo.	12.0	3.1	7.0	2.6		Hydrocephalus with myelocele
1552/65	R. K.	3 mo.	1 y. 7 mo.	7.0	3.3	5.0	2.2		Occlusive hydroc. after meningitis
1598/65	E. M.	5 mo.	1 y. 3 mo.	14.0	3.8	12.0	2.6		Hydrocephalus with myelocele
1603/65	E. W.	6 mo.	1 y. 6 mo.	14.0	4.5	5.0	2.2		Communicating hydrocephalus
1615/65	J. M.	3 mo.	4 y. 3 mo.	—	—	6.0	2.2	Brain mantle only 15 mm in thickness at operation	Communicating hydrocephalus

Echogram No.	Name	Age at Operation	Age at Control	Echo-Encephalographic Finding: Before Operation, Third Ventricle (mm)	Before Operation, BMI	At Control, Third Ventricle (mm)	At Control, BMI	Remarks	Type of Lesion
1627/65	B. L.	3 mo.	1 y. 5 mo.	—	7.1	9.0	3.0		Communicating hydrocephalus
1676/65	A. H.	1 y. 9 mo.	2 y.	12.0	4.3	10.0	3.5	After op. occipital meningocele disappeared	Hydrocephalus with meningocele
1688/65	B. B.	3 mo.	2 y.	—	3.2	—	3.1	Valve fills only slowly	Communicating hydrocephalus
1765/65	R. D.	4 mo.	1 y. 6 mo.	17.0	5.2	17.0	4.0		Communicating hydrocephalus
1759/65	B. L.	1 mo.	1 y. 4 mo.	9.0	3.5	10.5	2.4	Transient increase of BMI up to 4.5; insertion of a new ventricular catheter	Hydrocephalus with myelocele
1788/65	F. Z.	1 mo.	1 y. 4 mo.	—	3.8	—	3.0		Occlusive hydroc.
1827/65	H. S.	3 y.	6 y.	—	—	9.5	3.0	Distinct improvement in comparison to operative finding	Occlusive hydroc.
1900/65	O. H.	1 y.	1 y. 7 mo.	—	4.0	23.0	4.1	No improvement despite good valve function; brain atrophy?	Hydrocephalus following subdural effusions
1955/65	J. G.	7 mo.	1 y. 3 mo.	22.0	5.5	14.5	3.7		Communicating hydrocephalus
2012/65	J. K.	1 mo.	2 mo.	10.0	3.0	7.0	2.6		Hydrocephalus with myelocele
2039/66	R. Z.	9 mo.	1 y. 3 mo.	10.0	3.1	7.5	2.3		Hydrocephalus with myelocele
2369/66	M. E.	2 mo.	4 y. 5 mo.	—	—	18.0	4.0	No improvement despite good valve function; brain atrophy?	Hydrocephalus following perinatal intracranial hemorrhage
2384/66	P. W.	3 mo.	8 mo.	—	7.7	17.5	5.6		Occlusive hydroc.
2385/66	A. S.	5 mo.	9 mo.	12.5	4.2	9.5	2.9	Valve fills only slowly	Occlusive hydroc.
2392/66	B. J.	5 mo.	6 mo.	8.5	2.6	9.0	2.6	Valve fills only slowly	Communicating hydrocephalus
2435/66	P. H.	4 mo.	9 mo.	16.0	4.9	15.5	3.4		Postmeningitic occlusive hydroc.
2518/66	M. S.	7 mo.	9 mo.	25.0	6.1	16.0	4.7		Communicating hydrocephalus
2576/66	M. S.	18 mo.	1 y. 8 mo.	26.0	7.6	24.0	6.8		Hydrocephalus of unknown origin
2621/66	M. U.	$2^1/_2$ mo.	3 mo.	16.5	5.0	10.0	3.2		Hydrocephalus with myelocele

F. Other Applications of Echo-Encephalography

1. Subdural Effusions and Hematomas in Infants and Children

Although LITHANDER regarded the use of echo-encephalography in the diagnosis of subdural effusions and hematomas in small children as useless as late as 1961, GELETNEKY (1965 and 1966) as well as JACOBI, KAZNER and WOLLENSAK (1966) were able to describe certain characteristics of the echo-encephalograms in this group of lesions which facilitate the diagnosis of subdural effusions and hematomas in numerous cases.

According to LITHANDER and GELETNEKY *even larger unilateral hematomas do not lead to midline echo shifts;* on the other hand we found slight to moderate shifts of the midline echo which could be confirmed by dye studies in all 7 cases of unilateral fluid accumulations in a group of 45 children. 3 of these children were under one year of age. In 38 children, the hematomas or effusions were bilateral. 37 times the midline echo was at the normal site but in one child the midline structures were displaced by 3.5 mm. This finding can be explained by the difference in the amount of fluid accumulation on the two sides.

While the determination of the midline structures generally does not give any diagnostic pointers in infants with subdural effusions, another echo-encephalographic sign deserves increasing attention: *the width of the third ventricle.* This was measured echo-encephalographically in 31 children by means of the characteristic double echo. Particularly remarkable here was the fact that the average diameter of the third ventricle which was 9.0 mm (range of variation 6 to 15 mm) in 26 cases, was considerably larger than the age norm. However, the amount of ventricular dilatation usually did not reach the values obtained in cases of infantile hydrocephalus. Furthermore, the ventricular echoes originating from the inferior horn could be recorded with great regularity.

Table 27. *Subdural hematomas and hydromas in infancy* (PIA and GELETNEKY, 1968)

Type of Lesion	No. of Cases	Midline Echo Displaced	Hematoma Echo	Third Ventricle Dilated
Unilateral	12	8	9	2
Bilateral	30	3	24 [a]	6
Small hydromas without membranes	15	—	—	—
Total	57	11	33 (60%) 80%	8

[a] 4 unilateral

However, the most significant diagnostic sign in the echo-encephalogram is the *additional reflection developing at the interface between brain, hematoma membrane and fluid accumulation* which can be compared with the hematoma echo observed in cases of chronic subdural hematoma in adults. Just as in adults, however, a special technique of investigation is necessary to pick up these additional echoes. The probe is tipped upwards and directed towards the fluid accumulation (Fig. 157). In this way the interface of brain/fluid accumulation is hit perpendicularly — an absolute necessity for the development of useful reflections. In 33 of 45 children in our own investigation series, reflections from these interfaces were registered; these appear as needle-sharp echoes ahead of the end echo ("effusion echoes"). Figs. 182 to 185 show typical photographic examples. From the distance between

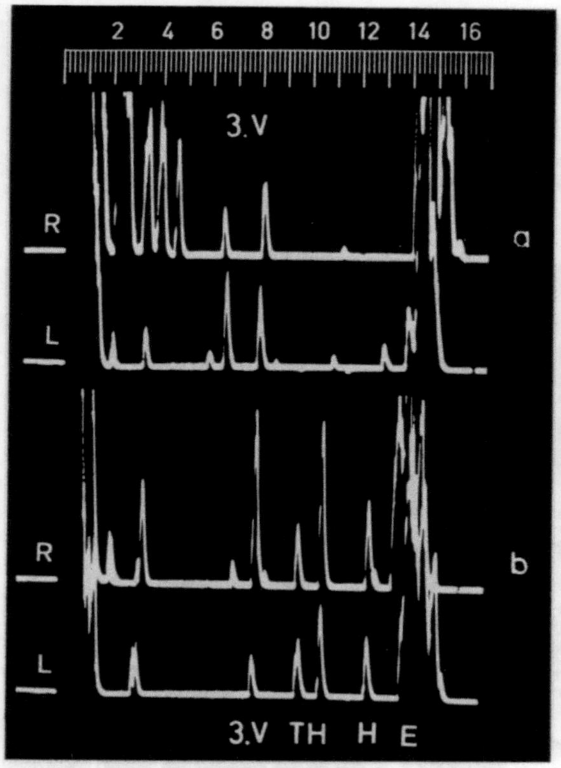

Fig. 182

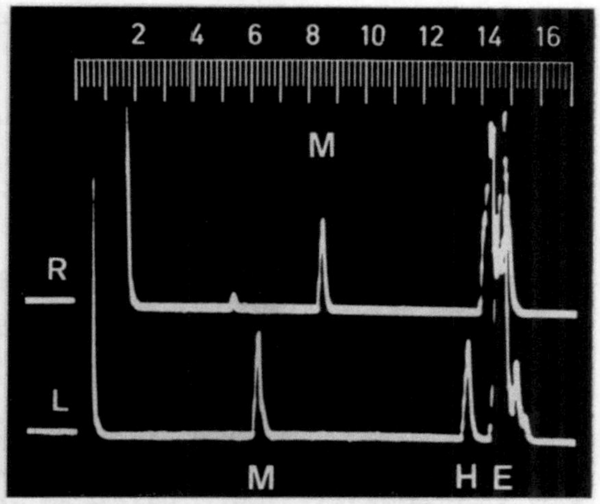

Fig. 183 a. Echo-encephalogram of a 6½-year-old boy with a subdural effusion over the right hemisphere. Displacement of the M-echo 10.0 mm towards the left. On sounding from the left, an additional reflection (H) appears ahead of the end echo probably originating from the internal wall of the encapsulated subdural hematoma. Pat. H. R., Echo-EG No. 1683/65

Fig. 182. Echo-encephalogram of an 18-month-old child with bilateral subdural effusions and pronounced internal hydrocephalus. a) (upper set of traces): bitemporal sounding technique reveals double echo from the walls of the markedly dilated third ventricle (3. V) 14—16 mm apart. b) (lower pair of traces, probe tipped 10° upwards): echo from the lateral wall of the third ventricle (3. V), double echo from the walls of the dilated temporal horn (TH), reflection from the hematoma membrane (H) and end echo (E). The subdural effusion was more marked on the right than on the left (14 mm to 10 mm). Pat. A. B., Echo-EG No. 806/1—2/63

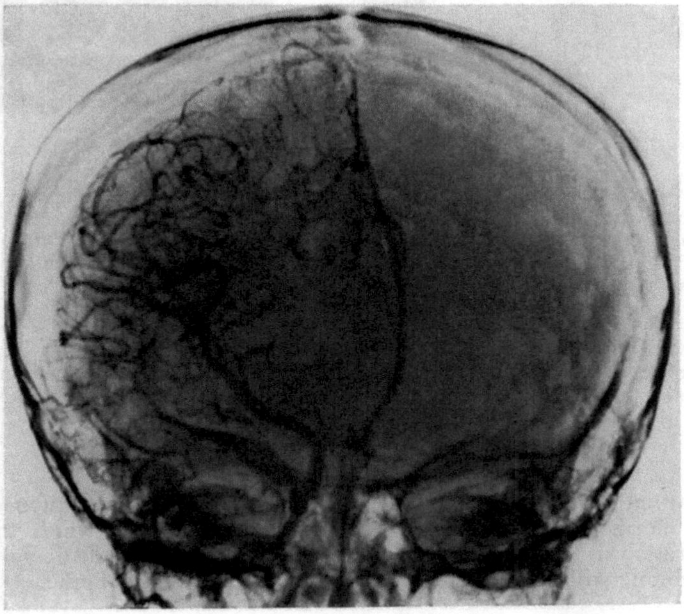

Fig. 183 b. Angiogram of the same patient as in Fig. 183 a. In addition to the sickle-shaped avascular space, corresponding to the subdural effusion, there is also an elevation of the middle cerebral artery which is produced by the coexistent arachnoid cyst in the temporal area

this additional echo and the end echo, conclusions can be drawn regarding the thickness of the effusion. If the fluid accumulations are unusually thick and extend far towards the temporal area, then "effusion echoes" appear not only ahead of the end echo but also in the first part of the echogram (Fig 184). Large effusions above the frontal pole can be detected by sounding from the front (Fig. 186). Pia and Geletneky (1968) have examined the largest series of subdural effusions by echo-encephalography. Their results are presented in Table 27.

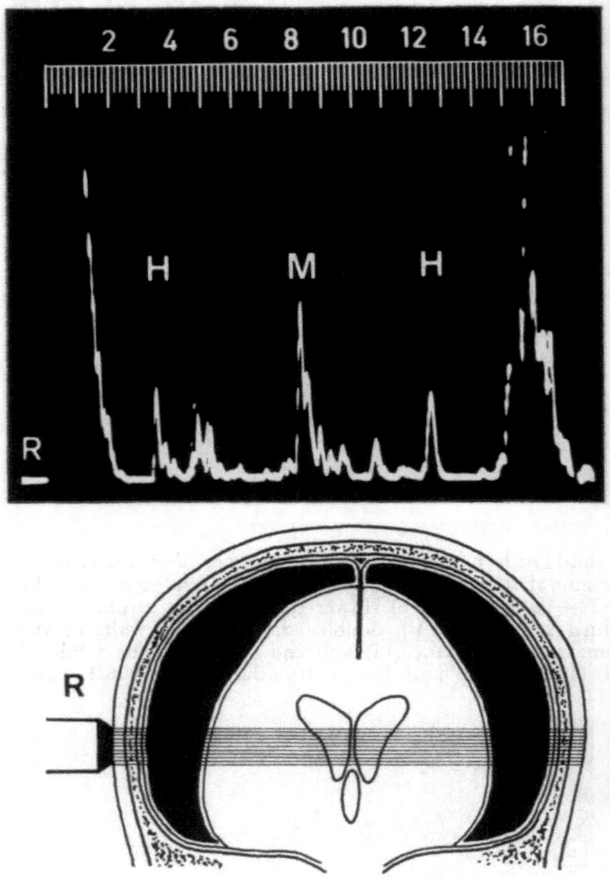

Fig. 184

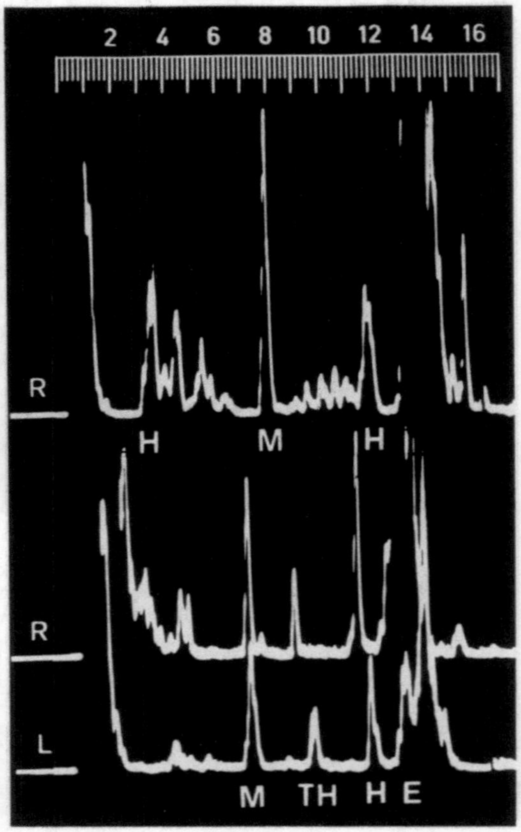

Fig. 185

Fig. 184. Echogram of a 6-year-old boy with a large bilateral subdural effusion. On both sides the boundary between brain and effusion is marked by a high reflection (H). In front of and behind these echoes, there are echo-free zones corresponding to the effusion. The brain "swims" in effusions. The diagrammatic representation will illustrate the development of such an echo-encephalogram. Pat. P. Ü., 6 years of age, Echo-EG No. 2784/67

Fig. 185. Echo-encephalogram of a 1-year-old boy with bilateral subdural effusions. Top: biparietal sounding shows remarkable similarity to the case shown in Fig. 184. Bottom: bitemporal sounding. Here an echo from the effusion membrane appears only in front of the end echo on both sides. The probe is tipped slightly upwards. Pat. T. W., Echo-EG No. 1876/2/66

Jacobi and Schuch (1966) report another method for the demonstration of "effusion echoes" in addition to the horizontal and diagonal sounding techniques, in which the probe is applied directly over the effusion in the precentral or parietal area. The reflections from the effusion or from the hematoma membrane will then become visible just behind the initial echo.

A special finding is a subdural effusion or hematoma together with an arachnoid cyst. Our own series of patients includes 3 children with such a combination (compare the case in Figs. 183 a and b). On sounding from the healthy side, one often sees individual or grouped reflections ahead of the end echo. The demonstration of the midline echo from the side of the lesion not infrequently presents difficulties making such echograms hard to interpret. The special echo-encephalographic findings with arachnoid cysts will be discussed in detail in the following section.

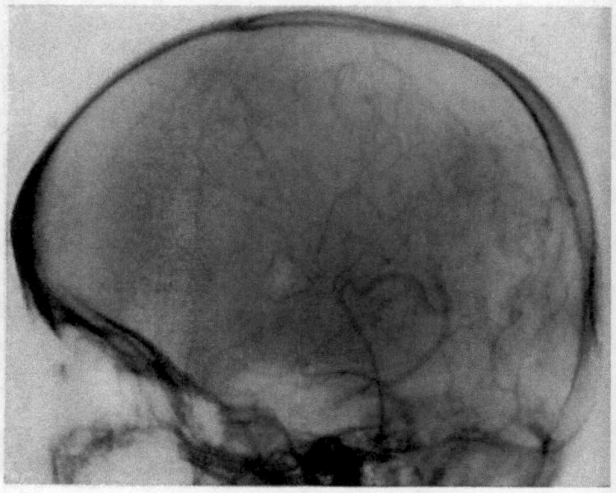

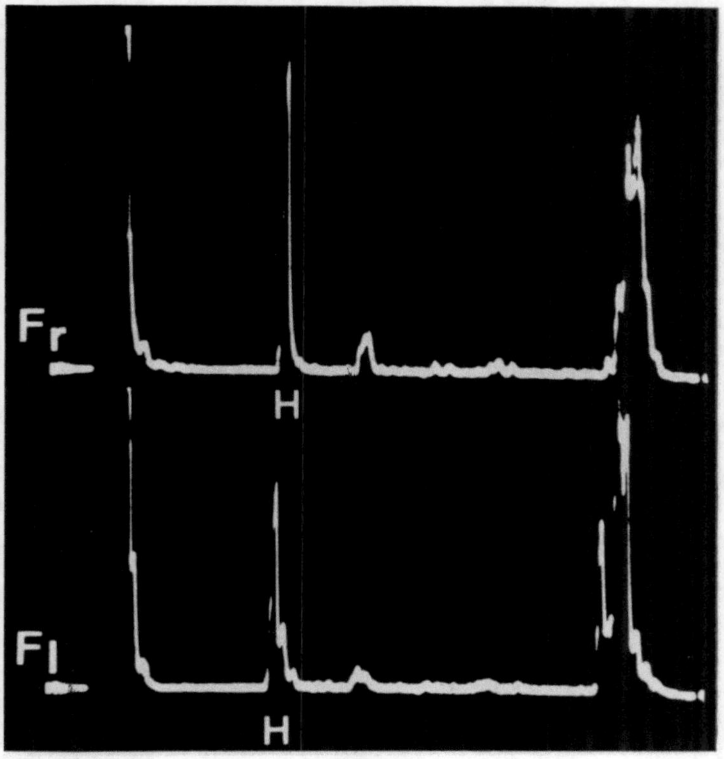

Fig. 186. Echo-encephalographic examination from a frontal probe application point in a 6-year-old boy with an unusually large accumulation of fluid. High echo (H) from the dense effusion membrane on both sides at the front. Top: the lateral angiogram reveals an avascular space, several cm wide which corresponds to the echo-encephalographic finding. Pat. P. Ü., 6 years of age. Echo-EG No. 2784/3/67

2. Infantile Brain Damage, Cerebral Malformations

Ventricular dilatations as a sequel of *infantile brain damage* are easy to detect by echo-encephalography. Naturally the ultrasonic picture does not permit a differentiation from hydrocephalus due to other causes. Especially in children with *unilateral atrophy of the hemispheres,* the echo-encephalogram not infrequently reveals a distinct *midline echo shift* in addition to the abnormal double echo from the walls of the third ventricle (Fig. 187). Geletneky (1965, 1966) has made a thorough echo-encephalographic study of such children and observed a displacement of the midline structures towards the atrophic side in 11 of 14 cases of hemiatrophy. The frequently variable height of the peaks on sounding from the left and from the right with the same pulse power, is due to differences in the thickness of the skull and scalp (see also Pia and Geletneky, 1968).

But even in adults, *a midline echo shift* can be a sequel of infantile brain damage. Fig. 188 shows the echogram and pneumoencephalogram of a 22-year-old patient who had been having epileptic seizures since childhood. The M-echo shift of 4 mm as well as the slight dilatation of the third ventricle is caused by a large occipital cyst which communicates with the ventricular system. The healthy hemisphere is shifted towards the atrophic side in such cases.

While cysts which are located so far posteriorly generally escape echo-encephalographic demonstration, *temporal arachnoid cysts* can be recognized by an *echo-free zone* which is bordered by an unusually high reflection from the medial cyst wall. Fig. 189 illustrates such a case. A tumor cyst can be distinguished from such a cyst only if there is a tumor echo complex present as well. The correct interpretation of these echograms can be extremely difficult. The final diagnosis of the clinical picture remains for contrast medium studies; the echo-encephalogram often reveals only a pathological finding. Attributing individual

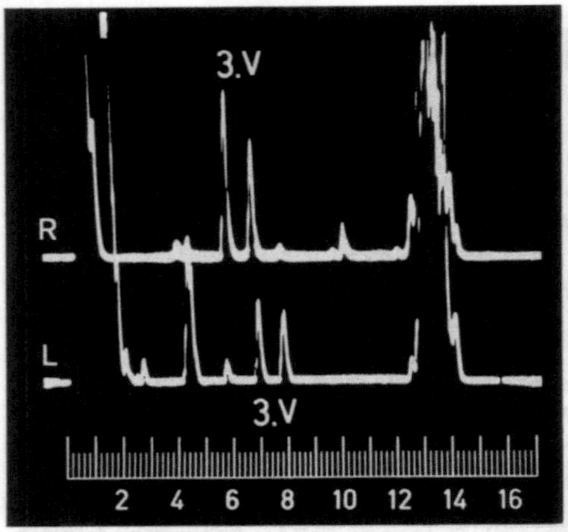

Fig. 187

Fig. 187. Echo-encephalogram of a $3^1/_2$-year-old girl with atrophy of one hemisphere. The double echo of the third ventricle which is dilated up to 9 mm has been shifted 5.5 mm towards the side of the lesion. Pat. H. Z., Echo-EG No. 1927/66

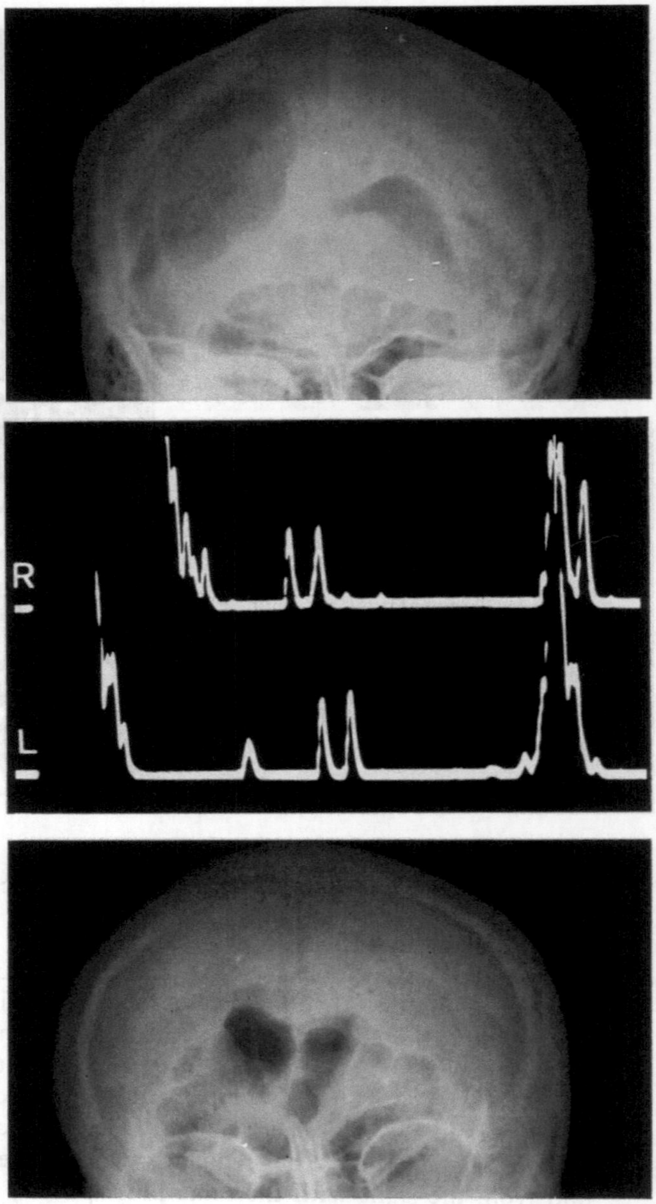

Fig. 188. Echo-encephalogram and pneumoencephalogram of a 22-year-old patient with a large occipital cyst. For further details, see text. Pat. F. B., Echo-EG No. 1322/64

Fig. 188

reflections of the ultrasonic picture to certain cerebral structures is probably only possible after pneumoencephalography has been performed (see also PIA and GELETNEKY, 1968). Echo-encephalography, however, is a suitable procedure for detecting such lesions on routine examination.

Echo-encephalography can also contribute to the diagnosis of *malformations of the brain.* (PIA and GELETNEKY, 1968; SJÖGREN et al., 1968.) Thus sounding an *hydranencephalic infant* in the usual direction reveals a total absence of reflections from intracranial structures. In cases of monoventricle (agenesis of the corpus callosum) the double echo from the third ventricle is absent (GELETNEKY, 1966); reflections are obtained only from the lateral walls of the large single ventricle (Fig. 190). SJÖGREN et al. (1968) have also reported such an observation.

Whether the demonstration of a double echo from the walls of a *septum pellucidum cyst* is always possible must remain an open question. It will be unusually difficult to distinguish such an abnormal double echo from the third ventricle echo (see also JACOBI, 1968). In 9 of our own cases of septum pellucidum cyst, the diagnosis was not made in a single case before pneumoencephalography had been performed. Multiple smaller cysts, as with porencephaly for example, will also produce lower echo peaks whose correct interpretation requires great experience, or is impossible. In cases of *complex malformations* such as the one which Fig. 191 shows, an interpretation of the echo-encephalographic trace is impossible and even after pneumoencephalograms are made, we are still unable to interpret the individual reflections accurately.

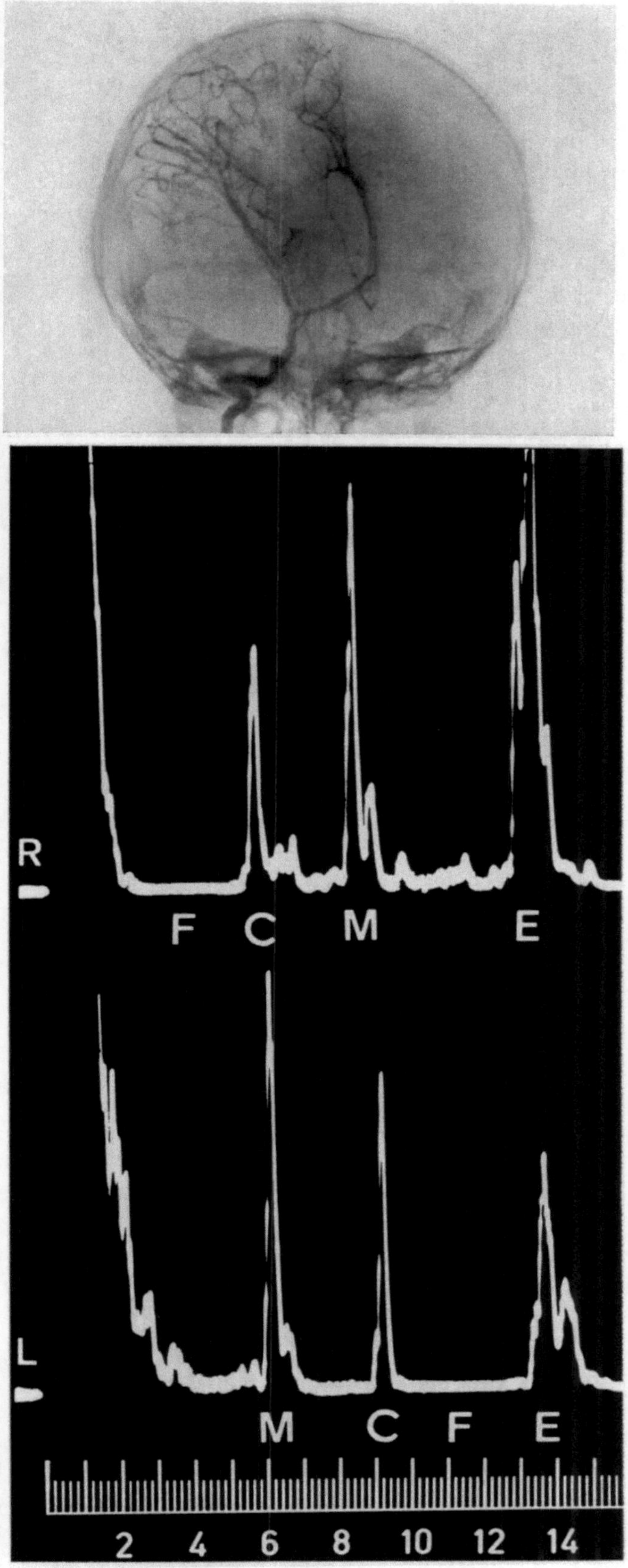

Fig. 189. Echo-encephalogram of a 3-year-old child with a large temporal arachnoid cyst. Considerable displacement of the midline echo (11.0 mm). Echo-free zone (F) is limited by a high echo (C). This reflection developed at the medial cyst wall. Complete agreement with the angiographic finding (top). Pat. J. A., Echo-EG No. 3017/67

a

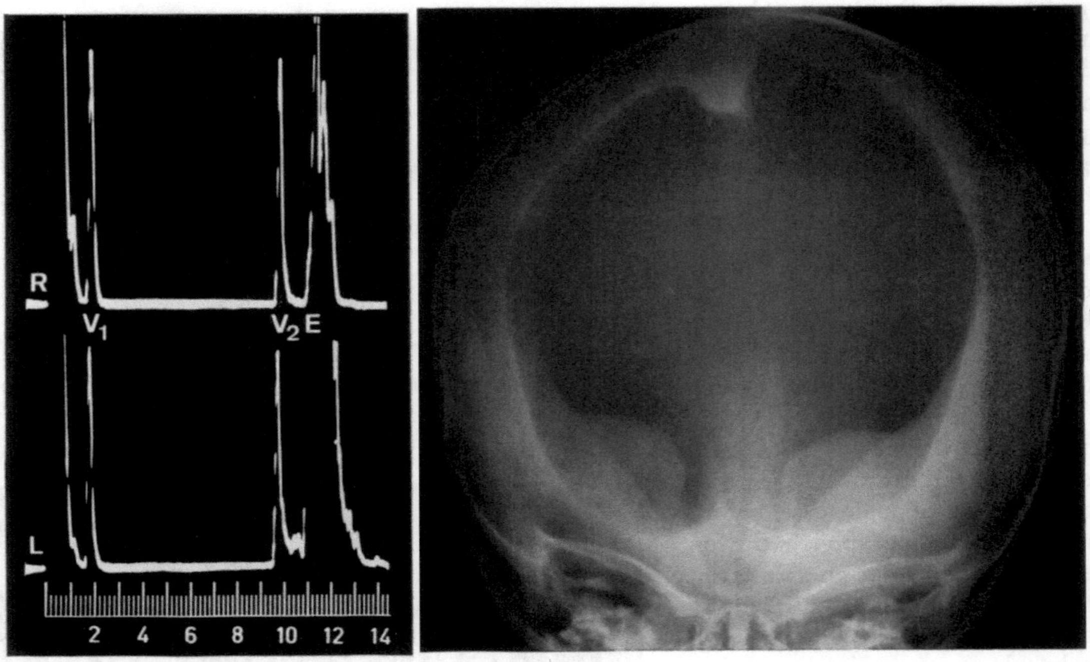

b

Fig. 190. a. Severe ventricular dilatation with agenesis of the corpus callosum diagnosed by echo-encephalography. On applying the probe high up in the temporal area, the midline echo is missing, the distance between the two ventricular wall reflections (V) remains echo-free even at maximum amplification. b. Confirmation of the echo-encephalographic diagnosis through subsequent pneumoencephalography. Pat. H. F., 6 weeks of age. Echo-EG No. 3626/68

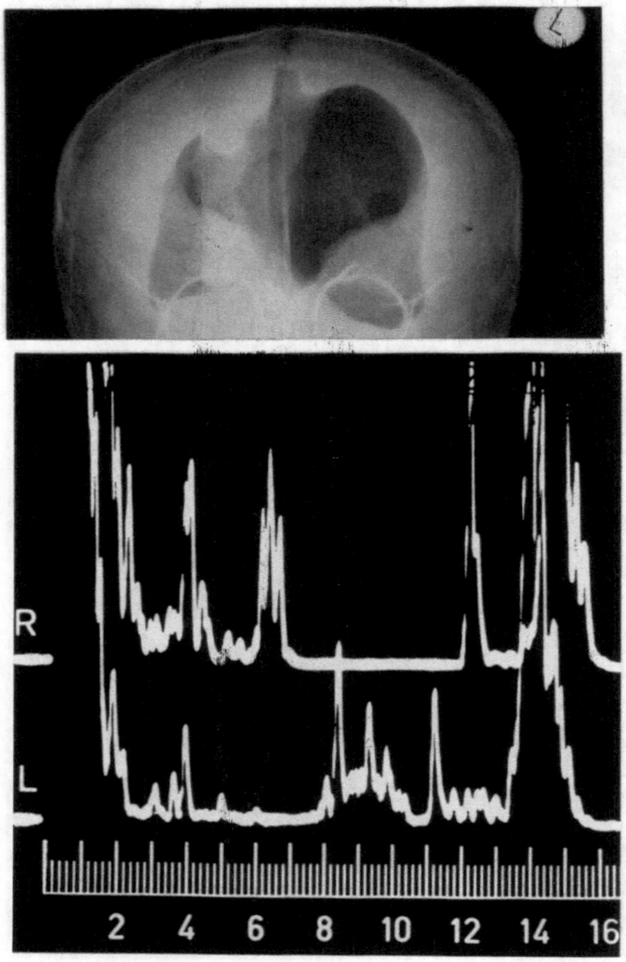

Fig. 191. Echo-encephalogram and pneumoencephalogram of a complex cerebral malformation in a 3-year-old child. The echo-encephalographic picture is nearly unintelligible even after seeing the pneumoencephalogram. Pat. A. Sch., Echo-EG No. 2208/66

In cases of *dysrhaphic malformations*, GELETNEKY (1965) believes it is possible to identify the various structures of the contents by means of ultrasound, so that meningoceles and myeloceles or encephaloceles and encephalo-meningoceles can be distinguished from each other. We have no experience in this area. However, in cases of this syndrome it presents no problems to detect the presence of a hydrocephalus or developing ventricular dilatation in addition to the malformation (see also p. 179).

3. Chronic Epilepsy

During the investigations on the measurability of the width of the third ventricle by echo-encephalography, we examined 183 patients with post-traumatic, symptomatic and idiopathic epilepsy. The midline echo on the echo-encephalogram was normal in all cases. Particularly remarkable was the *large number of epileptics with dilated third ventricles.* All in all, the width of this part of the ventricular system was determined in 173 of 183 patients (=94.5%) by means of the characteristic double echo. The more marked the dilatation, the easier was the measurement of the width of the third ventricle. 82 of the 173 epileptics, i. e. nearly half (47.4%) had a diameter of 7.0 mm and more, up to a maximum of 16.0 mm in one case.

The amount of ventricular dilatation is *clearly dependent on the duration and frequency of the seizures:* if the duration of the illness was 10 years, only every fifth patient revealed a ventricular dilatation on the echo-encephalogram; if the illness existed for 11 to 20 years, every second patient had a dilatation and in cases of more than 20 year's duration nearly every patient had one. Because of the good agreement between echo-encephalographic an pneumoencephalographic findings in other groups of patients, pneumoencephalography could be avoided in these cases, especially since most patients had had a previous clinical investigation where dye studies had also been performed.

4. Miscellaneous

In cases of *bacterial meningitis*, a slight increase in the width of the ventricles can often be observed within a few days. Circumscribed *encephalitis* even without the formation of an abscess occasionally leads to a displacement of the midline echo but rarely of greater proportions. Following the disappearance of the acute signs, the echo-encephalogram often reveals a dilatation of the third ventricle.

Tumors of the spinal cord which impair the reabsorption of CSF, occasionally produce a considerable dilatation of the third ventricle (see also ULBRICHT, 1966), The same is true for *basilar impressions* where a diameter of the third ventricle up to 14 mm has been observed.

An abnormally wide third ventricle was also seen in the echograms of 19 out of 32 individuals with idiopathic trigeminal neuralgia. Since most of the patients were between the ages of 60 and 80, however, a concurrent atrophic brain lesion must be assumed as the cause.

Dilatations of the third ventricle due to hypoxia of the blood and thus of the brain were observed in 17 of 39 children with *congenital heart defects,* mainly cases of tetralogy of FALLOT but also in the presence of isolated atrial and ventricular septal defects. The width of the third ventricle in these 17 children ranged from 7.0 to 10.5 mm which is certainly pathological.

Intracranial calcifications in cases of epiloia or Sturge-Weber syndrome cause irregular peak formations of great amplitude similar to those found with calcified brain tumors. The same is true for foreign bodies.

G. Possibilities and Limitations of One-Dimensional Echo-Encephalography (A-Scan)

If the *bitemporal sounding technique* of one-dimensional echo-encephalography is used, it attains a relatively high degree of accuracy. All authors report errors in measurement, but these make up only about 2 to 8% of all investigations. It is not always easy to discover the cause of such errors in retrospect. In the preceding chapters, the sources of error with cerebral disease processes have been repeatedly discussed in detail; however, a small number of errors in interpretation will always remain unexplained.

Since a large number of echoes appears on the screen of the Braun tube if the ultrasound diagnostic equipment is sensitive enough, it is understandable that it is not always possible to determine the origin of all these reflections with sufficient certainty. Therefore we must limit ourselves to those echoes which can be observed with greatest regularity or those that appear in a high percentage of cases of a certain clinical picture.

In the general section (p. 36 to 86), an attempt was made to survey those echoes which are of greatest diagnostic importance. However, it would be an error to assume that all these reflections can actually be recorded in all patients and their source correctly identified. The traces of one-dimensional echo-encephalography possess limited informative value and only on the basis of longer experiences is it possible to interpret reflections beyond the midline echo with certainty. For example, from the ultrasonic picture alone it is impossible to say whether the echo which precedes the end echo in Fig. 192 comes from a dilated temporal horn, from an epi- or subdural hematoma, a depressed fracture, a tumor, an area of calcification, a cyst or a brain abscess. Here the physical principles set a limit to the information provided by echo-encephalography. If there is an error in interpretation here, it is not an error which can be regarded as a failure of the method, rather it is an overdemand on the procedure.

The interpretation of the simple time-amplitude A-scan echogram requires great experience, a good visual imagination and a well-grounded knowledge of normal and pathological brain anatomy. The examining physician must also know the history and clinical findings of the patient. Only then can he associate certain reflections on the screen with the correct cerebral structures with adequate security. The photographed ultrasonic trace always represents only a momentary picture of the reflections from within the skull which can change considerably if the ultrasonic beam and the position of the probe are altered. Thus is depends solely on the investigator which momentary situation on the screen of the cathode ray tube he regards as characteristic and therefore documents photographically and which reflections or echo peaks he suppresses or amplifies by means of various manipulations.

Ranking authors have therefore pointed out the fact that the *results of A-scan echo-encephalography are subjective and easily influenced* (FISCHGOLD, 1964; WHITE and BLANCHARD, 1966), *and this is no doubt a great disadvantage of the method.* WHITE and BLANCHARD (1966) actually regard the echo-encephalographic technique as an art; however, it is one which it not exceptionally difficult to learn as JEFFERSON and HILL (1966) have said so well.

There has been no lack of efforts to compensate for the disadvantage of one-dimensional echo-encephalography in its present form, either by making the results more objective or even by delegating the entire examination to the domain of paramedical personnel. MCKINNEY (1964) for example, reported a method of determining the midline echo, called the B-mode determination which is thought to result in greater diagnostic accuracy. As in two-dimensional echo-encephalography, the echoes here appears as light spots on the screen of the Braun tube. The cathode ray beam wanders from top to bottom, over the screen in a certain period of time. In this interval, the probe is moved back and forth in the temporal area. The camera shutter remains open until the examination is over or a memory tube is employed. Instead of a single midline echo, a bright, more or less wide band, appears on the echogram; this is the sum of all reflections recorded during the procedure and is thought to simplify the exact determination of the midline echo.

As early as 1961, TER BRAAK, CREZÉE, GRANDIA and DE VLIEGER obtained a picture which showed the echoes originating from the midline structures more distinctly by using a long exposure time and moving the probe over a certain distance in the temporal area at the same time.

White and his collaborators (1964, 1965, 1966) have repeatedly reported on a technique of investigation which can even be carried out by a technician and which is thought to give greater certainty in the identification of the midline echo ("amplitude-averaging A-scan technique"). The probe is applied in the temporal area as usual, then the shutter of the camera is opened and all intracranial echoes are recorded and superimposed on each other over a time interval of one minute while the probe is being moved gently to and fro. Subsequently the same method of examination is undertaken from a corresponding point on the opposite side of the head. The polaroid photo obtained in this way usually shows the actual midline echo or the midline echo complex particularly well in contrast to the other echoes.

These different modifications of the normal technique of investigation may be an aid for the beginner in the determination of the midline echo but the practiced investigator has no need of them. Furthermore the results obtained in this manner are in no way superior to the normal method.

In summarizing the diagnostic possibilities of echo-encephalography, we must differentiate between indirect and direct signs of an intracranial lesion in the ultrasonic trace:

1. *indirect evidence of a brain disorder*

a) *by a shift of the midline echo:*
tumors of the cerebral hemispheres, unilateral intracranial hemorrhages, posttraumatic cerebral edema and encephalomalacia due to cerebrovascular disease:

b) *by abnormal echoes from the ventricular walls:*
cerebellar tumors, aqueduct stenoses or occlusions, atrophic brain lesions and infantile hydrocephalus.

2. *direct evidence of an intracranial lesion*

a) *by tumor echoes and reflections from cyst walls as well as intracranial calcifications:*
tumors of the cerebral hemispheres and oral brain stem

b) *hematoma reflections:*
intracerebral hematomas of any cause, epi- and subdural hematomas as well as subdural effusions in infants and small children.

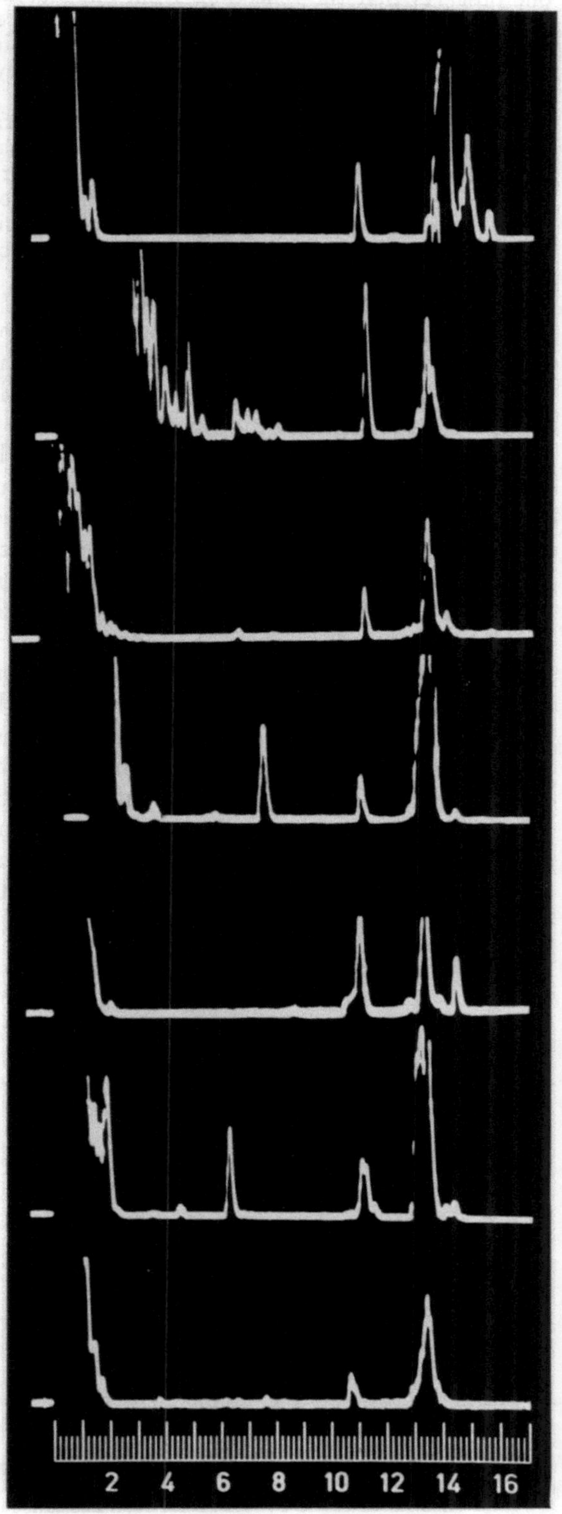

Fig. 192. Attributing the lateral reflections to the correct cerebral structures without knowing the history and clinical findings is usually impossible. The figure shows 7 different echograms, all of whom have an additional echo 2 cm ahead of the end echo. a) dilated temporal horn in a cerebellar tumor. Echo-EG No. 797/63); b) epidural hematoma (Echo-EG No. 1284/64); c) chronic subdural hematoma (Echo-EG No. 1281/64); d) depressed fracture (Echo-EG No. 1227/64); e) calcified astrocytoma (Echo-EG No. 1463/64); f) cystic astrocytoma (Echo-EG No. 999/64); g) brain abscess (Echo-EG No. 1035/64)

We have not included the diagnostic possiblities of *echo-pulsations* in this survey since the clinical application of this special investigation has not been clearly delineated as yet.

We have also not mentioned ultrasound absorption as another diagnostic aid here. Although it is true that the absorption of ultrasound is higher and in some cases lower in certain intracranial lesions, it is extremely difficult to recognize this distinctly in the ultrasonic trace. Not infrequently, one observes that the reflections from one side are higher than from the other. However, it seems somewhat daring to draw the conclusion that increased or decreased absorption of ultrasound in such cases is due to a hematoma or a tumor because the variations in the thickness of the bone can cause the very same effects. Also it would be necessary to know the echo-encephalogram of this patient before he became ill.

The fact remains however, that marked cerebral edema, following cerebral contusion for example, lowers the amplitude of all echoes while a normalization of this state restores the amplitude of the echoes to their original height. The opposite is true when the ventricles are dilated so that if there is a space-occupying lesion in the posterior fossa for example, the echoes from the ventricles will always be considerably higher. However, this observation can be explained on the basis of the enlarged ventricular walls which provide better conditions for reflecting ultrasonic energy.

Occasionally one encounters the opinion that echo-encephalography is in a position to make the entire diagnosis and to exclude the presence of an intracranial lesion with certainty. Naturally such a view will lead to disappointments. On the basis of the laws of mass shift with intracranial space-occupying lesions, it is to be expected that many of such lesions will escape echo-encephalographic demonstration because they do not lead to changes in those areas of the brain which can be scanned by ultrasound.

In evaluating echo-encephalographic findings, consideration must always be given to the fact that *this is usually only an ancillary investigation which frequently influences further steps decisively, however.* Therefore echo-encephalographic findings can be interpreted only in the light of the entire clinical picture. *It is not possible to exclude a space-occupying intracranial lesion by means of echo-encephalography with certainty.* In delineating the *limitations of this method* it may be stated that *midline displacements and ventricular dilatations can nearly always be demonstrated, hematoma interfaces in a high percentage of cases and tumors much more rarely.* However, similar limitations pertain to all other clinical methods of diagnosing space-occupying intracranial lesions.

To regard echo-encephalography as a complete replacement of dye studies would be misunderstanding its limitations and possibilities entirely. However, it is a valuable enrichment of the physician's diagnostic arsenal and it has earned a firm position in the early diagnosis of brain tumors and hydrocephalus, in the timely recognition of posttraumatic hemorrhages and in out-patient follow-up studies.

Appendix

Two-Dimensional Pulse-Echo Technique in the Diagnosis of Intracranial Lesions (B-Scan)

In one-dimensional echo-encephalography (A-scan), the distance of ultrasonic pulses between the piezoelectric crystal and a certain interface which reflects ultrasound is measured by the perpendicular deflections of the cathode ray. The amplitude of the echo traces corresponds to the amount of energy returning to the probe. The distance of the sound pulses can be converted to centimeters or millimeters of tissue, if the sound velocity of the examined structures is known and somewhat the same. In this way, the distance between the reflecting surface and the probe can be determined with sufficient accuracy.

Theoretically the ultrasono-tomographic procedure is naturally the ideal method for making the echo-encephalographic findings more easily legible and objective. Howry and Bliss successfully applied such a technique to soft body tissues in 1952 for the first time.

Instead of the perpendicular cathode ray deflections, light spots appear on the screen of the oscilloscope. The size, or more precisely the brightness of the individual light spots corresponds to the reflected ultrasound energy. Each line of a two-dimensional echogram may be imagined to be a one-dimensional ultrasonic picture which has been tipped over by 90° so that one is looking at the different echo peaks from the top so as to say. A two-dimensional echogram is composed of numerous such light spots produced by the movement of the probe in one level which transmit the impression of a cross-section of the part of the body being examined, comparable perhaps to an anatomic cross-section. Depending on the direction of the probe, the movement on the skull and the special technique used here, different B-scan-procedures were described (horizontal, linear, contact sector and compound scanning), all of which fall under the term *echo-tomography* or ultrasono-tomography. The first pictures obtained with such a procedure on the skull were published by Kikuchi, Uchida, Tanaka, and Wagai in 1957. These remind us of x-ray films. Subsequently Tanaka and his group reported on two-dimensional echo-encephalography again and again. These authors continue to apply the simple time-amplitude procedure as the first investigation, however.

Howry and Gordon (1964) questioned the very possibility of obtaining intelligible ultrasonic cross-sections of the head because of the physical conditions of the skull in most cases, since the sound beam is refracted by the bone if there is a deviation in the angle of incidence from the perpendicular. The experimental investigations of White and collaborators (1967, 1968) have demonstrated that differences in the thickness and irregularities of the skull bone result in considerable distortions and displacements in the B-scan display making it impossible to obtain an echo-tomogram corresponding to the actual anatomic relations. The only exceptions are pictures obtained by means of the compound-scan technique in hydrocephalic infants, who have an unusually thin skull bone which transmits ultrasound easily.

In recent years there has been no lack of attempts to develop usable echo-encephalographic cross-sections. The reader who is particularly interested in this field is referred to the work of de Vlieger, de Sterke, Molin and van der Ven (1963, 1966, 1967, 1968),

ADAPON, CHASE, KRICHEFF and BATTISTA (1965, 1966), GROSSMAN (1964, 1965, 1966), ITO, ISHIKAWA, WAGAI and UCHIDA (1965), BRINKER and TAVERAS (1966), DREESE et al. (1966, 1968), GALICICH, LOMBROSO and MATSON (1965), as well as MAKOW et al. (1966, 1967, 1968).

The only echo-tomograms showing the actual anatomical relations till now have been published by DENIER VAN DER GON, DUINHOUWER, MOLIN and DE VLIEGER (1966) and DE VLIEGER, DENIER VAN DER GON and MOLIN (1968). The reasons for the good quality of their pictures has been mentioned already (thin, infantile skull bone, dilated ventricles with large reflecting interfaces, compound-scan).

Relatively acceptable cross-sections can also be achieved with a technique where the probe is moved along a line running parallel to the connection line between the outher canthus and the external auditory meatus in the temporal region (so-called horizontal or linear scan (Fig. 193). By preparing the scalp with a jelly, sufficient contact can be maintained between

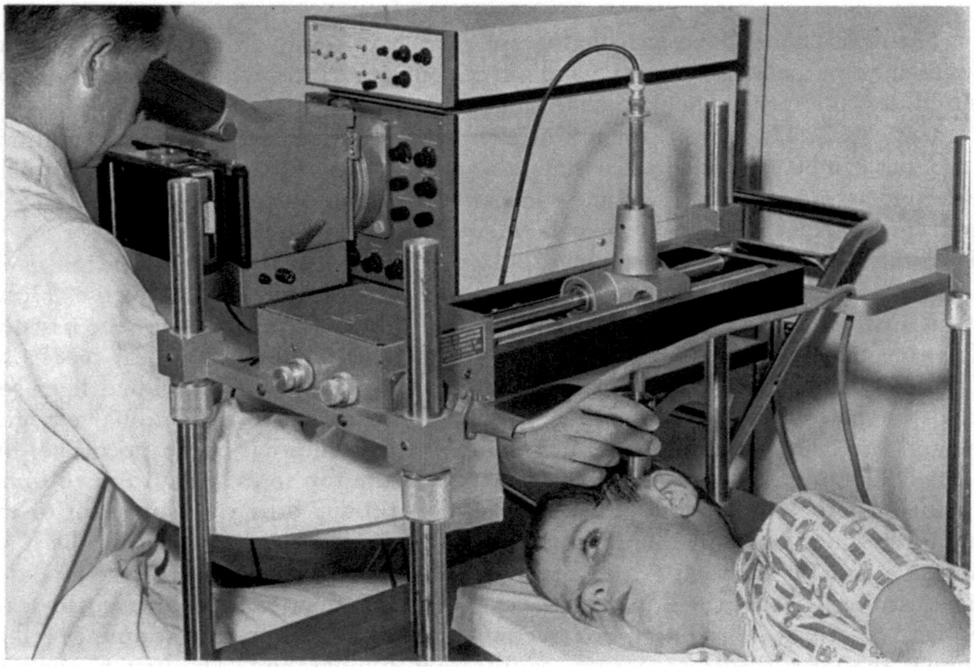

Fig. 193. Recording a two-dimensional echo-encephalogram (linear scanning) in a 6-year-old boy (HOFFREL 600 A Scanning System). The probe is automatically moved in a horizontal position. The investigator's hand follows the probe so that perfect contact with the scalp is assured

probe and scalp over a distance of approximately 10 cm. Especially GROSSMAN (1964, 1965, 1966), ADAPON et al. (1965, 1966) as well as DREESE et al. (1966, 1968) have published impressive tomograms demonstrating tumors directly. GALICICH, LOMBROSO and MATSON (1965) emphasize the value of the two-dimensional procedure in children with hydrocephalus and other types of infantile brain damage. Two other papers also come from the pediatric clinic in Boston; these deal with B-scan echo-encephalography in children and young people with different cerebral lesions (HOVIND, GALICICH and MATSON, 1967; LOMBROSO and ERBA, 1968). The authors arrive at the conclusion that "this method of examining the brain is a most useful adjunct to other more established procedures. This enables us to demonstrate the localized shift of the cerebral midline with great accuracy and to visualize enlarged ventricles, extracerebral hematomas, intracerebral cysts and intracerebral mass lesions, such as tumors, abscesses and hematomas. Ultrasonic scanning may have special value in detecting infiltrating gliomas and mass lesions of the thalamus, basal ganglia, parasagittal parietal lobe, and medial temporal lobe, areas notoriously difficult to examine by other methods". The investigations by LOMBROSO and ERBA (1968) however, also show that the probability of portraying certain cerebral structures in young people, decreases with increasing age.

ADAPON et al. (1966) investigated 127 patients using B-scan, 60 of whom had verified space-occupying intracranial lesions while the remainder failed to reveal a pathological finding in dye studies. Intracranial space-occupying lesions were demonstrated either directly

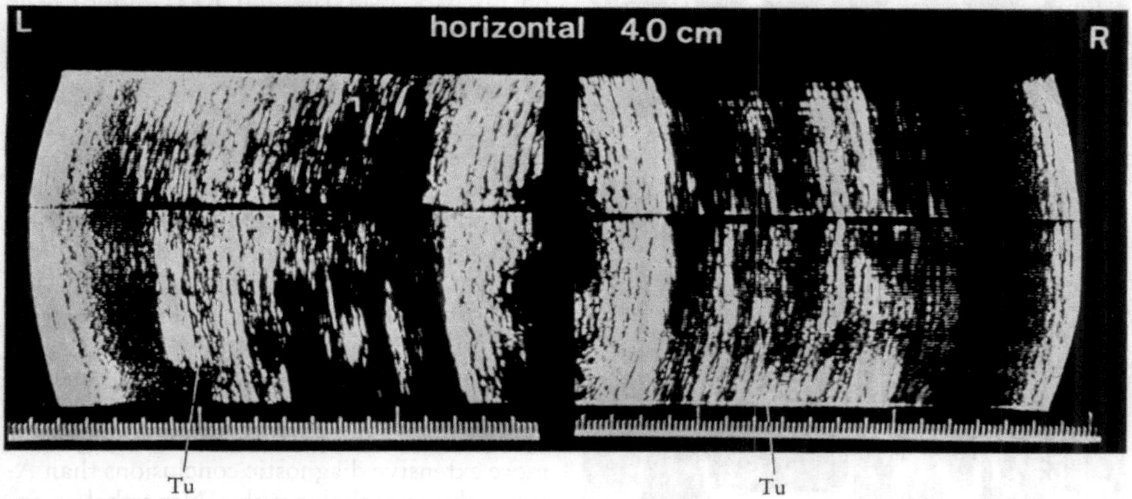

G. P. 7 years 18. 6. 68

Fig. 194. Two-dimensional echogram (horizontal B-scan) of a 7-year-old boy with a left-sided thalamic tumor. Increased echo activity on recording from the right and from the left in the left hemisphere near the midline. Note the marked distortion of the skull opposite the probe. A one-dimensional echogram of the same patient is seen in Fig. 98. Pat. G. P., Echo-EG No. 3529/68/20

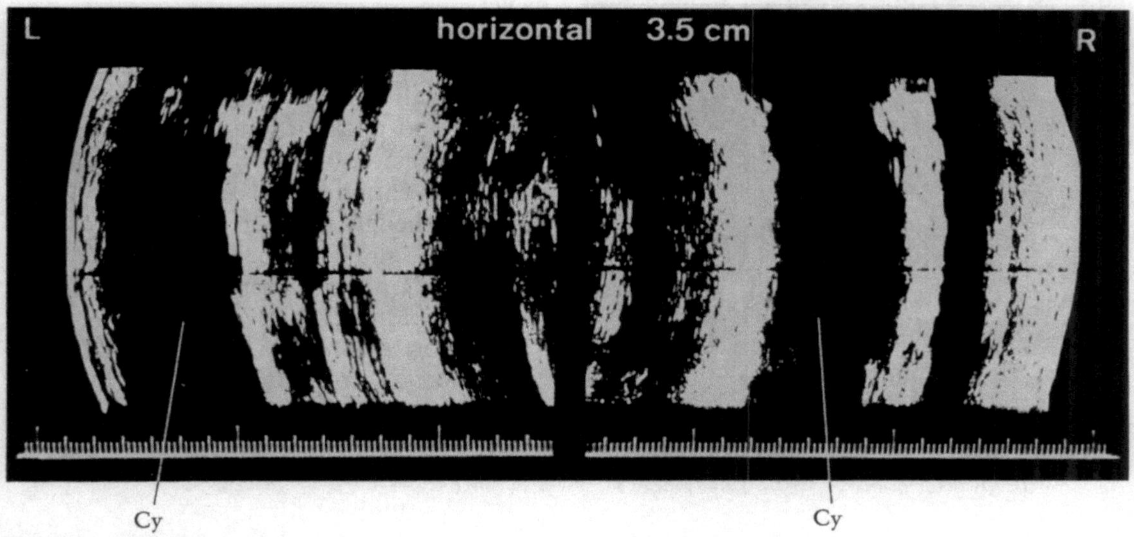

St. S. 2 months 25. 6. 68

Fig. 195 a. Two-dimensional echogram (horizontal B-scan) of a 2-month-old child with a large cyst following a perinatal hemorrhage in the left hemisphere. Large echo-free space in both records (Cy). Pat. St. S., Echo-EG No. 3711/68/20 (see also Fig. 195 b, which shows the A-scan of this child)

or indirectly in 44 patients (73.4%) using the B-scan technique. Thus 16 (26.6%) lesions escaped echo-tomographic investigation, although all these patients showed a displaced midline echo in A-scan echo-encephalograms.

Beyond this, new problems appear in two-dimensional echo-encephalography which make the investigation very much more difficult when the method is used on patients. It is often extremely hard to maintain equally good contact over a large part of the head unless the scalp is shaved. Another possibility is to use water as a coupling medium rather than applying the probe directly. This can be done either by using a water-filled plastic bag and laying it on the head or by immersing the patient's head in a container of water with the patient lying on his back (MAKOW and REAL, 1966). The probe then moves in the wather bath. Such a method of investigation, however, disposes of all the advantages of one-dimensional echo-encephalography at once: speed, little equipment, no bother for the patient. In addition, it necessitates the coorperation of the patient, thereby excluding unconscious or restless patients. The main field of application for echo-encephalography, namely the diagnostics of head injuries is lost here.

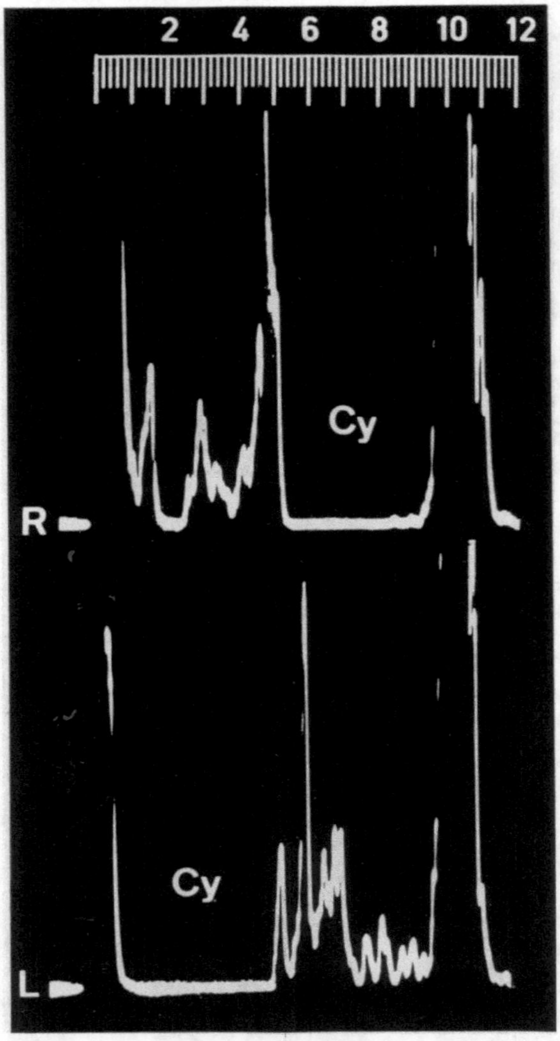

Fig. 195 b. One-dimensional echo-encephalogram of the same child as in Fig. 195 a. The diagnosis of a large left-sided cyst was easily possible here too. Pat. St. S., Echo-EG No. 3711/68

Thus we certainly cannot speak of the superiority of the two-dimensional method over the simple time-amplitude method. In special cases, the investigation may permit more extensive diagnostic conclusions than A-scan-echo-encephalography. Nevertheless, intensive studies will be required before the possibilities of B-scan can be surveyed as well as those of A-scan. Our own echo-tomographic investigations have not been very successful as yet.

However, it is hoped that the two-dimensional echo method will lead to better diagnostic results in the future, especially in children and possibly in adults with glioblastomas or tumors of the oral brain stem and of the third ventricle (Fig. 194 and 195). It is doubtful though whether the echo cross-sections of the skull and brain will be as exact as those obtained in obstetrics and gynecology (DONALD, MAC VICAR and BROWN, 1958, 1961; HOFMANN, HOLLÄNDER and WEISER, 1966; DONALD and ABDULLA, 1967; DONALD, 1968; KRATOCHWIL, 1968) because the physical conditions with the skull are quite different.

References

ABE, Y.: Diagnosis of intracranial hematoma by ultrasound. Arch. Jap. Chir. **36**, 63—83 (1967).

— Y. OHARA, K. SATO, K. SUGAWARA, K. ITO, and S. KIKUCHI: Study on the shift of midline echo in the intracranial diseases. Japan. Med. Ultrasonics **3**, 53 (1965).

— — K. SUGAWARA, K. ITO, and S. KIKUCHI: Diagnostic value of the third ventricle echo in the case of head injury. Med. Ultrasonics **4**, 36 (1966).

— K. SUGAWARA, M. ISHII, K. ITO, S. KIKUCHI, and K. TANAKA: Diagnosis of head injury by ultrasonic technique — Diagnostic values of hematoma echo. Med. Ultrasonics **4**, 58, (1966).

— K. TANAKA, T. WAGAI, and K. ITO: Diagnosis of intracranial hemorrhage using ultrasound. Acta radiol. **5**, diagnosis, 7th Symposium neuroradiol., New York 1964, 721—729 (1966).

ACHAR, V. S., R. P. K. COE, and J. MARSHALL: Echoencephalography in the differential diagnosis of cerebral haemorrhage and infarction. Lancet **1966, I,** 161—164.

ADAM, A.: Erste Versuche mit Ultraschallbehandlung im Kindesalter. Mschr. Kinderheilk. **97,** 391 (1949).

— Ultraschallbehandlung im Kindesalter. Ultraschall in Med. **1,** 274—277 (1949).

ADAMS, J.-H.: Echoencephalography (Letter to the editor). Lancet **1966, I,** 487—488.

ADAPON, B. D., N. E. CHASE, I. I. KRICHEFF, and A. F. BATTISTA: Cerebral ultrasonic tomography. Radiology **84,** 115—121 (1965).

— — — — 'B'scan encephalography. Acta radiol., **5,** diagnosis, 7th Symposium neuroradiol., New York 1964, 730—739 (1966).

AKERMAN, M., et A. BASSO: L'exploration cérébrale par une technique d'écho-B. Intérêt et limites. Neuro-chirurgie **13,** 527—536 (1967).

— et G. GUIOT: Le diagnostic des processus expansifs intra-craniens par l'association de la gamma- et de l'écho-encéphalographie (Etude statistique de 650 cas). Ann. radiol. **8,** 851—879 (1965).

ALVISI, C.: L'eco-encefalografia. Bologna: A. Caggi 1967.

AMBROSE, J.: Some clinical applications of ultrasound. Brit. J. Radiol. **36,** 302 (1963).

— Pulsed ultrasound. Illustrations of clinical applications. Brit. J. Radiol. **37,** 165—178 (1964).

ANDERSON, G. V., and J. W. NISWONGER: Cephalometry with ultrasound. Amer. J. Obstet. Gynec. **91,** 563—567 (1965).

ANDREW, D. S.: Ultrasonography in pregancy. An enquiry into its safety. Brit. J. Radiol. **37,** 185—186 (1964).

ASBERG, A.: Ultrasonic cinematography of the living heart. Ultrasonics **5,** 113—117 (1967).

AVANT, W. S. Jr.: Pulsatile echoencephalography. A review and the importance of the respiratory factor. Neurology (Minneap.) **16,** 1033—1040 (1966).

BALDES, E. J., J. F. HERRICK, and Ch. F. STROEBEL: Biologic effects of ultrasound. Amer. J. phys. Med. **37,** 111—121 (1958).

BALLANTINE, H. T., E. BELL, and J. MANLAPAZ: Progress and problems in the neurological applications of focused ultrasound. J. Neurosurg. **17,** 858—876 (1960).

— R. H. BOLT, T. F. HUETER, A. CAVALIERI, I. DYER, and E. DOZOIS: The detection of intracranial tumors by use of ultrasound. M. I. T. Acoust. Lab. Quarterly Progr. Report, Jan.—March 1951.

— — — and G. D. LUDWIG: On the detection of intracranial pathology by ultrasound. Science **112,** 525—528 (1950).

— T. F. HUETER, and R. H. BOLT: On the use of ultrasound for tumor detection. J. acoust. Soc. Amer. **26,** 581 (1954).

— — W. J. H. NAUTA, and D. M. SOSA: Focal destruction of nervous tissue by focused ultrasound. Biophysical factors influencing its application. J. exp. Med. **104,** 337—360 (1956).

— G. D. LUDWIG, R. H. BOLT, and T. F. HUETER: Ultrasonic localization of the cerebral ventricles. Trans. Amer. Neurol. Ass. **75,** 38—41 (1950).

BARROWS, H. S., P. DYCK, and T. KURZE: The diagnostic applications of ultrasound in neurological disease. Neurology (Minneap.) **15,** 361—365 (1965).

— — and TH. KURZE: The reliability and limitations of echo-encephalography in acute neurological conditions. In: Proceedings in echo-encephalography. Berlin-Heidelberg-New York: Springer 1968, pp. 114—116.

— and TH. KURZE: The value of ultrasonic encephalography as an adjunct to neurology and neurological surgery. In W. S. FIELDS: Neurological diagnostic techniques. Springfield, Ill.: Thomas 1966, pp. 1—29.

BARTH, G., u. H. A. BÜLOW: Zur Frage der Ultraschallschädigung jugendlicher Knochen. Strahlentherapie **79,** 271—280 (1949).

— J. PÄTZOLD u. F. WACHSMANN: Über den Wirkungsmechanismus biologischer Ultraschallreaktionen. I. Mitteilung. Möglichkeiten zur Klärung des Wirkungsmechanismus. Strahlentherapie **80,** 305—311 (1949).

— u. F. WACHSMANN: Biologische Grundlagen der Ultraschalltherapie. In: Ultraschalltherapie. Hsg. von R. POHLMAN. Bern: Huber 1951.

BARTHEL, R.: Sound velocity in some aqueous solutions as a function of concentration and temperature. J. acoust. Soc. Amer. **26,** 227—230 (1954).

BAUER, B. L., C. L. GELETNEKY, R. LORENZ, W. SEEGER, H. VOGELSANG u. W. WESEMANN: Differentialdiagnose und Behandlung des akuten Koma. Aus neurochirurgischer Sicht. Nervenarzt **37,** 540—545 (1966).

BAUM, G.: A comparison of the merits of scanned intensity modulated ultrasonography VS. Unscanned A scope ultrasonography. In: Diagnostic ultrasound. New York: Plenum Press 1966.

— and I. GREENWOOD: The applications of ultrasonics locating technique to ophthalmology. Theoretic considerations and acoustic properties of oculas media. Part I. Reflective properties. Amer. J. Ophthalmol. **46,** 319—329 (1958).

— — The application of ultrasonics locating techniques of ophthalmology. Arch. Ophthal. **60,** 263 (1958).

— — Ultrasound in ophthalmology. Amer. J. Ophthal. **49,** 249—261 (1960).

BAUM, G., and I. GREENWOOD: Ultrasonography — an aid in orbital tumor diagnosis. Arch Ophthal. **64,** 180—194 (1960).
— — Orbital lesion localization by three-dimensional ultrasonography. N. Y. St. J. Med. **61,** 4149—4157 (1961).
— — A critique of time-amplitude ultrasonography. Arch. Ophthal. **65,** 353—365 (1961).
BERANEK, L. L.: Acoustics. New York-Toronto-London: McGraw-Hill Book Company Inc. 1954.
BERGMANN, L.: Der Ultraschall und seine Anwendung in Wissenschaft und Technik. 6. Aufl. Stuttgart: S. Hirzel, 1954.
BERGSTRÖM, K., H. LODIN, and I. SJÖGREN: Echoencephalography and cerebral pneumography in infants and children. A comparative study with regard to ventricular size. 8e Symposium neuroradiol. Paris, 1967.
BETZ, H., and G. HUBER: The correlation between echo-encephalographic and pneumoencephalographic findings. In: Proceedings in echo-encephalography. Berlin-Heidelberg-New York: Springer 1968, pp. 25—26.
BINGEL, A.: Misinterpretations of echoencephalographic findings. Int. J. Neurol. Psychiat. **3,** 29—35 (1967).
BLATT, B.: Ekkoencefalografi med a-praesentation. Ugeskr. Laeg. **128,** 1052—1056 (1966).
BÖHMER, G., u. C. A. CARLSSON: Entscheidende Faktoren für den Behandlungserfolg beim Epiduralhämatom. Mschr. Unfallheilk. **67,** 414—428 (1964).
BOHN, E. R.: Über die Korrelation zwischen Pneumencephalographie und Echoencephalographie unter besonderer Berücksichtigung des 3. Ventrikels. Inauguraldissertation Bonn 1967.
BOLT, H., and T. F. HUETER: Some aspects of transcranial ultrasonic transmission. Ultraschall in Med. **6,** 202—203 (1953).
BOSTEM, F., A. THIBAUT, and E. COEURDEROI-GODART: Study of echo-encephalographic results correlated with electroencephalographic, anatomic and neuroradiologic data in 649 cases. In: Proceedings in echo-encephalography. Berlin-Heidelberg-New York: Springer 1968, pp. 164—170.
BRAAK, J. W. G. TER, P. CREZÉE, W. A. M. GRANDIA, and M. DE VLIEGER: The significance of some reflections in "Echo-Encephalography". Acta neurochir. (Wien) **9,** 382—397 (1961).
— W. A. M. GRANDIA, and M. DE VLIEGER: 'Echo-Encephalography' as an aid in the diagnosis of subdural and extradural haematomas. In: A. BIEMOND et al.: Recent neurological research, p. 37—45. Amsterdam: Elsevier 1959.
— and M. DE VLIEGER: Cerebral pulsations in echo-encephalography. Acta neurochir. (Wien) **12,** 678—694 (1965).
BRAUN, H., u. W. SCHMITT: Erfahrungen mit der Ultraschallkardiographie in der Diagnostik von Mitralvitien. Materia Medica Nordmark **14,** 533—539 (1962).
BRENNER, H.: Ist die Ultraschall-Echoencephalographie ein Fortschritt in der Diagnostik frischer Schädel-Hirnverletzungen? Wien. klin. Wschr. **77,** 267—269 (1965).
BRINKER, R. A.: Ultrasound brain scanning utilizing the contact method. In: C. C. GROSSMANN et al.: Diagnostic ultrasound, pp. 186—190. New York: Plenum Press 1966.
— Simultaneous presentation echoencephalography. Radiology **88,** 360—361 (1967).
— and B. L. HOLMAN: Compound sector echoencephalography. Anatomical correlations in the dog. Invest. Radiol. **2,** 160—164 (1967).
— D. L. KING, and J. M. TAVERAS: Echoencephalography. Amer. J. Roentgenol. **93,** 781—790 (1965).
— and J. M. TAVERAS: Ultrasound cross-sectional pictures of the head. Acta radiol., **5,** diagnosis, 7th Symposium neuroradiol., New York 1964, 745—753 (1966).
BROCKHOFF, V., E. KAZNER u. W. SCHIEFER: Die Kombination von Echo-Encephalographie und Kontrastmittelmethoden in der Diagnostik raumfordernder Prozesse in der hinteren Schädelgrube. 8e Symposium neuroradiol. Paris, 1967.
BRONSON, N. R.: Techniques of ultrasonic localization and extraction of intraocular and extraocular foreign bodies. Amer. J. Ophthal **60,** 596—603 (1965).
BROWN, R. E.: A standardized procedure for echoencephalography and for analyzing echoencephalograms. Can. med. Ass. J. **96,** 1349—1354 (1967).
BROWN, T. G.: Visualization of soft tissues in two and three dimensions — Limitations and development. Ultrasonics **5,** 118—124 (1967).
BRÜCKNER, H.: Die Bedeutung der Echo-Enzephalographie für Diagnose und Behandlung der subduralen Haematome. Beitr. Neurochir. **8,** 180—191 (1964).
— Die Echoencephalographie. Möglichkeiten und Grenzen der Methode. Helfen Heilen **1965,** 44—50.
— Der kindliche Hydrocephalus und seine Behandlung. Helfen Heilen **1965,** 52—55.
— Die postoperativen Komplikationen im Echoenzephalogramm. Acta neurochir. (Wien) **13,** 578—579 (1965).
— u. E. KAZNER: Die Diagnose intrakranieller Prozesse mit der Echoencephalografie. SRW-Nachr. **21,** 1—7 (1963).

Brüschke, G.: Tierexperimentelle Untersuchungen zur Frage der Schädigung von Testis, Ovar und gravidem Uterus durch Ultraschall. Z. ges. inn. Med. **10**, 895—899 (1955).
Brylski, J. R., and J. L. Izenstark: Anatomic localization of midline echo in sonograms of the brain. Amer. J. Roentgenol. **93**, 811—815 (1965).
Bull, J. W. D., J. Marshall, and D. A. Shaw: Cerebral angiography in the diagnosis of the acute stroke. Lancet **1960**, I, 562—565.
Buschmann, W.: Zur Diagnostik der Carotisthrombose. Albrecht v. Graefes Arch. Ophthal. **166**, 519—529 (1964).
— Probleme und Fortschritte der Ultraschalldiagnostik am Auge. Klin. Mbl. Augenheilk. **144**, 321—347 (1964).
— Acoustic illusions in ophthalmic ultrasonography. Amer. J. Ophthal. **57**, 461—466 (1964).
— Einführung in die ophthalmologische Ultraschalldiagnostik. Edition Leipzig 1966, 175 pp.
Busnel, R. G., P. Chauchard und H. Mazoué: Spezifische Wirkungen des Ultraschalls auf das neuromuskuläre System. Ultraschall in Med. **9**, 10—14 (1956).
Calatayud-Maldonado, V., C. Geletneky y R. Lorenz: Posibilidades de diagnóstico con el ecoencefalograma. Arch. Fac. Med. Zaragoza **13**, 61—83 (1965).
— — — El ecoencefalograma. Nuevo medio de diagnostico en neurocirurgia. Recio. Rev. Col. Med. Prov. Ciudadreal **3**, 25—36 (1966).
Callahan, B.: New dimensions in radiologic diagnosis. Hospital progress **45**, 50—54 (1964).
Campbell, J. B., and E. N. Weaver: The use of echo-encephalography in neurologic diagnosis. Sth. med. J. (Bgham, Ala.) **58**, 782—787 (1965).
Carlin, B.: Ultrasonics. New York: McGraw-Hill Book Company 1960.
Carstensen, E. L.: Measurement of the dispersion of velocity of sound in liquids. J. acoust. Soc. Amer. **26**, 858—861 (1954).
— K. Li, and H. P. Schwan: Determination of the acoustic properties of blood and its components. J. acoust. Soc. Amer. **25**, 286—289 (1953).
— and H. P. Schwan: Acoustic properties of blood and its components. In: Ultrasound in biology and medicine. Ed. E. Kelly. Washington: Amer. Inst. Biol. Scienc., 1957, pp. 1—14.
Cerbelle, J. C., J. Duhamel et J. Broussin: L'écho-encéphalographie chez les traumatisés crâniens. Soc. Franç. d'Electroradiol. Méd. Section Sud-Ouest, 29 Jan. 1966, 683—689.
Chadduck, W. M., and W. G. Crutchfield: Extracranial masses complicating the interpretation of echoencephalograms. A new technique for quantitative evaluation. J. Neurosurg. **21**, 699—703 (1964).
Ch'en, K.-P., and Y.-H. P'an: Intracerebral ultrasonic exploration. Chin. med. J. **83**, 506—510 (1964).
— — Preliminary report in the use of ventricular needle form lateral emission type ultrasonic probe for intracranial exploration. Zhong. Waike. Z. (Peking) **12**, 261—264 (1964).
Chesney, D. N.: Echoencephalography. Nursing Times **62**, 498—500 (1966).
Coe, R. P. K., and J. Marshall: Echoencephalography. Lancet **1966 I**, 430.
Colombati, S., e S. Petralia: Assorbimento di ultrasuoni in tessuti animali. Ricerca Scientifica **20**, 71—78 (1950).
Coraddu, M., et F. Napoleone: L'eco-encefalografia nella diagnosi delle lesioni compressive endocraniche. Rass. med Sarda **70**, 13—22 (1967).
Crawford, A. E.: Ultrasonic engineering. London: Butterworth 1955.
Dalsgaard-Nielsen, T.: Some clinical experience in the treatment of cerebral apoplexy (1000 cases). Acta psychiat. scand. Suppl. **108**, 101—119 (1956/57).
Davidoff, L. M., and C. G. Dyke: The normal encephalogram. Philadelphia: Lea & Febiger 1943.
Deftereos, T., und H. Wenker: Echo-Encephalographie. In: Chirurgie des Gehirns und Rückenmarks im Kindes- und Jugendalter. Hsgg. von K.-A. Bushe und P. Glees. Stuttgart: Hippokrates 1968, pp. 223—235.
Delius, P.: Ultraschall in der medizinischen Diagnostik. Elektromedizin **9**, 43—45 (1964).
Denier, A.: Les ultrasons; leurs applications au diagnostic: Ultrasonoscopie, et à la therapeutique: Ultrasonotherapie. J. Radiol Électrol. **27**, 481—487 (1946).
Denier van der Gon, J. J., J. C. Duinhouwer, C. E. Molin, and M. de Vlieger: Equipment for two-dimensional echoencephalography. Neurology (Minneap.) **16**, 927—933 (1966).
— C. E. Molin, and M. deVlieger: Comparison of scan techniques in two-dimensional echoencephalography. In: C. C. Grossmann et al.: Diagnostic ultrasound. New York; Plenum Press 1966, pp. 155—165.
Dilling, H.: Echoenzephalographie. Grundlagen, Anwendung und neuere Ergebnisse. Münch. med. Wschr. **109**, 2243—2251 (1967).
— u. W. Feuerlein: Altersbedingte Unterschiede in der echoencephalographischen Darstellbarkeit der medianen Hirnstrukturen. Arch. Psychiat. Nervenkr. **209**, 404—414 (1967).
Distel, L., and W. Kasper: The echo-encephalogram with hydrocephalus and subdural effusions in childhood. In: Proceedings in echo-encephalography. Berlin-Heidelberg-New York: Springer 1968, pp. 152—154.

Donald, I.: A new diagnostic echo-sounding technique in obstetrics and gynaecology. Proc. Roy. Soc. Med. **55**, 637—640 (1962).
— Use of ultrasonics in diagnosis of abdominal swellings. Brit. med. J. **1963, II**, 1154—1155.
— The interpretation of abdominal ultrasonograms. In: Diagnostic ultrasound. New York: Plenum Press 1966.
— Ultrasonics in obstetrics. Brit. med. Bull. **24**, 71—75 (1968).
— and U. Adulla: Further advances in ultrasonic diagnosis. Ultrasonics **6**, 8—12 (1967).
— and T. G. Brown: Demonstration of tissue interfaces within the body by ultrasonic echo sounding. Brit. J. Radiol. **34**, 539—546 (1961).
— J. MacVicar, and T. G. Brown: Investigation of abdominal masses by pulsed ultrasound. Lancet **1958, I**, 1188—1195.
Dreese, M. J., G. J. Hayes, and L. G. Kempe: Evaluating intracranial hematoma by echo EG. Acta radiol. **5**, diagnosis, 7th Symposium neuroradiol., New York 1964, 767—773 (1966).
— — — The value of an A-mode echoencephalogram display. Neurology (Minneap.) **16**, 355 (1966).
— F. E. McGee, and A. B. Harrelson: Correlative aids in B-scan echoencephalography. Neurology (Minneap.) **16**, 766—770 (1966).
— and L. C. McHenry: Studies of echo-EG correlations. Neurology (Minneap.) **14**, 271—272 (1964).
— and M. G. Netsky: The clinical use of Echo-Encephalography. Virginia med. Mth. **90**, 539—548 (1963).
— — Studies of lateral reflections in the echo-encephalogram. Neurology (Minneap.) **14**, 521—528 (1964).
— C. G. Suter, and L. E. Rennie: Two-dimensional echo-encephalography ("B-scan"): Description of a modified horizontal plane found clinically useful. In: Proceedings in echo-encephalography. Berlin-Heidelberg-New York: Springer 1968, pp. 210—215.
Dugdale, L. M.: Echo-encephalography: The use of pulsed ultrasound in neurological diagnosis. Aust. Radiol. **10**, 207—216 (1966).
Dussik, K. Th.: Über die Möglichkeit, hochfrequente mechanische Schwingungen als diagnostisches Hilfsmittel zu verwerten. Z. ges. Neurol. Psychiat. **174**, 153—168 (1942).
— Ultraschall-Diagnostik, insbesondere bei Gehirnerkrankungen, mittels Hyperphonographie. Z. phys. Ther. **1**, 140—145 (1948).
— Zum heutigen Stand der medizinischen Ultraschallforschung. Wien. klin. Wschr. **61**, 246 (1949).
— Ultraschallanwendung in der Diagnostik und Therapie der Erkrankungen des zentralen Nervensystems. In: K. Matthes und W. Rech: Der Ultraschall in der Medizin. Zürich: Hirzel, 1949, pp. 283—287.
— Weitere Ergebnisse der Ultraschalluntersuchung bei Gehirnerkrankungen. Acta neurochir. (Wien) **2**, 379—401 (1952).
— Probleme der Ultraschallanwendung in der Therapie der Erkrankungen des Zentralnervensystems. Ultraschall in Med. **6**, 159—162 (1953).
— The ultrasonic field as a medical tool. Amer. J. phys. Med. **33**, 5—20 (1954).
— F. Dussik und L. Wyt: Auf dem Wege zur Hyperphonographie des Gehirns. Wien. med. Wschr. **97**, 425—429 (1947).
Dyck, P., H. S. Barrows, and Th. Kurze: The intra-operative utilization of ultrasound in the localization of cerebral mass lesions. In: Proceedings in echo-encephalography. Berlin-Heidelberg-New York: Springer 1968, pp. 76—79.
— T. Kurze, and H. S. Barrows: Intra-operative ultrasonic encephalography of cerebral mass lesions. Bull. Los Angeles neurol. Soc. **31**, 114—124 (1966).
Eberle, H.: Das Echoenzephalogramm bei Schädel-Hirn-Trauma. Praxis **53**, 1710—1714 (1964).
Edler, I.: The diagnostic use of ultrasound in heart disease. Acta med. scand., Suppl. **308**, 32—36 (1955).
— The use of ultrasound as a diagnostic aid and its effects on biological tissues. Continuous recording of the movements of various heart-structures using an ultrasound echo-method. Acta med. scand. Suppl. **170**, 9—65 (1961).
— Ultrasoundcardiography in mitral valve stenosis. Amer. J. Cardiol. **19**, 18—31 (1967).
— Ultrasound cardiography. Ultrasonics **5**, 72—79 (1967).
— and C. H. Hertz: The use of ultrasonic reflectoscope for continuous recording of the movements of heart walls. Kung. Fysiogr. Sällskapets i Lund Förhandlingar, Bd. **24**, 1—19 (1954).
— and A. Gustafson: Ultrasonic cardiogram in mitral stenosis. Acta med. scand. **159**, 85—90 (1957).
Effert, S.: Pre- and postoperative evaluation of mitral stenosis by ultrasound. Amer. J. Cardiol. **19**, 59—65 (1967).
— F. Deupmann, and Karytsiotis: Registration of movement of cardiac walls using an ultrasonic reflection procedure. In: Ultrasonic energy, ed. by E. Kelly, Urbana: University of Illinois Press, 1965, p. 327.
— H. Erkens u. F. Grosse-Brockhoff: Über die Anwendung des Ultraschall-Echoverfahrens in der Herzdiagnostik. Dtsch. med. Wschr. **82**, 1253—1257 (1957).

Effert, S., u. E. Domanig: Diagnostik intraaurikulärer Tumoren und großer Thromben mit dem Ultraschall-Echoverfahren. Dtsch. med. Wschr. **84**, 6—8 (1959).

Elizondo-Martel, G., and J. Gershon-Cohen: Medical Ultrasonics. Essentials of echoencephalography. Amer. J. Roentgenol. **93**, 791—802 (1965).

Emery, J. L.: The examination of the normal and hydrocephalic infant brain using ultrasound. Develop. Med. child. Neurol. Suppl. **13**, 87—95 (1967).

Engeset, A., and Lönnum, A.: Third ventricles of 12 mm width or more. Preliminary report. Acta radiol. (Stockh.) **50**, 5—11 (1958).

— and E. Skraastad: Methods of measurement in encephalography. Neurology (Minneap.) **14**, 381—385 (1964).

Erba, G., and C. T. Lombroso: Detection of ventricular landmarks by two dimensional ultrasonography. J. Neurol. Neurosurg. Psychiat., **31**, 232—244 (1968).

Esche, R.: Untersuchungen zur Ultraschallabsorption in tierischen Geweben und Kunststoffen. Acustica **1—2**, Beih. 2, 71—74 (1952).

Evans, G. C., J. St. Lehmann, B. L. Segal, W. Likoff, M. Ziskin, and B. Kingsley: Echoaortography. Amer. J. Cardiol. **19**, 91—96 (1967).

Falk, B., Kirsten, L., S. Löfstedt, and B. Silferskiöld: Pneumo-encephalographic investigations in epilepsy. Acta psychiat. scand. **33**, 440—451 (1958).

Falls, H. F., R. C. Basset, and A. E. Lamberts: Ocular complications encountered in intracranial arteriography. Arch. Ophthal. (Chicago) **45**, 623—626 (1951).

Faust, C.: Die psychischen Störungen nach Hirntraumen. Akute traumatische Psychosen und psychische Spätfolgen nach Hirnverletzungen. In: Psychiatrie der Gegenwart **II**, Berlin-Göttingen-Heidelberg: Springer 1960, pp. 552—645.

Feigenbaum, H.: Diagnostic ultrasound. Ann. Int. Med. **65**, 185—189 (1966).

— J. A. Waldhausen, and L. P. Hyde: Ultrasound diagnosis of pericardial effusion. J. Amer. med. Ass. **191**, 711—714 (1965).

— A. Zaky, and J. A. Waldhausen: Use of ultrasound in the diagnosis of pericardial effusion. Ann. Int. Med. **65**, 443—452 (1966).

— — — Use of reflected ultrasound in detecting pericardial effusion. Amer. J. Cardiol. **19**, 84 (1967).

Ferrey, G.: Les applications diagnostiques des ultra-sons à la médecine. II. L'échoencéphalographie A. Ann. Radiol. **9**, 881—890 (1966).

Feuerlein, W.: Echoenzephalographie in der Chirurgie. Chir. Praxis **10**, 587—594 (1966).

— u. H. Dilling: Zur Bestimmung des Mittelechos in der Echoencephalographie. Nervenarzt **36**, 401—403 (1965).

— — Die Echoenzephalographie. Med. Klinik **61**, 1061—1064 (1966).

— — Das Echo-Encephalogramm des 3. Ventrikels in verschiedenen Lebensaltern. Arch. Psychiat. Nervenkr. **209**, 137—147 (1967).

— — The echo-encephalogram of the third ventricle in different age groups. In: Proceedings in echo-encephalography. Berlin-Heidelberg-New York: Springer 1968, pp. 132—135.

Fincher, E. F., G. J. Strewler, and H. S. Swanson: The Torkildsen-Procedure. A report of 19 cases. J. Neurosurg. **5**, 213—219 (1948).

Firestone, F. A.: Flaw detecting device and measuring instrument. USA-P. 2 280 226 (1940).

— The supersonic reflectoscope, an instrument for inspecting the interior of solid parts by means of sound waves. J. acoust. Soc. Amer. **17**, 287—299 (1945).

Firnhaber, W., and M. Djawdan: The diagnostic value of echo-encephalographic evidence. In: Proceedings in echo-encephalography. Berlin-Heidelberg-New York: Springer 1968, pp. 171—174.

Fischer, G., P. Gerin, M. Revol et J. Courjon: Place de l'écho-encéphalographie dans le diagnostic des lésions intracrâniennes. J. Méd. Lyon **43**, 85—102 (1965).

— A. Morin, P. Gerin et J. Courjon: Aspects comparatifs des données écho- et électroencéphalographiques (A propos de 100 cas). Ann. méd. psychol. **122**, 434 (1964).

— M. Revol, F. Munier, P. Gerin, J. Courjon et P. Wertheimer: L'écho-encéphalographie. Valeur diagnostique et résultats. Neurochirurgia **10**, 45—58 (1967).

Fischgold, H.: Die Schädelunfall-Diagnostik. Dtsch. Röntgenkongreß 1965. Stuttgart: G. Thieme 1966, pp. 126—130.

— Le diagnostic technique des tumeurs intracraniennes avant et après 1954. Neuro-chirurgie **13**, 11—18 (1967).

— Développement de l'Écho-Encéphalographie. Presse méd. **76**, 780 (1968).

— M. F. Strauss, and P. Hazemann: A-scan echo-encephalography in acute and chronic head injuries. In: Proceedings in echo-encephalography. Berlin-Heidelberg-New York: Springer 1968, pp. 80—87.

Fisher, T. R.: Source of the midline echo and its implications in echoencephalography. J. Neurol. Neurosurg. Psychiat. **29**, 379—382 (1966).

Ford, R. M.: Echo-encephalography: A method of estimating frontal midline displacement. In: Proceedings in echo-encephalography. Berlin-Heidelberg-New York: Springer 1968, pp. 48—55.

FORD, R., and J. AMBROSE: Echoencephalography. The measurement of the position of mid-line structures in the skull with high frequency pulsed ultrasound. Brain **86**, 189—196 (1963).
— and D. L. McRAE: Echoencephalography — A standardized technique for the measurement of the width of the third and lateral ventricles. In: Diagnostic ultrasound. New York: Plenum Press 1966, pp. 117—129.
FOSTER, J. J.: Ultrasonic encephalography. An aid in diagnosis of head trauma. J. int. Coll. Surg. **44**, 523—529 (1965).
FOX, J. L., J. R. RICHARDS, and A. J. LUESSENHOP: Method for detecting intracranial arteriovenous malformation with an ultrasonic transducer using the Doppler phenomenon. J. Neurosurg. **26**, 322—326 (1967).
FRAY, W. W.: Methods for determining pineal position with analysis of their errors. Amer. J. Roentgenol. **42**, 490—497 (1939).
FRENCH, L. A., J. J. WILD, and D. NEAL: Detection of cerebral tumours by ultrasonic pulses. Pilot studies on post-mortem material. Cancer **3**, 705—708 (1950).
— — — The experimental application of ultrasonics to localization of brain tumours. Preliminary report. J. Neurosurg. **8**, 198—203 (1951).
FREUND, H.-J.: Ultraschallregistrierung der Pulsationen einzelner intrakranieller Arterien zur Diagnostik von Gefäßverschlüssen. Arch. Psychiat. Nervenkr. **207**, 247—253 (1965).
— und H. KAPP: Ultraschall-Echo-Registrierung arterieller Pulsationen in tiefen Körperregionen. Pflügers Arch. ges. Physiol. **289**, 96 (1966).
— — Eine Methode zur Registrierung arterieller Pulsationen mittels Ultraschall. Pflügers Arch. ges. Physiol. **291**, 268—276 (1966).
— — and K. KENDEL: Recording arterial pulse curves with ultrasound. — Experimental investigations and diagnostic possibilities. In: Proceedings in echo-encephalography. Berlin-Heidelberg-New York: Springer 1968, pp. 192—196.
FRIEDEL, B.: Die Bestimmung des „Sollechos" zur besseren Identifizierung des Mittelechos. Nervenarzt **39**, 75—80 (1968).
FRIEDMANN, G., u. F. THUN: Zuverlässigkeit und Fehlermöglichkeiten der Echo-Enzephalographie bei supratentoriellen raumfordernden Prozessen. Med. Welt **50**, 689—696 (1964).
— und O. WILCKE: Vergleichende Untersuchungen der Kontrastmittel- und Isotopendiagnostik bei Hirntumoren. Dtsch. med. Wochenschr. **87**, 2311—2314 (1962).
FRIEDRICH, W., P. MIFKA, V. MAY u. L. NAVRATIL: Zur Frage des progredienten posttraumatischen Hydrocephalus internus. Wien. klin. Wschr. **75**, 826—828 (1963).
FRUCHT, A. H.: Die Schallgeschwindigkeit in menschlichen und tierischen Geweben. Z. ges. exp. Med. **120**, 526—557 (1952/1953).
FRY, W. J.: Ultrasound in neurology. Neurology (Minneap.) **6**, 693—704 (1956).
— Use of intense ultrasound in neurological research. Amer. J. phys. Med. **37**, 143—147 (1958).
— Present and future applications of ultrasonics in biomedicine. Proc. Inst. Radio Engrs. **50**, 1393—1404 (1962).
FUCHS, P.: Die Bedeutung der Echo-Encephalographie bei der Beurteilung cerebraler Komplikationen Kieferverletzter. Fortschr. Kiefer- u. Gesichtschirurg. **11**, 112—115 (1966).
FUJIMOTO, Y., A. OKA, R. OMOTO, and M. HIROSE: Ultrasonic scanning of the thyreoid gland as a new diagnostic approach. Ultrasonics **5**, 177—180 (1967).
GALICICH, J. H., C. T. LOMBROSO, and D. D. MATSON: Ultrasonic B-scanning of the brain. J. Neurosurg. **22**, 499—510 (1965).
GARG, A. G., and A. R. TAYLOR: An investigation into the effect of pulsed ultrasound on the brain. Ultrasonics **6**, 208—212 (1967).
— — A-scan echoencephalography in measurement of the cerebral ventricles. J. Neurol. Neurosurg. Psychiat. **31**, 245—249 (1968).
GELETNEKY, C.-L.: Echoenzephalographie im Säuglings- und Kindesalter. Acta neurochir. (Wien) **13**, 579—581 (1965).
— Echoencephalographie bei cerebralen Erkrankungen im Säuglings- und Kindesalter. Acta radiol. **5**, diagnosis, 7th Symposium neuroradiol., New York 1964, 779—785 (1966).
— Die Bedeutung der Echo-Encephalographie für die Diagnostik und Verlaufskontrolle bei Tumoren der hinteren Schädelgrube. Deutscher Röntgenkongreß 1967, Stuttgart: Thieme 1968 Teil A, pp. 114—116.
— and E. KAZNER: Echo-encephalography in the diagnosis of ventricular dilatation. In: Proceedings in echo-encephalography. Berlin-Heidelberg-New York: Springer 1968, pp. 122—132.
GERLACH, J., H.-P. JENSEN, W. KOOS und H. KRAUS: Pädiatrische Neurochirurgie. Stuttgart: Thieme 1967.
GERNET, H.: Ultraschall-Biometrie des Auges. Klin. Mbl. Augenheilk. **151**, 853—871 (1967).
GÖGLER, E.: Unfallopfer im Straßenverkehr. Documenta Geigy, Series chirurgica No. 5, Basel 1961.
GÖLLNITZ, G.: Über das normale Encephalogramm im Kindesalter. Nervenarzt **22**, 101—107 (1951).
GOHR, H., u. TH. WEDEKIND: Der Ultraschall in der Medizin. Klin. Wschr. **19**, 25—29 (1940).

GOKALP, H., R. K. JAKOBY, and J. W. WATTS: Echoencephalography: An evaluation based on its use in 300 individuals. Med. Ann. Distr. Columbia **35**, 641—645 (1966).
GOLDMAN, D. E., and T. F. HUETER: Tabular data of the velocity and absorption of high-frequency sound in mammalian tissues. J. acoust. Soc. Amer. **28**, 35—37 (1956).
— — Data for high frequency sound in mammalian tissues (Errata). J. acoust. Soc. Amer. **29**, 655 (1957).
GORDON, D.: The use of ultrasonic rays in diagnostic radiology. Paper presented at 4th Symposium neuroradiol., London, 1955. Publ. in: Ultrasound as a diagnostic and surgical tool. Edinburgh & London: Livingstone 1964.
— Echoencéphalographie. Rev. neurol. **99**, 652—653 (1958).
— Echo-encephalography. Ultrasonic rays in diagnostic radiology. Brit. med. J. **1959, I**, 1500—1504.
— Echo-encephalography. In: Medical Electronics, proceedings of the second international conference on medical electronics, Paris, 24—27 June 1959. London: Iliffe 1960, pp. 380—390.
— Ultrasonic rays in diagnosis and surgery. Trans. med. Soc. Lond. **79**, 173—180 (1963).
— Ultrasound as a diagnostic and surgical tool. Edinburgh & London: Livingstone 1964.
— The limitations and uses of ultrasound in localizing cerebral lesions. Proc. roy. Soc. Med. **58**, 1053—1058 (1965).
— Two-channel echo-encephaloscope (Letter to the editor). Ultrasonics **3**, 154 (1965).
— Ultrasonic surgery of the central nervous system. Acta radiol. **5**, diagnosis, 7th Symposium neuroradiol., New York 1964, 796—803 (1966).
— Three-dimensional echo-encephalography in stereotaxic surgery. In: Proceedings in echo-encephalography. Berlin-Heidelberg-New York: Springer 1968, pp. 238—241.
GOTTESFELD, K. R.: The practical applications of ultrasound in obstetrics and gynecology. In: Diagnostic ultrasound. New York: Plenum Press 1966.
GREATOREX, C. A., and H. J. D. IRELAND: An experimental scanner for the use with ultrasound. Brit. J. Radiol. **37**, 179—184 (1964).
GREEBE, H. M.: De waarde van de echo-encefalografie als klinische en poliklinische methode van onderzoek. Ned. T. Geneesk. **108**, 10—13 (1964).
GROSSMAN, C. C.: Clinical diagnostic application of ultrasound in brain disorders (Sono-Encephalography). Dis. nerv. Syst. **25**, 403—411 (1964).
— The normal sonoencephalogram (SEG). Ultrasonic echoes and brain structures. Dis. nerv Syst. **25**, 717—723 (1964).
— Experience with combined EEG and ultrasound techniques in neurology. Neurology (Minneap.) **14**, 272 (1964).
— Acoustic phenomena in ultrasonic detection of brain tumors. Ultrasonics **3**, 22—24 (1965).
— EEG-SEG or sonar. Ultrasonics **3**, 100—101 (1965).
— Ultrasonic cross section technique for visualization of intracranial masses. Neurology (Minneap.) **15**, 277 (1965).
— The use of diagnostic ultrasound in brain disorders. Springfield, Ill.: Thomas 1966.
— A and B scan sonoencephalography (SEG) — A new dimension in neurology. In: Diagnostic ultrasound. New York: Plenum Press 1966, pp. 130—141.
— J. H. HOLMES, C. JOYNER, and E. W. PURNELL: Diagnostic ultrasound. Proceedings of the First International Conference, University of Pittsburgh, 1965. New York: Plenum Press 1966.
GÜTTNER, W.: Betrachtungen zur Möglichkeit der US-Abbildung am menschlichen Schädel. Ultraschall in Med. **6**, 199—202 (1953).
— Die Energieverteilung im menschlichen Körper bei Ultraschall-Einstrahlung. Acustica **4**, 547—554 (1954).
— G. FIEDLER u. J. PÄTZOLD: Über Ultraschallabbildungen am menschlichen Schädel. Acustica **2**, 148—156 (1952).
GUSTAFSON, A.: Correlation between ultrasoundcardiography, hemodynamics and surgical findings in mitral stenosis. Amer. J. Cardiol. **19**, 32—41 (1967).
HACHE, H., u. P. SCHEID: Das epidurale Hämatom. Pädiat. Prax. **5**, 415—422 (1966).
HANSEN, L. H., and J. H. HOLMES: Echoencephalography in a large city hospital. Rocky Mount. med. J. **63**, 34—39 (1966).
HATA, H., H. MARUTA, J. YOSHIDA, K. ISHIDA, and Y. INABA: Observation of hydrocephalus by ultrasound — Evaluation of effect of ventriculoauriculostomy. Japan. Med. Ultrasonics **2**, 41—43 (1964).
HAZEMANN, P.: L'écho-encéphalographie de la ligne mediane. Comparaison avec les résultats fournis par l'E.E.G. en neuro-chirurgie. Paris: Foulon 1965.
HEMMER, R.: Erfahrungen mit der modernen operativen Hydrozephalus-Behandlung. Beih. Arch. Kinderheilk., 51. Heft (1964).
— Über atypische posttraumatische epidurale Blutungen und die Ursachen ihrer Fehldeutung. Dtsch. med. Wschr. **90**, 1945—1948 (1965).
HENKE, G.: Ultraschall in der medizinischen Diagnostik heute. Elektromedizin **9**, 68—72 (1964).
HERTZ, C. H.: Continuous registration of movement of heart structures by reflectoscope techniques. In: Ultrasonic energy, ed by E. KELLY. Urbana, University of Illiois Press, 1965, p. 294.

HERTZ, C. H.: Ultrasonic engineering in heart diagnosis. Amer. J. Cardiol. **19**, 6—17 (1967).
HOCKEY, A. A.: Echoencephalography. J. Indiana med. Ass. **59**, 342—346 (1966).
HÖPKER, W.: Die biologischen Wirkungen des Ultraschalls auf das Gehirn. Ultraschall in Med. **5**, 178—185 (1952).
HOFMANN, D., H.-J. HOLLÄNDER u. P. WEISER: Neue Möglichkeiten der Ultraschalldiagnostik in der Gynäkologie und Geburtshilfe. Fortschr. Med. **84**, 689—693 (1966).
— — — Über die geburtshilfliche Bedeutung der Ultraschalldiagnostik. Gynaecologica **164**, 24—36 (1967).
— H. MAST u. H.-J. HOLLÄNDER: Die Bedeutung der Plazentalokalisation mittels Ultraschall für die Amniozentese. Geburtsh. u. Frauenheilk. **27**, 1199—1209 (1967).
HOLMES, J. H.: The present status of ultrasonic diagnostic techniques in internal medicine. Ultrasonics **5**, 60—66 (1967).
— and D. H. HOWRY: Ultrasonic diagnosis of abdominal disease. Amer. J. dig. Dis. **8**, 12—32 (1963).
HOLYST, J.: Zastosowanie ultradźwięków do badania organicznych zmian śródczaszkowych (echoencefalografia). Pol. Tyg. Lek. **18**, 1109—1111 (1963).
— Echoencefalografia (Ultrasonoencefalografia). Pol. Tyg. Lek. **18**, 1749—1752 (1963).
HORENSTEIN, R., A. PERSSON, A. WENNBERG, and L. WIDÉN: The path of the ultrasonic beam in clinical echoencephalography. Acta radiol. (Stockholm), diagnosis **7**, 305—313 (1968).
HOTERMANS, J. M., et S. THIRY: Applications cliniques de l'échoencéphalographie en neurologie et en neurochirurgie. Rev. Méd. Liège **21**, 477—486 (1966).
— — Etude critique de 600 échoencéphalogrammes pré- et postopératoires. Acta chir. belg. **65**, 823—856 (1966).
— — L'échoencéphalographie post-opératoire. Expérience personelle basée sur 500 cas. Neurochirurgie **13**, 827—840 (1967).
— — L'échoencéphalographie — Examen préparatoire à l'angiographie carotidienne dans le diagnostic et la surveillance des accidents vasculaires cérébraux. 8e Symposium neuroradiol. Paris, 1967.
HOVIND, K. H., J. H. GALICICH, and D. D. MATSON: Normal and pathological intracranial anatomy, revealed by two-dimensional echoencephalography. Neurology (Minneap.) **17**, 252—267 (1967).
HOWRY, D. H.: Techniques used in ultrasonic visualization of soft tissue structures of the body. IRE, Convention Record **9**, 75—88 (1955).
— and R. W. BLISS: Ultrasonic visualization of soft tissue structures of the body. J. Lab. clin. Med. **40**, 579—592 (1952).
— and D. GORDON: Ultrasonic tomography. In: D. GORDON: Ultrasound as a diagnostic and surgical tool. Edinburgh & London: Livingstone 1964, pp. 103—123.
— D. A. STOTT, and W. R. BLISS: The ultrasonic visualization of carcinoma of the breast and other soft-tissue structures. Cancer **7**, 354—358 (1954).
HUBER, G.: Das Pneumencephalogramm bei Psychopathien und psychoreaktiven Störungen. Ein Beitrag zur Frage des „normalen" Encephalogramms. Arch. Psychiat. Nervenkr. **202**, 234—243 (1961).
— H. BETZ u. I. KLEINÖDER: Echoencephalographische Untersuchungen der 3. Hirnkammer bei einer männlichen Normalbevölkerung. Nervenarzt **39**, 82—84 (1968).
— u. C. PATIRI: Das Echoencephalogramm des 3. Ventrikels bei einer weiblichen Normalbevölkerung. Arch. Psychiat. Nervenkr. **210**, 61—67 (1967).
HUETER, T. F.: Messung der Ultraschallabsorption in tierischen Geweben und ihre Abhängigkeit von der Frequenz. Naturwissenschaften **35**, 285—287 (1948).
— Messung der Ultraschallabsorption im menschlichen Schädelknochen und ihre Abhängigkeit von der Frequenz. Naturwissenschaften **39**, 21—22 (1952).
— and R. H. BOLT: An ultrasonic method for outlining the cerebral ventricles. J. acoust. Soc. Amer. **23**, 160—167 (1951).
— — Sonics. Techniques for use of sound and ultrasound in engineering and science. New York—London: Wiley & Sons 1955.
HUSSMAN, L. H., H. W. KIENAST, and L. D. BOSHES: Determination of compressibility by ultrasound and its diagnostic significance for pathological indurations, degenerations and pulsations. In: Proceedings in echo-encephalography. Berlin-Heidelberg-New York: Springer 1968, pp. 34—37.
ICHIHARA, A.: Studies on diagnostic application of ultrasound. II. Echo-encephalography. Med. J. Hiroshima Univ. **14**, 129—140 (1966).
— Y. MASAOKA, and M. OTANI: Examination of bone echo in normal skull using simple stereotaxic apparatus. Japan. Med. Ultrasonics **2**, 43—45 (1964).
IIZUKA, J. H.: Correlation between neuroradiological and echoencephalographical findings. 8e Symposium neuroradiol. Paris, 1967.
— The physical properties of ultrasound. In: Proceedings in echo-encephalography. Berlin-Heidelberg-New York: Springer 1968, pp. 7—12.
INABA, Y., H. MARUTA, and H. HATA: A study of cerebral edema by ultrasound. Japan. Med. Ultrasonics **1**, Proc. 3rd Meeting of Japan. Soc. of Ultrasonics in Medicine, pp. 13—14, Tokyo, Juntendo University, 1963.

INUGAMI, K., HACHIYA, H. SASAKI, A. HAYASHI, T. WAGAI, I. ISHIKAWA, and K. YUKISHITA: Ultrasonic attenuation in epilepsy. Japan. Med. Ultrasonics **2**, 57—58 (1964).

ISHII, M.: An experimental study of stereotaxic destruction on the cat brain by intense focused ultrasound — The improvement of the apparatus for experiment. Arch. jap. Chir. **34**, 619—633 (1965).

ISHIKAWA, S.: Attenuation of ultrasound in brain tissue. Arch. jap. Chir. **33**, 923—940 (1964).

— H. NONAKA, and K. YUKISHITA: Ultrasonic attenuation in brain tissue (4th report). Japan. Med. Ultrasonics **2**, 60—61 (1964).

— K. YUKISHITA, and K. ITO: Ultrasonic attenuation in brain tissue (5th report). Attenuation of ultrasound in brain tumor. Japan. Med. Ultrasonics **3**, 33 (1965).

— — K. SATO, K. ITO, and T. WAGAI: Ultrasonic attenuation in brain tissue (7th report). Relationship of ultrasonic attenuation and histological structure of tumor tissue. Japan. Med. Ultrasonics **3**, 48 (1965).

ITO, K.: The clinical application of ultrasonics to the localization of brain tumors. Arch jap. Chir. **33**, 770—799 (1964).

— Y. ABE, and S. KIKUCHI: Diagnostic application of ultrasound in intracerebral hemorrhage (4th report). Japan. Med. Ultrasonics 2, 49—52 (1964).

— S. IJSHIKAWA, T. WAGAI, and R. UCHIDA: Ultrasono-tomography of brain (the 2nd report) — contact-sector-scanning-tomography —. Japan. Med. Ultrasonics **3**, 30 (1965).

— and K. NAKAMURA: The effects of ultrasonic wave on the growth of young bones. Japan. Med. Ultrasonics **1**, Proc. 3rd Meeting Japan Soc. of Ultrasonics in Medicine, pp. 1. Tokyo: Juntendo University 1963.

— H. NONAKA, Y. OHARA, and K. TANAKA: Diagnosis of brain tumor by ultrasound through the dura mater. Japan. Med. Ultrasonics **3**, 52 (1965).

— K. SUGAWARA, Y. OHARA, Y. ABE, and S. KIKUCHI: Ultrasonics in the diagnosis of head injury — "Hematoma echoes" as an aid in the diagnosis of subdural and epidural hematoma. Japan. Med. Ultrasonics **4**, 39 (1966).

— K. TANAKA, and Y. ABE: Ultrasound in the diagnosis of head injury. In: Proceedings in echo-encephalography. Berlin-Heidelberg-New York: Springer 1968, pp. 90—97.

IWATA, K., S. WATANABE, and M. TOMIYASU: A diagnostic method of brain tumors using A-scope. Japan. Med. Ultrasonics 2, 45—47 (1964).

JACKSON, F. E., M. HUSSEY, and D. RELYEA: Utilization of pulsed sonic beams (echoencephalogram) for detection of fragments of bone indriven into the brain. Milit. Med. **130**, 1107—1109 (1965).

JACOBI, J., P. KAMM u. W. SCHWARZ: Bedeutung der Ultraschallkardiographie in der Herzdiagnostik. Materia Medica Nordmark **14**, 251—257 (1962).

JACOBI, G.: Sources of error in echo-encephalograms of children with brain atrophy. In: Proceedings in echo-encephalography. Berlin-Heidelberg-New York: Springer 1968, pp. 155—157.

— E. KAZNER u. J. WOLLENSAK: Subdurale Ergüsse und Hämatome bei Säuglingen und Kindern. Betrachtungen zur Pathogenese, Klinik, Therapie und Prognose. Z. Kinderheilk. **96**, 199—227 (1966).

— u. P. SCHUCH: Echo-Enzephalographie und ihre Ergebnisse bei Kindern. Pädiat. Prax. **5**, 433—443 (1966).

— — Echo-Encephalographie und ihre Ergebnisse bei Kindern. Chir. Praxis **11**, 363—373 (1967).

— u. U. STEPHAN: Über die Anwendungsmöglichkeit der Echoencephalographie bei Kindern. Mschr. Kinderheilk. **113**, 344—345 (1965).

JACOBI, M.: Anwendungsmöglichkeiten und Ergebnisse der Echoencephalographie im Kindesalter. Inaug.-Diss. Erlangen 1966.

JACQUINOT, B., et B. GEUDELIN: Place de l'écho-encéphalographie parmi les examens complémentaires en neurologie. Concours med. **89**, 7227—7236 (1967).

JAMES, T. G., and E. A. TURNER: Traumatic intracranial haematoma. Lancet **1951, II**, 45—50.

JANSSON, F., and E. SUNDMARK: Determination of the velocity of ultrasound in ocular tissue at different temperatures. Acta Ophthal. **39**, 899—910 (1961).

JEFFERSON, A.: Some experiences with echo-encephalography. J. Neurol. Psychiat. **22**, 83—84 (1959).

— Clinical experiences with echo-encephalography. Acta neurochir. (Wien) 10, 392—409 (1962).

— and A. I. HILL: Echo-Encephalography. Progr. neurol. Surg. **1**, Basel-New York: Karger 1966, pp. 64—93.

— — The importance of the non-midline echoes in A-scan echo-encephalography with a commentary on their relevance to the reliability of the method. In: Proceedings in echo-encephalography. Berlin-Heidelberg-New York: Springer 1968, pp. 66—70.

JEPPSSON, ST.: Echoencephalography. III. Further studies on the sources of the midline echo and a clinical evalution. Acta chir. scand. **119**, 445—462 (1960).

— Echoencephalography. IV. The midline echo; an evaluation of its usefullness for diagnosing intracranial expansivities and investigation into its sources. Acta chir. scand. Suppl. **272**, 1—151 (1961).

— Echoencephalography V. A method for recording the intracranial pressure with the aid of the echoencephalographic technique. A preliminary report. Acta chir. scand. **128**, 218—224 (1964).

JEPPSSON, ST.: The use of the M-Echo in clinical echoencephalography. Acta neurol. scand. **41**, Suppl. 13 (1965) pp. 7—12.

— The use of the M-echo in clinical echoencephalography. In: Technical Progress in neurological diagnostics, ed. by O. GILLAND, Amsterdam: Elsevier 1965.

— A method for recording the intracranial pressure with the aid of the echo-encephalographic technique. In: Proceedings in echo-encephalography. Berlin-Heidelberg-New York: Springer 1968, pp. 186—189.

JOHANNSON, C.: The central veins and deep dural sinuses of the brain. Acta radiol. (Stockh.), Supplementum **107** (1957).

JOYNER, C. R., E. B. HEY, J. JOHNSON, and J. M. REID: Reflected ultrasound in the diagnosis of tricuspid stenosis. Amer. J. Cardiol. **19**, 66—73 (1967).

— J. M. REID, and J. P. BOND: Reflected ultrasound in the assessment of mitral valve disease. Circulation **27**, 503—511 (1963).

KAMM, P. J.: Diagnostische Möglichkeiten der Ultraschallkardiographie. Dtsch. med. Wschr. **93**, 1022—1025 (1968).

KANAYA, H., H. YAMASAKI, and I. SAIKI: Study of echoencephalography of intracerebral hematomas. Japan. Med. Ultrasonics **1**, 15—16 (1963).

— — — Echoencephalography of apoplexy, with special reference to the differential diagnosis of acute hypertensive intracerebral hemorrhage and brain softening. Brain Nerve (Tokyo), **16**, 769—778 (1964).

— — — Ultrasonic diagnosis of intracranial disease. J. Ther. (Tokyo), **46**, 753—760 (1964).

— — — and K. FURUKAWA: The use of echoencephalography to differentiate intracerebral hemorrhage and brain softening. J. Neurosurg. **28**, 539—543 (1968).

KATANUMA, M., S. AZAMI, and Y. TSUTSUMI: Pen-writing echo-encephalograph. Japan. Med. Ultrasonics **2**, 52—54 (1964).

KAUTZKY, R., u. K. J. ZÜLCH: Neurologisch-Neurochirurgische Röntgendiagnostik und andere Methoden zur Erkennung intrakranialer Erkrankungen. Berlin-Göttingen-Heidelberg: Springer 1955.

KAZNER, E.: Die Echoenzephalographie beim epiduralen Haematom. Acta neurochir. (Wien) **13**, 582 (1965).

— Recognition and differential diagnosis of intracranial complications following head injuries by means of echo-encephalography. In: Proceedings in echo-encephalography. Berlin-Heidelberg-New York: Springer 1968, pp. 98—107.

— Echo-encephalographische Diagnostik bei raumfordernden intrakraniellen Prozessen im Kindesalter. Verlag Wiener Med. Akademie (in press).

— ST. KUBICKI, ST. KUNZE, W. SCHIEFER u. S. WENDE: Die Bedeutung der klinischen Zusatzuntersuchungen für die Differentialdiagnose zerebrale Massenblutung — Hirninfarkt. Fortschr. Neurol. Psychiat. (in press).

— and ST. KUNZE: Echo-encephalographic findings with midline tumors and tumors of the base of the skull. In: Proceedings in echo-encephalography. Berlin-Heidelberg-New York: Springer 1968, pp. 71—75.

— — Die Echo-Encephalographie als ergänzende Untersuchung bei Mittelhirntumoren. Der Radiologe (in press).

— — u. W. SCHIEFER: Die Bedeutung der Echoencephalographie für die Erkennung epiduraler Haematome. Langenbecks Arch. klin. Chir. **310**, 267—291 (1965).

— — — Echoencephalography as an aid to the diagnosis of space-occupying lesions in the posterior fossa by measuring the size of the third and lateral ventricles. J. Neurosurg. **26**, 511—520 (1967).

— u. W. SCHIEFER: Das Ultraschall-Echo-Verfahren (Echoencephalographie), eine Methode zur frühzeitigen Erkennung raumfordernder intrakranieller Prozesse. Med. Mschr. **18**, 27—31 (1964).

— — Operative Entfernung eines Zystizerkus aus dem IV. Ventrikel. Med. Bild-Dienst Roche, Heft 3, 1965, 18—23.

— — Echoencephalographische Untersuchungsergebnisse bei Schädel-Hirnverletzungen. Acta radiol. **5**, diagnosis, 7th Symposium neuroradiol., New York 1964, 832—842 (1966).

— — Die Echoenzephalographie bei raumfordernden Prozessen der hinteren Schädelgrube. Acta neurochir. (Wien) **14**, 177—196 (1966).

— — The diagnosis of epidural haematoma on the basis of echo-encephalography. Excerpta medica, International Congress Series **139**, (1967) No. 173.

— — and K. J. ZÜLCH: Proceedings in Echo-Encephalography. International Symposium on Echo-Encephalography Erlangen, April 14th and 15th, 1967. New York: Springer 1968.

KEIDEL, W.-D.: Über die Verwendung des Ultraschalls in der klinischen Diagnostik. Ärztl. Forschg. (Wörishofen) **1**, 349—357 (1947).

KELLY, E.: Ultrasonic energy. Biological investigations and medical applications. Urbana: University of Illinois Press 1965.

KERNOHAN, J. W., and H. W. WOLTMAN: Incisure of the crus due to contralateral brain tumor. Arch. Neurol. Psychiat. (Chicago) **21**, 274—287 (1929).

KESSLER, G. B.: Echoencephalography. Its use as a diagnostic aid in neurological disease; plus a comparison with EEG foci. Bull. Los Angeles neurol. Soc. 30, 21—26 (1965).

KIENAST, H. W., and L. BOSHES: Echoencephalographic studies in epilepsy. Proceedings 8th Intern. Congress Neurol., Vienna, 1965, 354.

KIKUCHI, S., and K. ITO: Ultrasonic diagnosis of intracranial disease. Rinsho Geka (Clin. Surg., Tokyo) 18, 376—381 (1963).

— — and Y. ABE: Diagnosis of apoplexy by ultrasonic pulses (2nd report). Japan. Med. Ultrasonics 2, 53—56 (1963).

— R. UCHIDA, K. TANAKA, and T. WAGAI: Early cancer diagnosis through ultrasonics. J. acoust. Soc. Amer. 29, 824—833 (1957).

KINGSLEY, B., G. B. FLINT, G. T. RABER, and B. L. SEGAL: Another look at echocardiography. Concepts in biomedical engineering. Amer. J. Cardiol. 19, 108—116 (1967).

KLINGLER, D.: Die diametrale bitemporale Echoencephalographie im A-Scope unter Abstimmung auf die ideale Medianebene. Communicationes, 6. Intern. Kongreß Elektroenc. u. klin. Neurophysiol. Wien 1965, pp. 597—600.

KLINGLER, M.: Die Bedeutung angiographischer Schrägaufnahmen für die Diagnose epiduraler Hämatome. Acta neurochir. (Wien) 12, 695—701 (1965).

KOSLOWSKI, L., u. W. THIES: Bericht über 5900 Schädel-Hirn-Traumen. Mschr. Unfallheilk. 67, 97—103 (1964).

— — Erfahrungen mit der Echo-Encephalographie bei frischen Schädel-Hirn-Traumen. Hefte Unfallheilk. 87, 170—172 (1966).

KOSSOFF, G., D. E. ROBINSON, and W. J. GARRETT: Two dimensional ultrasonography in obstetrics. In: Diagnostic ultrasound. New York: Plenum Press 1966.

— — C. N. LIU, and W. J. GARRETT: Design criteria for ultrasonic visualization system. Ultrasonics 2, 29—38 (1964).

KRAMER, G.: Zur Diagnostik raumfordernder intrakranieller Prozesse mit Hilfe des Echoimpulsverfahrens. Chirurg 34, 290—298 (1963).

— Die Echoenzephalographie. Dtsch. med. Wschr. 89, 564—567 (1964).

— Vierjährige Erfahrungen mit der Echoencephalographie bei Schädelhirntraumen. Mschr. Unfallheilk. 69, 145—150 (1966).

— Difficulties in the interpretation of echo-encephalographic findings with head injuries. In: Proceedings in echo-encephalography. Berlin-Heidelberg-New York: Springer 1968, pp. 107—110.

KRATOCHWIL, A.: Die diagnostische Anwendung des Ultraschalls in der Geburtshilfe und Gynäkologie. Zbl. Gynäk. 88, 1032 (1966).

— Ultraschalldiagnostik in Geburtshilfe und Gynäkologie. Lehrbuch und Atlas. Stuttgart: Thieme 1968, pp. 116.

KRAULAND, W.: Über die Quellen des akuten und chronischen subduralen Hämatoms. Stuttgart: Thieme 1961.

KRAUSE, W., u. R. SOLDNER: Ultraschallbildverfahren (B-scan) mit hoher Bildfrequenz für medizinische Diagnostik. Electromedica 4, 1—5 (1967).

KRAUTKRÄMER, J., u. H. KRAUTKRÄMER: Werkstoffprüfung mit Ultraschall. 2. Auflage. Berlin-Heidelberg-New York: Springer 1966.

KRESSE, H.: The physical principles in the use of the ultrasound echo method. In: Proceedings in echo-encephalography. Berlin-Heidelberg-New York: Springer 1968, pp. 3—7.

KRÜGER, H., V. ZUMPE u. A. VELTIN: Echoencephalographische Untersuchungen der III. Hirnkammer bei Gesunden verschiedenen Lebensalters. Arch. Psychiat. Nervenkr. 210, 161—168 (1967).

— — — Das Echoencephalogramm des III. Ventrikels bei Gesunden und Schizophrenen. Nervenarzt 38, 412—414 (1967).

— — — Zur Validität der Echoencephalographie der inneren Liquorräume (Eine Untersuchung zur Frage der Übereinstimmung von Echo- und PEG-Meßwerten des III. Ventrikels). Nervenarzt 39, 80—82 (1968).

KUGLER, J.: Elektroencephalographie in Klinik und Praxis. Stuttgart: Georg Thieme 1963.

KUNZE, ST.: Die operative Behandlung intracerebraler Blutungen unter besonderer Berücksichtigung des Spontanhämatoms. Inauguraldissertation, Erlangen 1964.

— Contribution of echo-encephalography to the differential diagnosis of cerebral hemorrhage and brain softening. In: Proceedings in echo-encephalography. Berlin-Heidelberg-New York: Springer 1968, pp. 159—164.

KURZE, TH., P. DYCK, and H. S. BARROWS: Neurosurgical evaluation of ultrasonic encephalograph. J. Neurosurg. 22, 437—440 (1965).

LANG, W., u. H. W. PIA: Prognose und Indikation zur Ventrikulocisternostomie (Torkildsen-Operation). Dtsch. Z. Nervenheilk. 182, 176—184 (1961).

LANGEVIN, M. P.: Les ondes ultrasonores. Rev. gen. electr. 23, 626—634 (1928).

Lapayowker, M. S., and G. E. Christen: Echoencephalography in general hospital practice. Amer. J. Roentgenol. **93**, 803—810 (1965).
—, F. Murtagh, and J. A. Kirkpatrick: Echoencephalography in the evaluation of hydrocephalus. Radiology **86**, 1052—1055 (1966).
Lauber, H.: Das Pneumencephalogramm. Meßverfahren bei Erwachsenen. München: Joh. Ambrosius Barth 1965.
Lausberg, G., und C. L. Geletneky: Frühdiagnose, Therapie und Verlaufsformen des Hydrocephalus bei cystischen Dysrhaphien. Zbl. Neurochir. **28**, 161—168 (1967).
Leary, G. A.: Basic techniques applying ultrasonics to ophthalmic measurement and diagnosis. Ultrasonics **5**, 84—87 (1967).
Lee, R. G., and T. Morley: The routine use of echoencephalography in the diagnosis of intracranial masses. Canad. med. Ass. J. **91**, 987—990 (1964).
Lehman, J. St., G. C. Evans, and L. W. Brady: Ultrasound exploration of the spleen. In: Diagnostic ultrasound. New York: Plenum Press 1966.
Leksell, L.: Kirurgisk behandling av skallskador. Paper read at the meeting of Svenska Läkarsällskapet, Stockholm, Dec. 7, 1954. Nord. Med. **54**, 1444 (1955).
— Echo-encephalography. I. Detection of intracranial complications following head injury. Acta chir. scand. **110**, 301—315 (1955/56).
— Echoencephalography. II. Midline echo from the pineal body as an index of pineal displacement. Acta chir. scand. **115**, 255—259 (1958).
LeMay, M. J., and R. G. Ojemann: Angiographic changes associated with a dilated temporal horn. Acta radiol. (Stockh.), Diagnosis **1**, 358—366 (1963).
Lepintre, J., J. Lefebvre, G. Debrun et J. Legal: Transillumination et échoencéphalographie. Examens physiques simples dans diagnostic et dans le surveillance des collections liquidiennes sousdurales chez l'enfant. Neuro-chirurgie **11**, 285—293 (1965).
Lepoire, J., et Cl. Lapras: L'échoventriculographie. Neuro-chirurgie **13**, 256—258 (1967).
Lhotsky, J., and F. Weiss: Echo-encephalographic experience with surgical patients. In: Proceedings in echo-encephalography. Berlin-Heidelberg-New York: Springer 1968, pp. 174—175.
Lithander, B.: A control method for echo-encephalography. Acta psychiat. scand. **35**, 235—240 (1960).
— The clinical use of echo-encephalography. Acta psychiat. scand. **35**, 241—244 (1960).
— Origin of echoes in the echoencephalogram. J. Neurol. Psychiat. **24**, 22—31 (1961).
— Clinical and experimental studies in echo-encephalography. Acta psychiat. scand. **36**, Suppl. 159 (1961).
— The true displacement of the midline brain structures in relation to midline echo displacement. Acta psychiat. scand. **36**, Suppl. 159, 9—12 (1961).
— Estimation of the experimental error and normal individual variation in echo-encephalography. Acta psychiat. scand. **36**, Suppl. 159, 13—17 (1961).
— Echo-encephalography and extracerebral haematomas. Comparison between echo-encephalography and other methods of diagnosing extracerebral haematomas. Acta psychiat. scand. **36**, Suppl. 159, 31—36 (1961).
— Echo-encephalography in children. Acta psychiat. scand. **36**, Suppl. 159, 37—50 (1961).
— Echoencephalography. A general survey. Acta psychiat. scand. **41**, Suppl. 13, 1—6 (1965).
— Echoencephalography. A general survey. In: Technical progress in neurological diagnostic. Ed. by O. Gilland. Amsterdam: Elsevier 1965.
— and O. Marions: Echo-encephalography in adults. Acta psychiat. scand. **36**, Suppl. 159, 18—30 (1961).
Lombroso, C. T., and G. Erba: Two-dimensional ultrasonography for the visualization of ventricular landmarks. In: Proceedings in echo-encephalography. Berlin-Heidelberg-New York: Springer 1968, pp. 215—230.
Löpping, B.: Klinische und experimentelle Ultraschalluntersuchungen bei Augen mit Netzhautablösung und intraokularen Tumoren. Inaugural-Diss., Köln, 1962.
Ludwig, G. D.: The velocity of sound through tissues and the acoustic impedance of tissues. J. acoust. Soc. Amer. **22**, 862—866 (1950).
— R. H. Bolt, T. F. Hueter, and H. T. Ballantine: Factors influencing the use of ultrasound as a diagnostic aid. Trans. Amer. neurol. Ass. **75**, 225—228 (1950).
Lugt, P. J. M. van der, and C. E. Molin: Factors influencing cerebral pulsations in Echo Encephalography. Acta neurochir. (Wien) **18**, 246—255 (1968).
Lynn, J. G., and T. J. Putnam: Histology of cerebral lesions produced by focused ultrasound. Amer. J. Path. **20**, 637—649 (1944).
Makow, D. M.: Délimitation des contours ventriculaires par les ultrasons. Neuro-chirurgie **10**, 467 to 469 (1964).
— u. D. L. McRae: Echo-Tomographie des Schädels. Image Roche, Heft 4, 1966, 15—22.
— — Horizontal scanning of the head with ultrasound. Med. biol. Engng. **5**, 33—40 (1967).

Makow, D. M., u. D. L. McRae: Two dimensional echo-encephalography using immersion scanning: recent results. In: Proceedings in echo-encephalography. Berlin-Heidelberg-New York: Springer 1968, pp. 197—201.
— and R. R. Real: Immersion ultrasonic brain examination with 360° Scan. Ultrasonics 3, 75—80 (1965).
— — Development of a 360° compound immersion head scanner. In: Diagnostic ultrasound. New York: Plenum Press 1966, pp. 166—185.
— W. Wyslouzil, D. N. White, and J. Blanchard: Novel immersion scanner and display system for ultrasonic brain tomography. Acta radiol. **5**, diagnosis, 7th Symposium neuroradiol., New York 1964, 855—864 (1966).
Marková, J., a F. Pokorny: Echoencefalografie a diagnostické metody založené na využití ultrazvuku. Lékařská věda v zahraničí **6**, 112—117 (1966).
Maruta, H., H. Hata, K. Yamaoka, and Y. Inaba: A study of the ultrasonic attenuation of the brain during anaesthesia — The ultrasonic monitoring of anaesthesia. Japan. Med. Ultrasonics **2**, 54—57 (1964).
Matthes, K., u. W. Rech: Der Ultraschall in der Medizin. Kongreßbericht der Erlanger Ultraschall-Tagung 1949. Zürich: Hirzel, 1949.
Mattsson, O.: Ultrasonics as a complement to roentgen diagnosis. A preliminary communication. Acta radiol. (Stockholm), diagnosis **7**, 361—368 (1968).
Mayer, K.: Differentialdiagnose — Alkoholrausch und posttraumatische Bewußtseinsstörung —. Med. Welt **18**, 2213—2219 (1967).
McGinnis, K. D., and C. J. Zylak: Echoencephalography as an adjuvant to the routine examination of the skull. Radiology **85**, 291—297 (1965).
McKinney, W. M.: The value of B-mode determination of midline in echo-encephalography. Neurology (Minneap.) **14**, 259 (1964).
— W. S. Avant, F. L. Thurstone, and Bette Pou: Intracranial pulsations as measured by ultrasound. Engin. Med. Biol. Proc. of the 18th Ann. Conf. p. 68 (1965).
— F. L. Thurstone, W. S. Avant, and W. K. Wallace: The significance of intracranial echo pulsations. In: Diagnostic ultrasound. New York: Plenum Press 1966, pp. 114—116.
— J. F. Toole, W. T. Sharp, E. A. MacMillan, J. W. Gibson, and B. D. South: Evaluation of five hundred psychiatric inpatients by midline echoencephalography. Acta radiol. **5**, diagnosis, 7th Symposium neuroradiol., New York 1964, 865—870 (1966).
McLaurin, R. L., and F. Helmer: Angiographic observations following removal of subdural hematomas. Acta radiol. **5**, diagnosis. 7th Symposium neuroradiol., New York 1964, 381—387 (1966).
McRae, D. L., and D. M. Makow: Horizontal laminography of the head with ultrasound. J. Canad. Ass. Radiol. **17**, 75—80 (1966).
— — Horizontal laminography of the head with ultrasound. Further results. 8e Symposium neuroradiol. Paris, 1967.
Metzel, E., u. W. Umbach: Akute posttraumatische Blutungen des Schädelinneren. Actuelle Chir. **1**, 145—152 (1966).
Micsky, L. I. von: Ultrasonic tomography in obstetrics and gynecology. In: Diagnostic ultrasound. New York: Plenum Press 1966.
Mikkelsen, B.: Ekkoencefalografi. Nord. Med. **77**, 347—352 (1967).
Mikol, F.: L'écho-encéphalographie. Presse méd. **73**, 729—730 (1965).
Milbled, G., F. Wyremblewski et T. Ramez: L'écho-encéphalographie. Exploration cérébrale par les ultrasons. Lille méd. **9**, 712—715 (1964).
Miller, L. D., C. R. Joyner, S. J. Dudrick, and D. J. Eskin: Clinical use of ultrasound in the early diagnosis of pulmonary embolism. Ann. Surg. **166**, 381—393 (1967).
Mitsuno, T., H. Kanaya, S. Shirakata, K. Ohsawa, and Y. Ishikawa: Surgical treatment of hypertensive intracerebral hemorrhage. J. Neurosurg. **24**, 70—76 (1966).
Mizuno, Sh., H. Takeuchi, and K. Nakano: Diagnostic application of ultrasound in obstetrics and gynecology. In: Diagnostic ultrasound. New York: Plenum Press 1966.
Mühlhäuser, O.: Verfahren zur Zustandsbestimmung von Werkstoffen, besonders zur Ermittlung von Fehlern darin. DRP 569 598 (1931).
Müller, E.: Das epidurale Hämatom und die Schwierigkeit einer klinischen Diagnosestellung. Untersuchung an Hand eines gerichtsmedizinischen Obduktionsgutes. Inaug.-Diss. Köln 1965.
Müller, H. R.: Zur Echoencephalographie; ihre Indikation beim Schädel-Hirn-Trauma. Schweiz. med. Wschr. **94**, 119—124 (1964).
— Diagnose des epiduralen Hämatoms im präklinischen Stadium. Zum Wert der routinemäßigen Echoencephalographie bei Schädelkalottenfrakturen. Schweiz. med. Wschr. **97**, 1055—1057 (1967).
— and U. Blauenstein: Monitoring of midline-echo by the use of a new probe-holder. Europ. Neurol. **1**, 28—30 (1968).

Müller, H. R., and M. Klingler: Problems with echo-encephalography in the postoperative phase. In: Proceedings in echo-encephalography. Berlin-Heidelberg-New York: Springer 1968, pp. 116—120.
— and A. Lévy: A simple method of two-dimensional intraoperative sonencephalography, employing the A-scan technique. Europ. Neurol. **1**, 31—40 (1968).
Mundt jr., G. H., and W. F. Hughes jr.: Ultrasonics in ocular diagnosis. Amer. J. Ophthal. **41**, 488—498 (1956).
Nagai, H., K. Sakurai, M. Hayashi, M. Furuse, and K. Okamura: Echoencephalogram following experimental intracerebral hemorrhage and infarction. Proc. 3rd Meeting Japan. Soc. Ultrasonics in Medicine, p. 15. Tokyo: Juntendo University 1963.
— — — — — A. Shintani, and T. Kobyashi: Ultrasonic findings in intracranial diseases — Analysis of 114 cases. Japan. Med. Ultrasonics **2**, 47—49 (1964).
Nadjmi, M.: Relationships between radiological anatomy and echo-encephalography. In: Proceedings in echo-encephalography. Berlin-Heidelberg-New York: Springer 1968, pp. 27—31.
Newell, J. A.: The use of ultrasonics in medical diagnosis. Proc. 3rd Int. Conf. Med. Electron., Institution of Electrical Engineers, London, 422—424 (1961).
— Ultrasonic localization. Brit. J. Radiol. **34**, 546—550 (1961).
— Ultrasonics in medicine. Phys. in Med. Biol. **8**, 241—264 (1963).
Nichols, R. A., J. P. Whisnant, and H. L. Baker, Jr.: A-mode echoencephalography: Its value and limitations. Mayo Clin. Proc. **43**, 36—53 (1968).
Nicholson, A. M.: Piezo-electric effect in composite Rochelle salt crystal. Proc. Amer. Inst. Elect. Eng. **38**, 1467—1493 (1919).
Nonaka, H., K. Yukishita, S. Ishikawa, and K. Ito: Experimental brain edema by ultrasound. Japan. Med. Ultrasonics **2**, 9—11 (1964).
Nover, A.: Klinische Ultraschalluntersuchungen bei Netzhautablösung und intraokularen Tumoren. Klin. Mbl. Augenheilk. **142**, 176—194 (1963).
— u. G. Kunde: Ultraschalluntersuchungen am Glaskörper gesunder und kranker Augen. Klin. Mbl. Augenheilk. **152**, 639—648 (1968).
— u. H. Stallkamp: Über die Möglichkeiten und Grenzen der Ultraschalldiagnostik intraokularer Fremdkörper. Bericht über die Zusammenkunft d. Deutsch. Ophthalm. Ges., Heidelberg 1960. München: Bergmann 1960, pp. 251—255.
— — Experimentelle Ultraschalluntersuchungen bei Augen mit intraokularen Fremdkörpern. Albrecht v. Graefes Arch. Ophthal. **164**, 517—523 (1962).
Nürnberger, S., u. G. Schaltenbrand: Messungen am Encephalogramm. Ein Beitrag zum Begriff des „normalen Encephalogramms". Dtsch. Z. Nervenweilk. **174**, 1—14 (1955).
Oberschulte-Beckmann, D., u. D. Otto: Nachweise künstlicher epiduraler Hämatome durch Echo-Encephalographie. Chirurg. **38**, 370—373 (1967).
— — The anatomical basis of the end echo in echo-encephalography from an experimental point of view. In: Proceedings in echo-encephalography. Berlin-Heidelberg-New York: Springer 1968, pp. 32—34.
— — Echo-encephalographic investigations of cadaver skulls with artificial epidural hematomas. In: Proceedings in echo-encephalography. Berlin-Heidelberg-New York: Springer 1968, pp. 120—122.
Özgür, T.: Kafa travmalarinin EEG yaninda ECHO ile birlikte incelenmesinin diyagnostik, olayin gelişimi ve prognostiğin tayini yönünden değeri. Bornova: Ege Üniversitesi Matbaasi 1967.
Ohara, Y., Y. Abe, S. Ishikawa, K. Ito, and K. Tanaka: Ultrasonotomography of brain (3rd report). Med. Ultrasonics **4**, 61 (1966).
Oka, M.: Experiences of cranial test by measuring the ultrasonic loss in traumatic brain. Proc. 3rd Meeting Japan. Soc. of Ultrasonics in Medicine, p. 14. Tokyo: Juntendo University 1963.
— Y. Naito, H. Sakamoto, Y. Kijiraoka, and K. Kiyoshima: Study on the dynamic test of head injury by measuring the ultrasonic attenuation of traumatized brains — The ultrasonic attenuation of injured brains. Japan. Med. Ultrasonics **2**, 58—60 (1964).
Oksala, A., u. A. Lehtinen: Über die diagnostische Verwendung von Ultraschall in der Augenheilkunde. Ophthalmologica **134**, 387—395 (1957).
— — Diagnosis of detachment of the retina by means of ultrasound. Acta ophthal. **35**, 461—467 (1957).
— — Diagnostics of rupture of the scalera by means of ultrasound. Acta ophthal. **36**, 37—40 (1958).
— — Measurement of the velocity of sound in some parts of the eye. Acta ophthal. **36**, 633—639 (1958).
— — Das Echogramm in der Diagnostik von Augenkrankheiten. Klin. Mbl. Augenheilk. **137**, 72—87 (1960).
— — Experimental observations on acoustic biomicroscopy of various parts of the eye. Acta ophthal. **38**, 599—605 (1960).
— — Acoustic biomicroscopy of intraocular tumors. Amer. J. Ophthal. **51**, 1203—1211 (1961).
Oldendorf, W. H.: Ultrasound in neurologic diagnosis. Trans. Bio-Med. Engin. B.M.E. II, 118—122 (1964).

OLDENKOTT, P.: Die eindimensionale Echo-Enzephalographie — Ein diagnostisches Hilfsmittel bei Erkrankungen und Verletzungen des Hirnschädels. Med. Welt. **18**, 2906—2908 (1967).

OLIVIER, L., M.-F. STRAUSS, M. DAVID et H. FISCHGOLD: Aspects positifs et négatifs de l'echo-encéphalographie. Remarques à propos des traumatismes crâniens. Presse med. **75**, 2213—2216 (1967).

OMOTO, R.: Ultrasonic tomography of the heart: An intracardiac scan method. Ultrasonics **5**, 80—83 (1967).

OSSOINIG, K.: Zur Ultraschalldiagnostik intraocularer Fremdkörper. Bericht d. Ophthalm. Ges. **67**, 288—294. München: J. F. Bergmann 1965.

— Erste Erfahrungen mit der kombinierten A-Bild- und B-Bildechographie orbitaler Tumoren. Wien. klin. Wschr. **80**, 72—74 (1968).

— Die Ultraschalldiagnostik in der Augenheilkunde. Wien. med. Wschr. **118**, 362—368 (1968).

— u. K. SEHER: Ergebnisse der Ultraschalldiagnostik orbitaler Tumoren. Klin. Mbl. Augenheilk. **151**, 519—524 (1967).

— et E. VALENÇAK: L'échotomographie dans le diagnostic des tumeurs orbitaires. Neuro-chirurgie **13**, 899—905 (1967).

OVERTON, M. C., J. E. COE, and S. R. SNODGRASS: Echoencephalography: a clinical appraisal. Tex. St. J. Med. **61**, 823—829 (1965).

P'AN, Y.-H., H.-H. WU, K.-P. CH'EN, and M.-F. SU: Ultrasonics in the diagnosis of intracranial space-occupying lesions. Zhong. Waike Z. (Peking) **11**, 536—541 (1963).

PANTSCHENKO, D. I., and E. L. MATSCHERET: Diagnostic possibilities of echoencephalography. Electromedica **1967**, 3, 24—26.

— — The diagnostic possibilities of echo-encephalography. In: Proceedings in echo-encephalography. Berlin-Heidelberg-New York: Springer 1968, pp. 157—159.

PARKINSON, D., and H. CHOCHINOV: Subdural hematomas — some observations on their post-operative course. J. Neurosurg. **17**, 901—904 (1960).

PATEISKY, K., D. KLINGLER u. R. GERSTNER: Echoenzephalographie. Darstellung im A-Skope und Erfahrungen mit einer symmetrisch auf die ideale Mitte abgestimmten Technik. Wien. Z. Nervenheilk. **25**, 341—372 (1967).

PATIRI, C.: Zur Frage der Echoencephalographie des 3. Ventrikels bei weiblichen Durchschnittspersonen. Inauguraldissertation Bonn 1967.

PELL, R. L.: Ultrasound for routine clinical investigations. Ultrasonics **2**, 87—89 (1964).

PENZHOLZ, H.: Zur Diagnose traumatischer intrakranieller Blutungen unter besonderer Berücksichtigung von Carotisangiographie und Echoencephalographie. Forschung, Praxis, Fortbildung **11**, 390—397 (1966).

PEPIN, B., B. JAQUINOT et B. GENDELIN: L'écho-encéphalographie. Méd. int. **1**, 35—40 (1966).

PERRET, G., and H. NISHIOKA: Cerebral angiography. An analysis of the diagnostic value and complications of carotid and vertebral angiography in 5484 patients. J. Neurosurg. **25**, 98—114 (1966).

PERSSON, A. W., A. WENNBERG, et L. WIDÉN: Valeur diagnostique de l'écho-encéphalographie comme complément de l'EEG. Resultats préliminaires. Rev. neurol. **108**, 106 (1963).

PETERS, G.: Ultraschallwirkung auf das Nervensystem. Kongreßbericht der Erlanger Ultraschalltagung. Zürich: Hirzel 1949, p. 166.

— Morphologische Untersuchungen über die Wirkung von Ultraschallwellen auf das Zentralnervensystem. Fortschr. Neurol. Psychiat. **17**, 85—94 (1949).

— Die Wirkung der Ultraschallwellen auf das Zentralnervensystem. Strahlentherapie **79**, 653—658 (1949).

— Experimentelle Ultraschallschäden am Gehirn. Ultraschall in Med. **4**, 60—62 (1952).

— Schädigungen des Zentralnervensystems durch Ultraschall. In: Handb. spez. path. Anatomie u. Histologie. Ed. by W. SCHOLZ. Vol. XIII/3. Berlin: Springer 1955.

PEZOLD, F. A.: Zur Frage des Ultraschallschadens. Ultraschall in Med. **4**, 1—28 (1952).

PIA, H. W.: Zur Frage der Umgehungsdrainage nach Torkildsen. Zbl. Neurochir. **13**, 102—107 (1953).

— Hirnverletzungen bei Kindern und ihre akuten Komplikationen. Münch. med. Wschr. **108**, 760—768 (1966).

— Die Echoenzephalographie in der Hirngeschwulstdiagnostik. Dtsch. med. Wschr. **92**, 999—1002 (1967).

— Klinik und operative Therapie der intrakraniellen Aneurysmen. Dtsch. Ärztebl. **64**, 2265—2272 (1967).

— Die Echoenzephalographie in der Hirngeschwulstdiagnostik. In: Diagnostische und therapeutische Fragen bei Hirngeschwülsten. Stuttgart: Thieme 1968, pp. 24—32.

— Die Bedeutung der Echoencephalographie für die Hirntumordiagnostik. Nervenarzt **39**, 322 to 327 (1968).

— u. C.-L. GELETNEKY: Echoenzephalographie. Stuttgart: Thieme 1968.

— — Echo-encephalography with infantile encephalopathies and their sequelae. In: Proceedings in echo-encephalography. Berlin-Heidelberg-New York: Springer 1968, pp. 145—151.

PLANIOL, TH.: La gamma-encéphalographie dans le diagnostic des lésions expansives intra-crâniennes. Confin. neurol. **21**, 137—158 (1961).
— L'echo-encéphalographie dans la détection des néoformations intracraniennes. Neurochirurgie **13**, 19—22 (1967).
— Results of four years experience with echo-encephalography of brain tumors. In: Proceedings in echo-encephalography. Berlin-Heidelberg-New York: Springer 1968, pp. 44—48.
— et O. BETTI: Diagnostic des méningiomes intracraniens par la gamma-encéphalographie. Rev. neurol. **105**, 420—429 (1961).
— H. FISCHGOLD, M. DAVID, M. SACHS et B. COURSON: Sur les examens physiques ambulatoires du cerveau. Intérêt de l'écho-encéphalographie médiane dans une observation d'empyème sous-dural. Neuro-chirurgie **11**, 273—284 (1965).
— J. METZGER, M. DAVID et H. FISCHGOLD: Explorations techniques ambulatoires dans le diagnostic des tumeurs cérébrales hémispheriques. Acta radiol., **5**, diagnosis, 7th Symposium neuroradiol., New York 1964, 906—914 (1966).
— F. MIKOL, J. CHARPENTIER et J. BUISSON: L'écho-encéphalographie. Premiers résultats après six mois d'investigations en neurochirurgie. Rev. neurol. **110**, 489—505 (1964).
— and J. M. POTTIER: Registration of cerebral echo-pulsations and comparison with rheo-encephalographic oscillations. In: Proceedings in echo-encephalography. Berlin-Heidelberg-New York: Springer 1968, pp. 189—192.
POHLMAN, R.: Die Ultraschalltherapie. Praktische Anwendung des Ultraschalls in der Medizin. Bern: Huber 1951.
— Über die erforderlichen Voraussetzungen einer brauchbaren Ultraschall-Diagnostik. Rehabilitation **15**, 135 (1962).
PORRAS, C. L.: La ecoencefalografia en el diagnostico de los procesos intracraneales. Rev. clin. esp. **89**, 29—32 (1963).
POTTHOFF, P. C.: Der heutige Stand der Neurochirurgie des Kleinhirns und der Kleinhirnerkrankungen. Nervenarzt **39**, 299—310 (1968).
POZO, L. DEL, J. V. ANASTASIO y ALIJARDE: Ecos ventriculares en la exploración ultrasónica cerebral. Rev. clin. esp. **100**, 48—49 (1966).
PRIDIE, R. B., and V. BRODIE: The shift of the mid-line demonstrated ultrasonically in cerebral atrophy. 8e Symposium neuroradiol. Paris, 1967.
PROLER, M. L., R. R. NORTH, and M. D. LOW: Echoencephalography — The measurement of the position of the midline structures in the skull by pulsed ultrasound. Med. Rec. (Houston) **57**, 416—417 (1964).
RADDA, H., u. E. SCHIMA: Wert und Grenzen der Echoencephalographie bei frischen Schädel-Hirnverletzungen. Wien. med. Wschr. **115**, 881—883 (1965).
RASKIND, R.: Clinical studies in echoencephalography. Preliminary report. J. int. Coll. Surg. **42**, 512—515 (1964).
— Clinical studies in echoencephalography: Report of 600 scans. J. int. Coll. Surg. **44**, 153—158 (1965).
— Diagnostic ultrasonography. J. Amer. Geriat. Soc. **13**, 887—892 (1965).
RAUSCH, F., W. SCHIEFER u. G. STRUCK: Über den Wert der zerebralen Angiographie für die Diagnose arteriosklerotischer Gefäßprozesse. Fortschr. Neurol. Psychiat. **24**, 512—520 (1956).
REID, J. M., and J. J. WILD: Ultrasonic ranging for cancer diagnosis. Electronics (N. Y.), July 1952.
— — Ultrasonic echo-ranging for tissue diagnostic studies. IRE, Convention Record **9**, 68—74 (1955).
RICHARDSON, A.: The management of primary intracranial hemorrhage. Modern trends in neurology (Third series). London: Butterworth and Co. 1962, pp 89—107.
— and J. AMBROSE: Echoencephalography in cerebral hemorrhage and infarction. Lancet **1966 I**, 319. (Letter to the editor).
RICHARDSON, E. G.: Ultrasonic physics. Sec. Ed. Amsterdam: Elsevier Publ. Co. 1962.
RICHTER, H.: Zur Problematik der frischen, schwersten Schädel-Hirn-Verletzungen. Bruns' Beitr. klin. Chir. **214**, 298—304 (1967).
— Die Echoenzephalographie beim frischen Schädel-Hirn-Trauma. Dtsch. med. Wschr. **93**, 1144 to 1146 (1968).
— L. KOSLOWSKI, and G. LAU: The value of echo-encephalography with acute life-threatening closed head injuries. In: Proceedings in echo-encephalography. Berlin-Heidelberg-New York: Springer 1968, pp. 110—113.
— u. G. LAU: Zur Konkurrenz der Echoencephalographie und der Carotisangiographie beim gedeckten, schweren Schädel-Hirn-Trauma. Langenbecks Arch. klin. Chir. **319**, 601 (1967).
ROBERTSON, E. G.: Pneumoencephalography. Springfield: Thomas 1957.
ROBINSON, D. E., and G. KOSSOFF: The C.A.L. two-channel echo-encephaloscope. Ultrasonics **3**, 69—74 (1965).
— — An ultrasonic echo-encephaloscope for the examination of the human brain. Proc. I.R.E.E. Australia **1966**, 39—44.

ROTHMAN, J., N. E. CHASE, I. I. KRICHEFF, R. MAYORAL, and E. R. BERANBAUM: Ultrasonic diagnosis of pericardial effusion. Circulation **35**, 358—364 (1967).

ROZSIVAL, V., PETR, R., J. ČIHÁK a I. KALISTOVÁ: Echoencephalografické vyšetření u mozkových expansí. Sbor. Věd. Pr. Lék. Fak. Hr. Králové **10**, 391—398 (1967).

RUSSO, G. L., and J. G. ARNOLD: Neurosurgical experience with echoencephalography. M. Ann. D. C. **33**, 260—264 (1964).

SAIER, J., J. PAILLAS, R. VIGOUROUX et H. GASTAUT: Intérêt de l'échoencéphalographie pour le diagnostic neuro-chirurgical. Rev. neurol. **111**, 520—522 (1964).

SANDEN, K. VON: Schalleistungsmessung in der Therapie und die Grenzen ihrer dosimetrischen Anwendbarkeit. Ultraschall in Med. **1**, 45—52 (1949).

SCHÁB, R., és J. KOPA: Az echoencephalographiáról (About echo-encephalography). Orv. Hetil. **106**, 1745—1752 (1965).

SCHEID, W.: Lehrbuch der Neurologie. Stuttgart: Thieme 1963.

SCHERZER, E.: Die Echoenzephalographie als Ergänzung des EEGs und ihre Grenzen. Wien. med. Wschr. **115**, 957—959 (1965).

— Das normale Echoenzephalogramm und seine Bedeutung in der Traumatologie. Wien. klin. Wschr. **78**, 301—304 (1966).

SCHIEFER, W.: Verbesserte Indikationsstellung zur Kontrastmitteluntersuchung durch vorhergehende Echoenzephalographie. Dtsch. Röntgenkongreß 1964. Stuttgart: G. Thieme 1965, pp. 326—330.

— Echo-Enzephalographie. Acta neurochir. (Wien) **13**, 576—578 (1965).

— Raumbeschränkende Prozesse im Schädelinnern. Therapiewoche **16**, 1471—1478 (1966).

— Spezielle Hämatomdiagnostik. Echoencephalographie, Angiographie. Langenbecks Arch. klin. Chir. **319**, 566—576 (1967).

— Echo-encephalography with tumors of the cerebral hemispheres. In: Proceedings in echo-encephalography. Berlin-Heidelberg-New York: Springer 1968, pp. 55—66.

— Reliability and limitations of A-scan echo-encephalography. In: Proceedings in echo-encephalography. Berlin-Heidelberg-New York: Springer 1968, pp. 181—185.

— u. E. KAZNER: Die Echo-Enzephalographie. Diagnostische Möglichkeiten. Dtsch. med. Wschr. **89**, 1394—1400 (1964).

— — Methodik und diagnostische Möglichkeiten der Echoencephalographie. Fortschr. Med. **84**, 151—154 (1966).

— — Der heutige Stand der Echoenzephalographie. Beitr. Neurochir. **13**, 93—108 (1966).

— — Klinische Echo-Encephalographie. Mit einer Einführung in die akustischen Grundlagen von W. GÜTTNER. Berlin-Heidelberg-New York: Springer 1967.

— — and H. BRÜCKNER: The diagnostic possibilities offered by echo-encephalography. Excerpta medica, International Congress Series **60** (1963), No. 58.

— — — Die Echoenzephalographie, ihre Anwendungsweise und klinischen Ergebnisse. Fortschr. Neurol. Psychiat. **31**, 457—491 (1963).

— — u. ST. KUNZE: Ergebnisse der Echoenzephalographie bei supratentoriellen Geschwülsten. Zbl. Neurochir. **26**, 281—295 (1965).

— — — Möglichkeiten und Grenzen der Echoenzephalographie beim Schädel-Hirntrauma. Dtsch. Ärztebl. **63**, 2645—2648, 2719—2722, 2777—2780 u. 2793—2974 (1966).

— M. LEWKE u. E. KAZNER: Der hämorrhagische Schock als Leitsymptom für die Erkennung posttraumatischer intrakranieller Hämatome bei Säuglingen und Kleinkindern. Zbl. Neurochir. **29**, 131—138 (1968).

— u. H. W. STEINMANN: Über Kreislaufwirkungen und bioelektrische Veränderungen bei Anwendung verschied. Röntgenkontrastmittel zur cerebralen Angiographie. Zbl. Neurochir. **18**, 173–188 (1958).

SCHIERSMANN, O.: Einführung in die Enzephalographie. Leipzig: Georg Thieme 1942.

SCHIFFER, K. H.: Zur Auswertung von Ventrikelbildern am Encephalogramm. Fortschr. Röntgenstr. **75**, 50—54 (1951).

SCHLAGENHAUFF, R. E., J. MAZUROWSKI, and B. H. SMITH: Echoencephalography in neurologic diagnosis. Critical evaluation of a-scan technique (time-amplitude detection of fixed echoes). N. Y. St. J. Med. **67**, 1035—1041 (1967).

SCHLEGEL, J. U., P. DIGGDON, and J. CUELLAR: The use of ultrasound for localizing renal calculi. J. Urol. **86**, 367—369 (1961).

SCHMIEDER, F.: Das Encephalogramm nach Fleckfieber. Klin. Wschr. **26**, 14—19 (1948).

SCHMITT, W., u. H. BRAUN: Mitteilung der mittels Ultraschall-Kardiographie gewonnenen Ergebnisse bei Mitralvitien und Herzgesunden. Z. Kreisl.-Forsch. **49**, 214—222 (1960).

SCHUM, U.: Ultraschalldiagnostik in der Ophthalmologie. Münch. med. Wschr. **110**, 1369—1375 (1968).

SEGAL, B. L.: Echocardiography—Ultrasoundcardiography. Amer. J. Cardiol. **19**, 1—5 (1967).

— W. LIKOFF, and B. KINGSLEY: Echocardiography; Clinical application in mitralstenosis. J. Amer. med. Ass. **195**, 161—166 (1966).

SEGAL, B. L., W. LIKOFF, and B. KINGSLEY: Echocardiography. Clinical application in combined mitral stenosis and mitral regurgitation. Amer. J. Cardiol. **19**, 42—49 (1967).
— — — Echocardiography. Clinical application in mitral regurgitation. Amer. J. Cardiol. **19**, 50—58 (1967).
SIMON, MIKOL et HAZEMANN: Cited from H. FISCHGOLD: Die Schädelunfall-Diagnostik. Dtsch. Röntgenkongreß 1965. Stuttgart: G. Thieme 1966, pp. 126—130.
SJÖGREN, I.: Hydrocephaluskonferens i Cambridge 1963. Svensk Läkartidn **60**, 3779 (1963).
— Echoencephalography in infantile hydrocephalus. Acta paed. scand. Suppl. **159**, 174 (1965).
— Echo-ventriculography in infantile hydrocephalus — Preliminary report. Acta neurol. scand. **41**, Suppl. 13, 13—18 (1965).
— Echo-ventriculography in infantile hydrocephalus. In: Technical progress in neurological diagnostics. Ed. by O. GILLAND, Amsterdam: Elsevier 1965.
— Echoencephalography in paediatric practice with special regard to measurement of the ventricular size. Acta paed. scand. Suppl. **178** (1967).
— Echoencephalography in paediatric practice with special regard to measurement of the ventricular size. Akademisk Avhandling, Uppsala 1968.
— Comparative studies of echo-ventriculography and cerebral pneumography in infantile hydrocephalus and cerebral malformations. In: Proceedings in echo-encephalography. Berlin-Heidelberg-New York: Springer 1968, pp. 135—142.
— K. BERGSTRÖM, and H. LODIN: Echoencephalography in infants and children. Comparison with cerebral pneumography in measuring ventricular size. Acta radiol. (Stockh.) Suppl. **278**, 1968.
SKUDRZYK, E.: Die physikalischen Ursachen für die mechanisch-chemisch-biologische Wirkung des Ultraschalls. Ultraschall in Med. **5**, 51—64 (1952).
SMYTH, M.: Echoencephalographic technique. Philadelphia: Smith Kline-Precision Company 1963.
— Echoencephalography: Supplementary information on importance of far-side group of echoes. Ekoline Bulletin No. 3. Philadelphia: Smith-Kline-Precision Company 1964.
SOKOLOW, S. J.: Zur Frage der Fortpflanzung ultraakustischer Schwingungen in verschiedenen Körpern. Elektr. Nachr. Techn. **6**, 454—461 (1929).
SOMER, J. C.: Instantaneous and continuous pictures obtained by a new two-dimensional scan technique with a stationary transducer. In: Proceedings in echo-encephalography. Berlin-Heidelberg-New York: Springer 1968, pp. 234—238.
SOULEN, R. L., M. S. LAPAYOWKER, and J. L. GIMENEZ: Echocardiography in the diagnosis of pericardial effusion. Radiology **86**, 1047—1050 (1966).
STALLKAMP, H.: Klinische und experimentelle Ultraschalluntersuchungen am gesunden Auge und bei intraokularen Fremdkörpern. Inaugural-Diss. Köln 1961.
— u. A. NOVER: Diagnostische Ultraschalluntersuchungen am gesunden Auge. Albrecht v. Graefes Arch. Ophthal. **164**, 399—410 (1962).
STOWELL, A., and J. E. KAUTH: The value of echo-encephalography in acute and chronic head injuries. J. Okla. med. Ass. **60**, 494—496 (1967).
STRIK, W. O.: Die diagnostische Anwendung des Ultraschalls. Med. Klin. **56**, 1817—1822 (1961).
SUGAR, O., and S. UEMATSU: The use of ultrasound in the diagnosis of intracranial lesions. Surg. Clin. N. Amer. **44**, 55—64 (1964).
SUNDÉN, B.: On the diagnostic value of ultrasound in obstetrics and gynecology. Acta obstet. gynec. scand. **43**, Suppl. 6 (1964).
— Obstetric diagnosis with ultrasound. Ultrasonics **5**, 67—71 (1967).
TAKEUCHI, T.: An experimental study of stereotaxic destruction of cat brain by focused ultrasound. Nihon Geka Hokan **34**, 634—650 (1965).
TANAKA, K.: Diagnosis of brain tumors using ultrasound. Nihon Igaku (Japan. Medic.), Supplementum III, 205—209 (1963).
— Diagnosis of brain disease by ultrasound. Clinical practice (Japanese). Shindan-To-Chiryo Sha, Co., Ltd. 1968.
— and K. ITO: Ultrasonic diagnosis of intracranial disease. Shinkei Kenkyû no Shimpo (Recent Advances Research Nervous System) **7**, 335—351 (1963).
— — Diagnosis of brain tumor using ultrasound. Acta radiol. **5**, diagnosis, 7th Symposium neuroradiol., New York 1964, 915—927 (1966).
— — and S. ISHIKAWA: Ultrasonic diagnosis of brain tumor. Sôgô Igaku (Medicine, Tokyo) **18**, 297—308 (1961).
— — — and T. WAGAI: Diagnosis of brain tumors using ultrasound. Nihon Rinshô (Japan. J. clin. Med.) **21**, 2195—2200 (1963).
— — and Y. OHARA: Ultrasonotomography of the brain. In: Proceedings in echo-encephalography. Berlin-Heidelberg-New York 1968, pp. 201—208.
— — — and Y. OHARA: Ultrasonics in the diagnosis of intracranial space occupying lesions. Excerpta Medica Foundation, Amsterdam 1966.

TANAKA, K., K. ITO, S. ISHIKAWA, and S. UEMATSU: Cerebral diagnosis in children by means of ultrasound. Shônika Shinryô **25**, 926—932 (1962).
— — — K. YUKISHITA, and Y. OHARA: Ultrasonics in the diagnosis of intracranial space occupying lesions. Proc. III Intern. Congr. Neurol. Surg., Excerpta medica Intern. Congr. Series **110**, 580 to 589 (1966).
— — and T. WAGAI: The localization of brain tumors by ultrasonic techniques. A clinical review of 111 cases. J. Neurosurg. **23**, 135—147 (1965).
— Y. KIKUCHI, and R. UCHIDA: Detection of intracranial anatomical abnormalities by ultrasound. J. acoust. Soc. Jap. **8**, 111—112 (1952).
— — — Ultrasonic diagnosis of brain tumor. Proc. 3rd int. Congr. Acoust. Stuttgart 1959. Amsterdam: Elsevier Publ. Co. pp. 1291—1294.
— — — Ultrasonic diagnosis of intracranial diseases. Proc. surg. Soc. Japan **53**, 242—243 (1963).
— and T. WAGAI: Ultrasonic diagnosis of brain disease in Japan. In: Ultrasound as a diagnostic and surgical tool. Ed. by D. GORDON, Edinburgh & London: Livingstone 1964, pp. 161—175.
— — K. ITO, S. UEMATSU, Y. ABE, S. ISHIKAWA, and Y. KIKUCHI: Diagnosis of intracranial hemorrhage. In: Advance of medical application of Ultrasound developed by the Ultrasonic Research Group in Juntendo University in 1963. Tokyo: Juntendo University School of Medicine, Hongo, 1964.
— — Y. KIKUCHI, R. UCHIDA, and S. UEMATSU: Ultrasonic diagnosis in Japan. In: Diagnostic ultrasound. New York: Plenum Press 1966, pp. 27—45.
TANK, T. M., and W. B. HAMBY: Echoencephalography: An aid in the diagnosis of intracranial lesions. Report of a case. Cleveland Clin. Quart. **31**, 101—105 (1964).
TAVERAS, J. M., and E. H. WOOD: Diagnostic neuroradiology. Baltimore: Williams and Wilkins Co. 1964.
TAYLOR, J. C., J. A. NEWELL, and P. KARVOUNIS: Ultrasonics in the diagnosis of intra-cranial space-occupying lesions. Lancet **1961, I**, 1197—1199.
THEISSMANN, H., u. F. PFANDER: Über die Durchlässigkeit des Knochens für Ultraschall. Strahlentherapie **80**, 607—610 (1949).
THOMPSON, H. E.: Studies of fetal growth by ultrasound. In: Diagnostic ultrasound. New York: Plenum Press 1966.
THUN, F.: Die Echoencephalographie bei postoperativen Komplikationen. Acta neurochir. (Wien) **12**, 652 (1965).
THURSTONE, F. L., N. I. KJOSNES, and W. M. MCKINNEY: Ultrasonic scanning of biologic tissue by a new technique. Science **149**, 302—303 (1965).
— and W. M. MCKINNEY: Focused transducer arrays in an ultrasonic scanning system for biologic tissue. In: Diagnostic ultrasound. New York: Plenum Press 1966, pp. 191—194.
TÖNNIS, W.: Die Chirurgie des Gehirns und seiner Häute. Bd. III. In: Die Chirurgie. Ed. by M. Kirschner u. O. Nordmann. Wien: Urban & Schwarzenberg 1948.
— Wie läßt sich die Frühdiagnose der Hirntumoren verbessern? Wien. med. Wschr. **103**, 835—840 (1953).
— Diagnostik der intrakraniellen Geschwülste. In: Handbuch der Neurochirurgie. Bd. IV, 3. Teil. Ed. by H. OLIVECRONA u. W. TÖNNIS. Berlin: Springer 1962.
— G. FRIEDMANN, E. SCHMIDT-WITTKAMP u. W. WALTER: Die traumatischen intrakraniellen Hämatome. Documenta Geigy, Series chir. **6**, 1—89 (1963).
— u. R. A. FROWEIN: Klinische Fehldiagnosen bei intrakraniellen Hämatomen. Dtsch. Röntgenkongreß 1965. Stuttgart: Thieme 1966, S. 137—142.
— — Probleme der neuzeitlichen Behandlung schwerer Schädel-Hirn-Verletzungen. Langenbecks Arch. klin. Chir. **316**, 323—329 (1966).
— — u. K. H. EULER: Zur Erkennung der akuten traumatischen intrakraniellen Hämatome. Chirurg **34**, 145—151 (1963).
— — F. LOEW, W. GROTE, R. HEMMER, W. KLUG u. H. FINKEMEYER: Organisation der Behandlung schwerer Schädel-Hirn-Verletzungen. Stuttgart: Thieme 1968.
— u. W. SCHIEFER: Klinik der raumbeengenden Prozesse des Occipitallappens. Dtsch. Z. Nervenheilk. **170**, 402—432 (1953).
— — Die Komplikationen bei Angiographie der Hirngefäße. Fortschr. Neurol. Psychiat. **26**, 265 to 300 (1958).
— — Zirkulationsstörungen des Gehirns im Serienangiogramm. Berlin-Göttingen-Heidelberg: Springer 1959.
TORKILDSEN, A.: Ventriculo-cisternostomy. A post-operative study. Acta chir. scand. **85**, 254—260 (1941).
— A follow-up study 14 to 20 years after ventriculocisternostomy. Acta psychiat. scand. **35**, 113—121 (1960).
UEMATSU, S.: Ultrasonic determination of the hemispheral thickness for the diagnosis of hydrocephalus. In: Diagnostic ultrasound. New York: Plenum Press 1966, pp. 148—154.

UEMATSU, S., O. SUGAR, and A. E. WALKER: Echoencephalography for head trauma. Proc. III Intern. Congr. Neurol. Surg., Excerpta medica Intern. Congr. Series **110**, 194—199 (1966).
— and A. E. WALKER: Ultrasonic determination of the size of cerebral ventricular system. Neurology (Minneap.) **17**, 81—86 (1967).
ULBRICHT, W.: Rückenmarkstumoren mit Stauungspapille. Acta neurochir. **15**, 138—149 (1966).
— u. V. DE SEIXAS: Echoencephalographie und 3. Ventrikel. Paper presented at Hungar. Neurol. Congr., Budapest, October 7—10, 1966.
ULBRICHT, W., and V. DE SEIXAS: Echo-encephalographic investigations with hydrocephalus of various origins, especially infantile hydrocephalus. In: Proceedings in echo-encephalography. Berlin-Heidelberg-New York: Springer 1968, pp. 143—145.
ULTAN, L. B., B. L. SEGAL, and W. LIKOFF: Echocardiography in congenital heart disease. Preliminary observations. Amer. J. Cardiol. **19**, 74—83 (1967).
UMBACH, W.: Echodiagnostik und Verlaufskontrollen beim kindlichen Hydrocephalus. Acta neurochir. (Wien) **13**, 581 (1965).
— u. M. KLEY: Untersuchungen mit Ultraschall zur Diagnose und Verlaufskontrolle des kindlichen Hydrocephalus. Dtsch. med. Wschr. **90**, 1313—1315 (1965).
UNGER, R. R., u. W.-D. SIEDSCHLAG: Echoenzephalographie und gedeckte Schädel-Hirn-Verletzungen. Mschr. Unfallheilk. **69**, 435—442 (1966).
— — Echoenzephalographie und Hirntumor. Dtsch. Gesundheitswesen **22**, 455—457 (1967).
U.S. Atomic Encergy Commission: Studies in methods in instruments to improve localization of radioactive materials in the body with special reference to the diagnosis of brain tumors and the use of ultrasonic techniques. AECU-3012, Minneapolis, Univ. Minnesota Press 1955.
VANÝSEK, J., u. J. PREISOVÁ: Über die Untersuchungsmöglichkeiten mit Ultraschall im Orbitalraum. Wiss. Z. Humboldt-Univ. Berlin, Math.-Nat. R. **14**, 179—184 (1965).
VASTINE, J. H., and K. K. KINNEY: The pineal shadow as an aid in the localization of brain tumours. Amer. J. Roentgenol. **17**, 320—324 (1927).
VIGOUROUX, R. P., and R. A. FROWEIN: Survey of the organization of services for acute head injury in european countries. Excerpta Medica Foundation, Amsterdam 1966.
VINCENT, J.-D.: L'échoencéphalographie. J. méd. Bordeaux **144**, 819—822 (1967).
— J. FAURE, M. GOT, B. SOUMIREU-MOURAT et P. ESCHAPASSE: Premiers résultats d'utilisation de l'échoencéphalographie en service d'exploration fonctionelle du système nerveux J. méd. Bordeaux **143**, 357—366 (1966).
VLIEGER, M. DE: Echo-encephalographie bei Hirnverletzungen. Acta neurochir. (Wien) **9**, 707 (1961).
— Diagnostische hulpmiddelen bij traumatische hersenletsels. Huisarts en Wetenschap **5**, 337 (1962).
— Echo-encefalografie als hulpmiddel bij de hersendiagnostiek. Ned. T. Geneesk. **108**, 5—10 (1964).
— Echo-encephalography and extra-cerebral haematomas. In: D. GORDON: Ultrasound as a diagnostic and surgical tool. Edinburgh & London: Livingstone 1964, pp. 145—160.
— De mogelijkheden van het twee-dimensionele echogram voor de algemene medische diagnostiek. Ned. T. Geneesk. **108**, 1513—1514 (1964).
— Echo-encephalografie en hersendiagnostiek. Hoofdlijnen **1**, 18—23 (1965).
— Echo-encephalography and cerebral circulation. Excerpta medica, International Congress Series **139** (1967), No. 21.
— Evolution of echo-encephalography in neurology — A review. Ultrasonics **6**, 91—97 (1967).
— Metingen met ultrageluid bij de mens. De Ingenieur **79**, 184—187 (1967).
— Problems in the differential diagnosis of hematoma echoes with posttraumatic intracranial hemorrhages. In: Proceedings in echo-encephalography. Berlin-Heidelberg-New York: Springer 1968, pp. 87—90.
— Two-dimensional echo-encephalography in the diagnosis of infantile hydrocephalus. In: Proceedings in echo-encephalography. Berlin-Heidelberg-New York: Springer 1968, pp. 208—210.
— J. J. DENIER VAN DER GON, P. J. M. VAN DER LUGT, and C. E. MOLIN: Pulsations in echo-encephalography. 6th Intern. Congr. Electroencephalogr. Clin. Neurophysiol. Wien, Sept. 5—10, 1965, 591—592.
— — — — Two-dimensional echoencephalography of the third ventricle in hydrocephalus. Neurology (Minneap.) **18**, 473—479 (1968).
— and H. J. RIDDER: Use of echoencephalography. Neurology (Minneap.) **9**, 216—223 (1959).
— A. DE STERKE, C. E. MOLIN, and C. VAN DER VEN: Ultrasound for two-dimensional echo-encephalography. Ultrasonics **1**, 148—151 (1963).
VOGELSANG, H.: Pneumencephalogramm-Verlaufsbeobachtungen bei Torkildsendrainagen. Deutscher Röntgenkongreß 1964. Stuttgart: Thieme 1965, S. 307—311.
WACHSMANN, F. F.: Schädigungsmöglichkeiten durch Ultraschall vom Standpunkt der physikalischen Vorgänge im Gewebe und der Applikationstechnik aus gesehen. Ultraschall in Med. **4**, 42—44 (1952).
WAGAI, T., R. MIYAZAWA, K. ITO, and Y. KIKUCHI: Ultrasonic diagnosis of intracranial disease, breast tumors and abdominal diseases. In: Ultrasonic Energy, Ed. by E. KELLY. Urbana: University of Illinois Press 1965, pp. 346—364.

WAGAI, T., M. TSUCHIDATE, K. ITO, M. ISHII, S. KATSUMI, S. UEMATSU, S. ISHIKAWA, T. TAKEUHI, H. OHASHI, H. NONAKA, K. TAKEUCHI, Y. ABE, M. ARIMA, T. OUCHI, Y. KIKUCHI, M. IDE, and R. UCHIDA: Recent Advance of medical application of Ultrasound developed by the ultrasonic research group in Juntendo University. Juntendo University School of Medicine, Hongo, Tokyo, Japan 1963.

WAKEFIELD, G.: Ultrasonic diagnosis in cerebral diseases. Univ. Coll. Hosp. Mag. **47**, 10 (1964).

WALKER, A. E., and S. UEMATSU: Dural echoencephalography. J. Neurosurg. **25**, 634—637 (1966).

WALLACE, W. K., W. S. AVANT Jr., W. M. MCKINNEY, and F. L. THURSTONE: Ultrasonic techniques for measuring intracranial pulsations. Research and clinical studies. Neurology (Minneap.) **16**, 380—382 (1966).

WEIDENBACH, W.: Zur Diagnose intrakranieller Hämatome. Langenbecks Arch. klin. Chir. **319**, 601—602 (1967).

WEIGELDT, H. D., and K. SCHRÖTER: Anatomical and technical causes of errors in echo-encephalography. In: Proceeding in echo-encephalography. Berlin-Heidelberg-New York: Springer 1968, pp. 179—181.

WEINLAND, W., and GERDA WEINLAND: Experience with echo-encephalography in a specialist neurological practice. In: Proceedings in echo-encephalography. Berlin-Heidelberg-New York: Springer 1968, pp. 176—179.

WELLS, P. N. T.: Resonance artifacts (Letter to the editor). Ultrasonics **3**, 154 (1965).

WERTHEIMER, P., G. FISCHER, J. COURJON, P. GERIN, M. REVOL et F. MUNIER: La valeur diagnostique de l'écho-encéphalographie. Etude d'une statistique de 1000 examens. Rev. neurol. **114**, 219—225 (1966).

WEST, K. A.: Eko-encefalografi. Suomen Laarkarilehti **19**, 2227—2234 (1964).

— Correlation between ultrasonic and roentgenological findings in infantile hydrocephalus. Acta paed. scand. **56**, 27—36 (1967).

WHITE, D. N.: A-scan echoencephalography. Canad. med. Ass. J. **94**, 180—189 (1966).

— Studies in ultrasonic echoencephalography. VI. A critical analysis of the amplitude-averaging, A-scan technique. Neurology (Minneap.) **16**, 858—866 (1966).

— Studies in ultrasonic echoencephalography (6 A). A critical analysis of the amplitude-averaged A-Scan. Trans. Amer. neurol. Ass. **1966**, 363—365.

— Accuracy of A scan determination of midline echo. In: C. C. GROSSMAN et al.: Diagnostic ultrasound. New York: Plenum Press 1966, pp. 142—147.

— The limitations of echo-encephalography. Ultrasonics **5**, 88—90 (1967).

— Results of the amplitude-averaged technique of echo-encephalography. 8e Symposium neuroradiol. Paris, 1967.

— and B. BLANCHARD: Studies in ultrasonic echoencephalography. IV. Results of an averaging technique to localize the cerebral midline structure. Neurology (Minneap.) **15**, 1041—1048 (1965).

— — Studies in ultrasonic echoencephalography II. An objective technique for the A-scan presentation of the cerebral midline structures. Acta radiol. **5**, diagnosis, 7th Symposium neuroradiol., New York 1964, 936—952 (1966).

— — and J. M. CLARK: Studies in ultrasonic echoencephalography-V. Echoes obtainable from a plastic skull. Brit. J. Radiol. **40**, 108—119 (1967).

— — and M. N. WHITE: Studies in ultrasonic echoencephalography. III. Limitations of a simple radial scan. Wiss. Z. Humboldt-Univ. Berlin, Math.-Nat. R. **14**, 23—28 (1965).

— J. N. CHESEBROUGH, and J. B. BLANCHARD: Studies in ultrasonic echoencephalography. I. A-Scan determination of the M-echo position in a group of patients. Neurology (Minneap.) **15**, 81—86 (1965).

— J. M. CLARK, J. N. CHESEBROUGH, M. N. WHITE, and D. J. CLARK: The effect of the skull in degrading resolution in echo-encephalographic B- and C-scans. In: Proceedings in echo-encephalography. Berlin-Heidelberg-New York: Springer 1968, pp. 230—234.

— — and M. N. WHITE: Studies in ultrasonic echoencephalography. VII: General principles of recording informations in ultrasonic B- and C-scanning and the effects of scatter, reflection and refraction by cadaver skull on this information. Med. biol. Engng. **5**, 3—14 (1967).

— — — Studies in ultrasonic echoencephalography. VIII: The effects on resolution of irregularities in attenuation of an ultrasonic beam traversing cadaver skull. Med. biol. Engng. **5**, 15—23 (1967).

WIDÉN, L., A. PERSSON, and A. WENNBERG: Echo-encephalography in the differential diagnosis between intracerebral haemorrhage and cerebral infraction. Thule Intern. Sympos., Stockholm: Nord. Bokhandelns Förlag, 1967, pp. 153—158.

WIEDAU, E., u. O. RÖHER: Ultraschall in der Medizin. Dresden u. Leipzig: Steinkopff 1963.

WILCKE, O.: Hirntumor-Diagnostik mit Positronenstrahlern. Acta neurochir. (Wien) **10**, 301—319 (1962).

— Möglichkeiten und Grenzen der Hirntumordiagnostik mit Positronenstrahlern (Cu^{64} und As^{74}). Neurochirurgia **7**, 33—41 (1964).

WILD, J. J.: The use of ultrasonic pulses for the measurement of biologic tissues and the detection of tissue density changes. Surgery **27**, 183—188 (1950).

WILD, J. J., and D. NEAL: Use of high frequency ultrasonic wave for detecting changes of texture in living tissues. Lancet **1951, I,** 655—657.

— and J. M. REID: Application of echo-ranging techniques to the determination of structure of biological tissues. Science **115,** 226—230 (1952).

— — Further pilot echographic studies on the histologic structure of tumor of the living intact human breast. Amer. J. Path. **28,** 839—861 (1952).

— — The effects on biological tissues of 15-Mc pulsed ultrasound. J. acoust. Soc. Amer. **25,** 270 to 280 (1953).

— — Fifteen megacycle pulsed ultrasonic reflection studies on biological tissues. Ultraschall in Med. **6,** 195—199 (1953).

— — Echographic visualization of lesions of the living human breast. Cancer Res. **14,** 277—283 (1954).

— — Diagnostic use of ultrasound. Brit. J. Phys. Med. **19,** 248—257 (1956).

WILLARD, G. W.: Improved methods for measuring ultrasonic velocity. J. acoust. Soc. Amer. **23,** 83—93 (1951).

WILLOCKS, J., I. DONALD, T. C. DUGGAN, and N. DAY: Foetal cephalometry by ultrasound. J. Obstet. Gynaec. Brit. Cwlth. **71,** 11—20 (1964).

WINTERS, W., J. GIMENEZ, and L. SOLOFF: Clinical application of ultrasound in the analysis of prosthetic ball valve function. Amer. J. Cardiol. **19,** 97—107 (1967).

WOEBER, K. H.: Der Ultraschall in der Medizin. Stuttgart: S. Hirzel 1949.

— Über das Auftreten von Schädigungen am Zentralnervensystem der Ratte durch Ultraschallwellen. Strahlentherapie **79,** 643—652 (1949).

WOJAHN, H.: Über die Letalität des epiduralen Hämatoms. Mschr. Unfallheilk. **67,** 150—163 (1964).

YAMAKAWA, K., and A. YOSHIOKA: Ultrasonic diagnostic methods. Sôgô Igaku (Medicine, Tokyo) **20,** 265—271 (1963).

YANG, C. L., and Y. L. TAN: Ultrasonic examination of 170 patients with intracranial hypertension. (Zhongua Shenjingjingshenke Zazhi). Science abstracts China Medicine **5,** 26 (1964).

YUKISHITA, K., Y. ABE, K. ITO, and T. WAGAI: Ultrasonics in the diagnosis of head injury. Japan. Med. Ultrasonics **3,** 29 (1965).

— K. EHARA, H. NONAKA, S. ISHIKAWA, K. ITO, and K. TANAKA: Ultrasonics in the diagnosis of brain tumor (3rd report). Med. Ultrasonics **4,** 59 (1966).

— S. ISHIKAWA, K. ITO, and K. TANAKA: Ultrasonics in the diagnosis of brain tumor. J. Japan. Surg. Soc. **67,** 1511 (1966).

— and K. TANAKA: Diagnosis of brain tumors by ultrasound. J. Japan. Surg. Soc. **66,** 80 (1965).

— — K. ITO, K. EHARA, and H. WATANABE: Ultrasonic diagnosis of brain tumors. In: Proceedings in echo-encephalography. Berlin-Heidelberg-New York: Springer 1968, pp. 38—44.

ZANDER, E., et G. FOROGLOU: L'apport de l'échoencéphalographie en neurochirurgie. Praxis **55,** 398—406 (1966).

ZICHNER, L.: Zur Bedeutung der Echoencephalographie in der klinischen Neurologie. Inaug.-Diss. Düsseldorf, 1967.

ZÜLCH, K. J.: Die Hirngeschwülste in biologischer und morphologischer Darstellung. 3. Auflage. Leipzig: J. Ambrosius Barth 1958.

— The morphologic basic of the abnormal echo-encephalogram. In: Proceedings in echo-encephalography. Berlin-Heidelberg-New York: Springer 1968, pp. 12—24.

Subject Index

Herstellung: Konrad Triltsch, Graphischer Betrieb, Würzburg

MIX
Papier aus verantwortungsvollen Quellen
Paper from responsible sources
FSC® C105338

If you have any concerns about our products,
you can contact us on
ProductSafety@springernature.com

In case Publisher is established outside the EU,
the EU authorized representative is:
Springer Nature Customer Service Center GmbH
Europaplatz 3, 69115 Heidelberg, Germany

Printed by Libri Plureos GmbH
in Hamburg, Germany